D1237853

X-RAY POLARIMETRY: A NEW WINDOW IN ASTROPHYSICS

Due to the advent of a new generation of detectors, X-ray polarimetry promises to join X-ray imaging, spectroscopy and timing as one of the main observational techniques in high energy astrophysics. This has renewed interest in the field, and indeed several polarimetric missions have recently been proposed. This volume provides a complete and up-to-date view of the subject for researchers in astrophysics. The contributors discuss the present status and perspectives of instruments, review current theoretical models, and examine future missions. As well as detailed papers, the book contains broad reviews that can be easily understood by astrophysicists new to the field.

RONALDO BELLAZZINI is Director of Research at the Istituto Nazionale di Fisica Nucleare (INFN), Italy. His main research interests are particle astrophysics, X-ray polarimetry, and the development of highly innovative particle detectors.

ENRICO COSTA is Director of Research at the Istituto di Astrofisica Spaziale e Fisica Cosmica-Roma of the Istituto Nazionale di Astrofisica (INAF). His areas of research are in X-ray and gamma-ray astrophysics and polarimetry.

GIORGIO MATT is a Professor at the Physics Department, Università Roma Tre, where he teaches both basic physics and astrophysics. His research activities are concerned mainly with the physics of accretion onto black holes, from both theoretical and observational points of view.

GIANPIERO TAGLIAFERRI is an Associate Astronomer at the Brera Astronomical Observatory of INAF. His research activities deal with the studies of gamma-ray bursts, active galactic nuclei, and with the development of new X-ray focussing telescopes.

X-RAY POLARIMETRY: A NEW WINDOW IN ASTROPHYSICS

Edited by

RONALDO BELLAZZINI
Instituto Nazionale di Fisica Nucleare (INFN)

ENRICO COSTA
Instituto di Astrofisica Spaziale e Fisica Cosmica-Roma, INAF

GIORGIO MATT
Università Roma Tre

and

GIANPIERO TAGLIAFERRI
Brera Astronomical Observatory, INAF

CAMBRIDGE UNIVERSITY PRESS
Cambridge, New York, Melbourne, Madrid, Cape Town, Singapore,
São Paulo, Delhi, Dubai, Tokyo

Cambridge University Press
The Edinburgh Building, Cambridge CB2 8RU, UK

Published in the United States of America by Cambridge University Press, New York

www.cambridge.org
Information on this title: www.cambridge.org/9780521191845

First published 2010

Printed in the United Kingdom at the University Press, Cambridge

A catalogue record for this publication is available from the British Library

ISBN 978-0-521-19184-5 Hardback

Contents

Contributors

N. Anabuki Osaka University, Department of Earth and Space Science, Machikaneyama, Toyonaka, Osaka 560-0043, Japan

A. Argan INAF – IASF – Roma, Via Fosso del Cavaliere 100, 00130 Rome, Italy

M. Axelsson Stockholm University and the Oskar Klein Centre, AlbaNova University Center, 106 91 Stockholm, Sweden

N. Barrière INAF – IASF – Roma, Via Fosso del Cavaliere 100, 00130 Rome, Italy

S. Basso INAF – Osservatorio Astronomico di Brera, Via Bianchi 46, 23807 Merate (LC), Italy

F. Bayer ECAP, University of Erlangen-Nürnberg, Erwin-Rommel-Str. 1, 91058 Erlangen, Germany

R. Beck Max Planck Institute for Radio Astronomy, Auf dem Huegel 69, 53121 Bonn, Germany

R. Bellazzini INFN – Sezione di Pisa, Largo Bruno Pontecorvo 3, 56127 Pisa, Italy

S. Bianchi Dipartimento di Fisica, Università degli Studi di Roma TRE, Via della Vasca Navale 84, 00146 Rome, Italy

K. Black GSFC/Rock Creek Scientific, Silver Spring, MD 20910, USA

P. F. Bloser University of New Hampshire, 8 College Road, Durham, NH 03824, USA

A. Brez INFN – Sezione di Pisa, Largo Bruno Pontecorvo 3, 56127 Pisa, Italy

N. Bucciantini University of California at Berkeley, 601 Campbell Hall, Berkeley, CA 94720, USA

M. Bursa Astronomical Institute of the Academy of Sciences of the Czech Republic, Bocni II 1401/1a, CZ-14131 Praha 4, Czech Republic

M. Chiaberge Space Telescope Science Institute, 3700 San Martin Drive, Baltimore, MD 21218, USA

E. M. Churazov MPA, Garching, Germany; IKI, Moscow, Russia

K. T. Chyży Astronomical Observatory of the Jagiellonian University, ul. Orla 171, 30-244 Krakòw, Poland

O. Citterio INAF – Osservatorio Astronomico di Brera, Via Bianchi 46, 23807 Merate (LC), Italy

T. P. Connor University of New Hampshire, 8 College Road, Durham, NH 03824, USA

E. Costa INAF – IASF – Roma, Via Fosso del Cavaliere 100, 00130 Rome, Italy

V. Cotroneo INAF – Osservatorio Astronomico di Brera, Via Bianchi 46, 23807 Merate (LC), Italy

S. Covino INAF – Osservatorio Astronomico di Brera, Via Bianchi 46, 23807 Merate (LC), Italy

R. Cowsik Washington University in St. Louis, Department of Physics, One Brookings Drive, St. Louis, MO 63130, USA

A. J. Dean University of Southampton, School of Physics and Astronomy, Southampton, SO17 1BJ, UK

P. Deines-Jones NASA – Goddard Space Flight Center, Greenbelt, MD 20771, USA

G. O. Depaola Facultad de Matemática, Astronomía y Física, Universidad Nacional de Córdoba, Medina Allende s/n, 5008 Córdoba, Argentina

S. Di Cosimo INAF – IASF – Roma, Via Fosso del Cavaliere 100, 00130 Rome, Italy

G. Di Persio INAF – IASF – Roma, Via Fosso del Cavaliere 100, 00130 Rome, Italy

K. Dolag Max Planck Institute for Astrophysics, Karl-Schwarzschild-Str. 1, Postfach 1317, 85741 Garching, Germany

M. Dovčiak Astronomical Institute of the Academy of Sciences of the Czech Republic, Bocni II 1401/1a, CZ-14131 Praha 4, Czech Republic

R. Duraichelvan Raman Research Institute, C.V. Raman Avenue, Sadashivanagar, Bangalore 560080, India

J. Durst ECAP, University of Erlangen-Nürnberg, Erwin-Rommel-Str. 1, 91058 Erlangen, Germany

M. Ehle XMM Science Operations Centre, European Space Astronomy Centre (ESAC) – ESA, P.O. Box 78, Villanueva de la Cañada, 28691 Madrid, Spain

S. Fabiani INAF – IASF – Roma, Via Fosso del Cavaliere 100, 00130 Rome, Italy

Y. Z. Fan Niels Bohr Institute, Blegdamsvej 17, DK-2100 Copenhagen, Denmark

M. Feroci INAF – IASF – Roma, Via Fosso del Cavaliere 100, 00130 Rome, Italy

W. Forman Harvard-Smithsonian Center for Astrophysics, 60 Garden Street, Cambridge, MA 02138, USA

M. Forot CEA Saclay, DSM/Irfu/Service d'Astrophysique, F-91191, Gif sur Yvette, France and Astrophysique Interactions Multi-échelles (AIM), CNRS

H. Fujimoto Kanazawa University, Kakuma, Kanazawa, Ishikawa 920-1192, Japan

N. Fujita Yamagata University, 1-4-12 Kojirakawa, Yamagata 990-8560, Japan

A. Fumi RIKEN, 2-1 Hirosawa, Wako, Saitama 351-0198, Japan

A. B. Garson Washington University in St. Louis, Department of Physics, One Brookings Drive, St. Louis, MO 63130, USA

P. Ghosh Tata Institute of Fundamental Research, Department of Astronomy and Astrophysics, Homi Bhabha Road, Mumbai 400 005, India

R. W. Goosmann Observatoire Astronomique de Strasbourg, 11 rue de l'Observatoire, F-67000 Strasbourg, France

D. Gotz CEA Saclay, DSM/Irfu/Service d'Astrophysique, F-91191, Gif sur Yvette, France and Astrophysique Interactions Multi-échelles (AIM), CNRS

J. Greiner Max Planck Institute for Extraterrestrial Physics, Giessenbachstrasse, 85748 Garching, Germany

M. Guainazzi XMM Science Operations Centre, European Space Astronomy Centre (ESAC) – ESA, P.O. Box 78, Villanueva de la Cañada, 28691 Madrid, Spain

S. Gunji Yamagata University, 1-4-12 Kojirakawa, Yamagata 990-8560, Japan

A. K. Harding NASA Goddard Space Flight Center, Greenbelt, MD 20771, USA

K. Hayashida Osaka University, Department of Earth and Space Science, Machikaneyama, Toyonaka, Osaka 560-0043, Japan

A. Hayato RIKEN, 2-1 Hirosawa, Wako, Saitama 351-0198, Japan

R. Heilmann MIT Kavli Institute for Astrophysics and Space Research, 77 Massachusetts Avenue, 37-287 Cambridge, MA 02139, USA

J. S. Heyl Department of Physics and Astronomy, University of British Columbia, Vancouver, British Columbia, Canada

W. C. G. Ho University of Southampton, School of Mathematics, Southampton SO17 1BJ, UK

J. Horák Astronomical Institute of the Academy of Sciences of the Czech Republic, Bocni II 1401/1a, CZ-14131 Praha 4, Czech Republic

M. L. Iparraguire Facultad de Matemática, Astronomía y Física, Universidad Nacional de Córdoba, Medina Allende s/n, 5008 Córdoba, Argentina

T. Iwahashi RIKEN, 2-1 Hirosawa, Wako, Saitama 351-0198, Japan

M. S. Jackson Royal Institute of Technology (KTH), Dept. of Physics, AlbaNova University Centre, SE 10691 Stockholm, Sweden

K. Jahoda NASA – Goddard Space Flight Center, Greenbelt, MD 20771, USA

J. Jakubek IEAP, Czech Technical University in Prague, Horská 3a/22, 128 00 Prague, Czech Republic

D. Jincy School of Pure and Applied Physics, Mahatma Gandhi University, Kottayam 686560, Kerala, India

P. Kaaret Department of Physics and Astronomy, University of Iowa, Iowa City, IA 52246, USA

T. Kallman NASA – Goddard Space Flight Center, Greenbelt, MD 20771, USA

G. Kanbach Max Planck Institute for Extraterrestrial Physics, Giessenbachstrasse, 85748 Garching, Germany

V. Karas Astronomical Institute of the Academy of Sciences of the Czech Republic, Bocni II 1401/1a, CZ-14131 Praha 4, Czech Republic

S. Kishimoto High Energy Accelerator Research Organization, KEK, 1-1 Oho, Tsukuba, Ibaraki 305-0801, Japan

Y. Kishimoto Yamagata University, 1-4-12 Kojirakawa, Yamagata 990-8560, Japan

M. Kiss Royal Institute of Technology (KTH), Dept. of Physics, AlbaNova University Centre, SE 10691 Stockholm, Sweden

M. Kohama ISAS, JAXA, 2-1-1 Sengen, Tsukuba, Ibaraki 305-8505, Japan

S. Konami RIKEN, 2-1 Hirosawa, Wako, Saitama 351-0198, Japan

H. Krawczynski Washington University in St. Louis, Department of Physics, One Brookings Drive, St. Louis, MO 63130, USA

J. H. Krolik Johns Hopkins University, 3400 N Charles St, Baltimore, MD 21218, USA

S. Kubo Clear Pulse Co., Ltd, 6-25-17 Chuoh, Ota-ku, Tokyo, 143, Japan

Z. Kuncic University of Sydney, School of Physics A28, University of Sydney, NSW 2006, Australia

D. Lai Cornell University, Department of Astronomy, Ithaca, NY 14853, USA

P. Laurent CEA Saclay, DSM/Irfu/Service d'Astrophysique, F-91191, Gif sur Yvette, France and Astroparticules et Cosmologie (APC), 10, rue Alice Domon et Léonie Duquet, F-75205, Paris Cedex 13, France

F. Lazzarotto INAF – IASF – Roma, Via Fosso del Cavaliere 100, 00130 Rome, Italy

D. Lazzati North Carolina State University, Dept. of Physics, Riddick Hall, Suite 421, Box 8202, 2401 Stinson Drive, Raleigh, NC 27695-8202, USA

F. Lebrun CEA Saclay, DSM/Irfu/Service d'Astrophysique, F-91191, Gif sur Yvette, France and Astroparticules et Cosmologie (APC), 10, rue Alice Domon et Léonie Duquet, F-75205, Paris Cedex 13, France

J. S. Legere University of New Hampshire, 8 College Road, Durham, NH 03824, USA

F. Longo INFN – Sezione di Trieste, Padriciano 99, I – 34012 Trieste, Italy

A. Marinucci Dipartimento di Fisica, Università degli Studi di Roma TRE, Via della Vasca Navale 84, 00146 Rome, Italy

H. L. Marshall MIT Kavli Institute for Astrophysics and Space Research, 77 Massachusetts Avenue, 37-287 Cambridge, MA 02139, USA

J. Marykutty School of Pure and Applied Physics, Mahatma Gandhi University, Kottayam 686560, Kerala, India

G. Matt Dipartimento di Fisica, Università degli Studi di Roma TRE, Via della Vasca Navale 84, 00146 Rome, Italy

M. L. McConnell University of New Hampshire, 8 College Road, Durham, NH 03824, USA

S. McGlynn Oskar Klein Centre, AlbaNova University Center, 106 91 Stockholm, Sweden

A. McNamara University of Sydney, School of Physics A28, University of Sydney, NSW 2006, Australia

T. Michel ECAP, University of Erlangen-Nürnberg, Erwin-Rommel-Str. 1, 91058 Erlangen, Germany

T. Mihara RIKEN, 2-1 Hirosawa, Wako, Saitama 351-0198, Japan

M. Minuti INFN – Sezione di Pisa, Largo Bruno Pontecorvo 3, 56127 Pisa, Italy

A. Moretti INAF – Osservatorio Astronomico di Brera, Via Bianchi 46, 23807 Merate (LC), Italy

F. Muleri INAF – IASF – Roma, Via Fosso del Cavaliere 100, 00130 Rome, Italy

T. Murakami Kanazawa University, Kakuma, Kanazawa, Ishikawa 920-1192, Japan

K. Murphy MIT Kavli Institute for Astrophysics and Space Research, 77 Massachusetts Avenue, 37-287 Cambridge, MA 02139, USA

L. Natalucci INAF – IASF – Roma, Via Fosso del Cavaliere 100, 00130 Rome, Italy

G. Pareschi INAF – Osservatorio Astronomico di Brera, Via Bianchi 46, 23807 Merate (LC), Italy

B. Paul Raman Research Institute, C.V. Raman Avenue, Sadashivanagar, Bangalore 560080, India

M. Pearce Royal Institute of Technology (KTH), Dept. of Physics, AlbaNova University Centre, SE 10691 Stockholm, Sweden

M. Pinchera INFN – Sezione di Pisa, Largo Bruno Pontecorvo 3, 56127 Pisa, Italy

J. Poutanen University of Oulu, Department of Physical Sciences, PO Box 3000, 90014 University of Oulu, Finland

N. Produit ISDC, Geneva Observatory, 16, Chemin d'Ecogia, CH-1290 Versoix, Switzerland

P. V. Rishin Raman Research Institute, C.V. Raman Avenue, Sadashivanagar, Bangalore 560080, India

A. Rubini INAF – IASF – Roma, Via Fosso del Cavaliere 100, 00130 Rome, Italy

J. M. Ryan University of New Hampshire, 8 College Road, Durham, NH 03824, USA

F. Ryde Oskar Klein Centre, AlbaNova University Center, 106 91 Stockholm, Sweden

Y. Saito ISAS, JAXA, 2-1-1 Sengen, Tsukuba 305-8505, Japan

T. Sakashita Kanazawa University, Kakuma, Kanazawa, Ishikawa 920-1192, Japan

H. Sakurai Yamagata University, 1-4-12 Kojirakawa, Yamagata 990-8560, Japan

S. Yu. Sazonov MPA, Garching, Germany; IKI, Moscow, Russia

J. D. Schnittman Johns Hopkins University, 3400 N Charles St, Baltimore, MD 21218, USA

H. Schnopper Deer Isle Observatory, Maine, USA

N. Schulz MIT Kavli Institute for Astrophysics and Space Research, 77 Massachusetts Avenue, 37-287 Cambridge, MA 02139, USA

E. Silver Harvard-Smithsonian Center for Astrophysics, 60 Garden Street, Cambridge, MA 02138, USA

P. Soffitta INAF – IASF – Roma, Via Fosso del Cavaliere 100, 00130 Rome, Italy

M. Soida Astronomical Observatory of the Jagiellonian University, ul. Orla 171, 30-244 Krakòw, Poland

G. Spandre INFN – Sezione di Pisa, Largo Bruno Pontecorvo 3, 56127 Pisa, Italy

R. A. Sunyaev MPA, Garching, Germany; IKI, Moscow, Russia

M. Suzuki ISAS, JAXA, 2-1-1 Sengen, Tsukuba, Ibaraki 305-8505, Japan

J. Swank NASA – Goddard Space Flight Center, Greenbelt, MD 20771, USA

G. Tagliaferri INAF – Osservatorio Astronomico di Brera, Via Bianchi 46, 23807 Merate (LC), Italy

H. Tajima SLAC National Accelerator Laboratory, 2575 Sand Hill Road M/S 29, Menlo Park, CA 94025, USA

S. Takeda ISAS, JAXA, 3-1-1 Yoshinodai, Sagamihara 229-8510, Japan

T. Tamagawa RIKEN, 2-1 Hirosawa, Wako, Saitama 351-0198, Japan

F. Tamborra Dipartimento di Fisica, Università degli Studi di Roma TRE, Via della Vasca Navale 84, 00146 Rome, Italy

Y. Tanaka Yamagata University, 1-4-12 Kojirakawa, Yamagata 990-8560, Japan

F. Tokanai Yamagata University, 1-4-12 Kojirakawa, Yamagata 990-8560, Japan

N. Toukairin Yamagata University, 1-4-12 Kojirakawa, Yamagata 990-8560, Japan

A. Trois INAF – IASF – Roma, Via Fosso del Cavaliere 100, 00130 Rome, Italy

H. Tsunemi Osaka University, Department of Earth and Space Science, Machikaneyama, Toyonaka, Osaka 560-0043, Japan

P. Ubertini INAF – IASF – Roma, Via Fosso del Cavaliere 100, 00130 Rome, Italy

M. Urbanik Astronomical Observatory of the Jagiellonian University, ul. Orla 171, 30-244 Krakòw, Poland

M. van Adelsberg Kavli Institute for Theoretical Physics, Kohn Hall, University of California, Santa Barbara, CA 93106, USA

C. Wang National Astronomical Observatories, Chinese Academy of Sciences, Beijing 100012, China and Department of Astronomy, Cornell University, Ithaca, NY 14853, USA

M. C. Weisskopf NASA – Marshall Space Flight Center, 320 Sparkmann Drive, Huntsville, AL 35805, USA

M. Weżgowiec Astronomical Observatory of the Jagiellonian University, ul. Orla 171, 30-244 Krakòw, Poland

B. Vollmer CDS, Observatoire astronomique de Strasbourg, 11 rue de l'université, 67000 Strasbourg, France

K. Wu Mullard Space Science Laboratory, University College London, Holmbury St. Mary, Dorking, Surrey RH5 6NT, UK

M. Yamauchi Osaka University, Department of Earth and Space Science, Machikaneyama, Toyonaka, Osaka 560-0043, Japan

D. Yonetoku Kanazawa University, Kakuma, Kanazawa, Ishikawa 920-1192, Japan

I. V. Zhuravleva Max Planck Institute for Astrophysics, Karl-Schwarzschild-Str. 1, Postfach 1317, 85741 Garching, Germany

A. Zoglauer Space Sciences Laboratory, University of California at Berkeley, 7 Gauss Way, Berkeley, CA 94720, USA

Preface

Advances in X-ray astronomy over the almost five decades since the first rockets were launched have been impressive in the domains of imaging, spectroscopy and timing. On the contrary, polarimetry has not progressed much since the historic results of the Columbia team headed by Bob Novick with rockets and with the OSO-8 satellite. The introduction, since *Einstein*, of X-ray telescopes and imaging detectors produced a dramatic jump in the sensitivity of X-ray missions. Polarimetry based on the conventional techniques of Bragg diffraction and Compton scattering has suffered from the increased mismatching in terms of sensitivity which resulted in the preclusion of the whole extragalactic sky. Moreover the shift from satellites stabilized on one axis to those stabilized on three axes made cumbersome the hosting of polarimeters, which needed the rotation of the whole instrument, in the focal plane of telescopes. As a consequence no polarimeter was included in the final design of *Einstein*, *Chandra* and *XMM-Newton*.

The advent of a new generation of detectors, to be combined with large area X-ray telescopes, has renewed interest in X-ray polarimetry, as demonstrated by the several polarimetric missions recently proposed. One of them, *GEMS*, has been recently selected by NASA within the SMEX program, with a launch due in 2014. There are discussions in Italy about the possibility of a national X-ray mission including polarimetry, to be launched in the same time frame. On a longer time frame and more ambitious scale, it should not be forgotten that a polarimeter is one of the focal plane instruments of the proposed ESA/JAXA/NASA International X-ray Observatory, *IXO*.

These new instruments, although based on traditional gas counters in the proportional-multiplication regime, benefit from the fantastic improvements, in terms of fine subdivision, introduced by VLSI electronics. It is worth noting that a similar evolution toward finely subdivided detectors has also rejuvenated the field of scattering polarimetry, both in the hard X-ray range and in the domain of future Compton telescopes. It is reasonable to expect that some of the proposed experiments will be launched in the next few years, along with *GEMS*, providing the scientific community with real data.

In parallel with this evolution of the experimental landscape, in recent years we have had significant progress in theoretical studies. The improved knowledge of X-ray emission in astrophysical sources, deriving from imaging, spectroscopic, and timing data, often makes it possible to predict in detail the polarization properties of such emission. For instance, the present knowledge on magnetars, on accretion onto black holes or on jets in blazars and in microquasars allows us to identify polarization measurements with the capability to confirm or disprove current scenarios.

We believed therefore that, five years after the Conference in Stanford, the time was ripe to organize a conference on X-ray polarimetry with the aim of discussing the present status and perspectives of instruments as well as reviewing and discussing theoretical models. Our aim was not only to gather the community actively involved in the field (both on the instrumental and theoretical sides), but also and foremost to stimulate the interest of a wider community, which so far suffered the lack of observational perspectives. 'The Coming of Age of X-ray Polarimetry' conference was held in Rome, at the Center for American Studies in Via Caetani 32, on 27-30 April 2009. This book collects the Proceedings of the conference.

It is a pleasure here to thank the Center for American Studies, chaired by Senator Professor Giuliano Amato and directed by Professor Karim Mezran, for allowing us to hold our conference in their beautiful library, even if our topics were a little bit off their humanistic mainstream. We appreciated very much the courtesy and willingness to help of the library staff; the participants were also stunned by the richness and beauty of the library rooms inside the historic building.

We also thank the representatives of organizing science institutions and of ASI who kindly attended the opening of the conference.

Gabriella Ardizzoia and Rachele Millul performed, with their usual competence and kindness, the never sufficiently appreciated secretarial work. Riccardo Campana, Ettore Del Monte, Yuri Evangelista and Fabio Muleri provided invaluable help in the organization of the conference and in the editing of this book. Andrea Marinucci and Francesco Tamborra also helped during the conference. Last but not least, we want to thank Sergio di Cosimo. Without him, neither the conference nor this book could ever have became a reality.

The success of a conference is eventually determined by the number of participants and the quality of their contributions, and by the liveliness of the scientific discussions, both during the formal sessions and at coffee breaks, lunches, dinners, etc. It is not appropriate for us to judge the success of the conference we organized. We can only say that at the end we felt extremely relieved and satisfied.

Therefore, we want to thank warmly all the participants for this, as well as for providing their contributions in written form, which the reader will find in the following pages.

Ronaldo Bellazzini
Enrico Costa
Giorgio Matt
Gianpiero Tagliaferri

1

X-ray polarimetry: historical remarks and other considerations

M. C. Weisskopf

NASA/Marshall Space Flight Center

We briefly discuss the history of X-ray polarimetry for astronomical applications including a guide to the appropriate statistics. We also provide an introduction to some of the new techniques discussed in more detail elsewhere in these proceedings. We conclude our discussion with our concerns over adequate ground calibration, especially with respect to unpolarized beams, and at the system level.

1.1 Introduction

Sensitive X-ray polarimetry promises to reveal unique and crucial information about physical processes in and structures of neutron stars, black holes, and ultimately all classes of X-ray sources. We do not review the astrophysical problems for which X-ray polarization measurements will provide new insights, as these will be discussed in some detail in many of the presentations at this conference.

Despite major progress in X-ray imaging, spectroscopy, and timing, there have been only modest attempts at X-ray polarimetry. The last such dedicated experiment, conducted by Bob Novick (Columbia University) over three decades ago, had such limited observing time (and sensitivity) that even ~10% degree of polarization would not have been detected from some of the brightest X-ray sources in the sky. Statistically significant X-ray polarization was detected in only one X-ray source, the Crab Nebula.

1.1.1 History

The first positive detection of X-ray polarization[11] was performed in a sounding-rocket experiment that viewed the Crab Nebula in 1971. Using the X-ray polarimeter

X-ray Polarimetry: A New Window in Astrophysics, eds. R. Bellazzini, E. Costa, G. Matt and G. Tagliaferri. Published by Cambridge University Press.

on the Orbiting Solar Observatory (OSO)-8, this result was confirmed[15] with a 19-σ detection ($P = 19.2\% \pm 1.0\%$), conclusively proving the synchrotron origin of the X-ray emission. Unfortunately, because of low sensitivity, only 99%-confidence upper limits were found for polarization from other bright X-ray sources (e.g. $\leq$13.5% and $\leq$60% for accreting X-ray pulsars Cen X-3 and Her X-1, respectively[13]). Since that time, although there have been several missions that had planned to include X-ray polarimeters – such as the original *Einstein* Observatory and *Spectrum-X* (v1) – no X-ray polarimeter has actually managed to be launched.

1.2 Instrumental approaches

There are a limited number of ways to measure linear polarization in the 0.1–50 keV band, sufficiently sensitive for astronomical sources. We discuss four techniques here, but see also G. Frazier's contribution for a discussion of other techniques. We emphasize that *meaningful X-ray polarimetry is difficult*:

(1) In general, we do *not* expect sources to be strongly ($\gg$10%) polarized. For example, the maximum polarization from scattering in an optically-thick, geometrically-thin, accretion disc is only about 10% at the most favorable (edge-on) viewing angle. Hence, most of the X-rays from such a source carry no polarization information and thus merely increase the background (noise) in the polarization measurement.

(2) With one notable exception – namely, the Bragg-crystal polarimeter – the modulation of the polarization signal in the detector, the signature of polarization, is much less than 100% (typically, 20%–40%) (and *energy-dependent*) even for a 100%-polarized source.

(3) The degree of linear polarization is positive definite, so that any polarimeter will always measure a (not necessarily statistically significant) polarization signal, even from an unpolarized source. Consequently, the statistical analysis is more unfamiliar to X-ray astronomers. For a detailed discussion of polarimeter statistics see [12]. The relevant equations are also summarized in slides 18–20 of our presentation.[1]

Concerning the statistics, one of the most important formulas is the minimum detectable polarization (MDP) at a certain confidence level. In the absence of any instrumental systematic effects, the 99%-confidence level MDP,

$$\mathrm{MDP}_{99} = \frac{4.29}{MR_S}\left[\frac{R_S + R_B}{T}\right]^{1/2}, \tag{1.1}$$

where the "modulation factor", M, is the degree of modulation expected in the absence of background for a 100%-polarized beam, R_S and R_B are, respectively, the source and background counting rates, and T is the observing time.

The MDP is *not* the uncertainty in the polarization measurement, but rather the degree of polarization which has, in this case, only a 1% probability of being equalled or exceeded by chance. One may form an analogy with the difference between measuring a handful of counts, say 9, with the Chandra X-Ray Observatory and thus having high confidence (many sigmas) that one has detected a source, yet understanding that the value of the flux is still highly uncertain – 30% at the 1-sigma level in this example. We emphasize this point because the MDP often serves as *the* figure of merit for polarimetry. While it is *a* figure of merit that is useful and meaningful, a polarimeter appropriate for attacking astrophysical problems must have an MDP significantly smaller than the degree of polarization to be measured, a point that is often overlooked.

As P. Kaaret noted during his summary (p. 251), consider an instrument with no background, a modulation factor of 0.5, and the desire to obtain an MDP of 1%: this requires detection of 10^6 counts! The statistics will be superb, but the understanding of the response function needs to be compatible. I know of no observatory where the response function is known so well that it may deal with a million-count spectrum.

1.2.1 Crystal polarimeters

The first successful X-ray polarimeter for astronomical application utilized the polarization dependence of Bragg reflection. In[14] we describe the first sounding-rocket experiment using a crystal polarimeter, the use of which Schnopper & Kalata[10] had first suggested. The principle of operation is summarized in slide 7 of the presentation and a photograph of the one of two crystal panels that focused the X-rays onto a proportional counter is shown in slide 8.

Only three crystal polarimeters (ignoring the crystal spectrometer on Ariel-5 which also served as a polarimeter) have ever been constructed for extra-solar X-ray applications. Only two – both using graphite crystals without X-ray telescopes – were ever flown (sounding rocket[14]; OSO-8 satellite[15]; *Spectrum-X* (v1) (not flown)[9]).

One of the virtues of the crystal polarimeter is that, for Bragg angles near 45 degrees, modulation of the reflected flux approaches 100%. From Equation (1.1) we see that this is very powerful, *all other things being equal*, since the MDP scales directly with the inverse of the modulation factor but only as the square root of the other variables. Obviously, a disadvantage is the narrow bandwidth for Bragg reflection.

1.2.2 Scattering polarimeters

There are two scattering processes from bound electrons: coherent and incoherent scattering. A comprehensive discussion of these processes may be found in many textbooks (see, e.g.,[8]).

Various factors dominate the consideration of the design of a scattering polarimeter. The most important are these:

(1) to scatter as large a fraction of the incident flux as possible while avoiding multiple scatterings;
(2) to achieve as large a modulation factor as possible;
(3) to collect as many of the scattered X-rays as possible; and
(4) to minimize the detector background.

The scattering competes with photoelectric absorption in the material, both on the way in and on the way out. The collection efficiency competes with the desire to minimize the background and to maximize the modulation factor.

Not counting more recent higher-energy payloads being developed for balloon and future satellite flights discussed elsewhere in these proceedings, only two polarimeters of this type have ever been constructed for extra-solar X-ray applications. The only ones ever flown were suborbital in 1968, see [1], in 1969 see [16], and in 1971 see, e.g., [11]. The scattering polarimeter[9] built for the *Spectrum-X* (v1) satellite was never flown.

The virtue of the scattering polarimeter is that it has reasonable efficiency over a moderately large energy bandwidth, thus facilitating energy-resolved polarimetry. The principal disadvantage is a modulation factor less than unity, since only for scattering into 90 degrees will the modulation approach 1.0 in the absence of background, and for a 100%-polarized beam. To obtain reasonable efficiency requires integrating over a range of scattering angles and realistic modulation factors are under 50%, unless the device is placed at the focus of a telescope. The modulation factor for the scattering polarimeter on *Spectrum-X* (v1) reached ~75%. At the focus it is feasible to make the scattering volume small which then limits the range of possible scattering angles.

1.3 New approaches

In this conference we will hear detailed presentations of a number of new approaches to X-ray (and higher-energy) polarimeters. We mention two of these approaches here.

1.3.1 Photoelectron tracking

The angular distribution (e.g. [7]) of the K-shell photoelectron emitted as a result of the photoelectric absorption process depends upon the polarization of the incident photon. The considerations for the design of a polarimeter that exploits this effect are analogous to those for the scattering polarimeter. In this case the competing effects are the desire for a high efficiency for converting the incident X-ray flux into photoelectrons and the desire for those photoelectrons to travel large distances before interacting with elements of the absorbing material.

Here we concentrate on polarimeters that use gas mixtures to convert the incident X-rays to photoelectrons. Currently there are three approaches to electron tracking polarimetry that use this effect.

To our knowledge, the first electron tracking polarimeter specifically designed to address polarization measurements for X-ray astronomy, and using a gas as the photoelectron-emitting material, was that designed by Austin and Ramsey at NASA/Marshall Space flight Center[2], (see also [3; 4]). They used the light emitted by the electron avalanches which occur after the release of the initial photoelectron in a parallel-plate proportional counter. The light was then focused and detected by a CCD camera.

Another gas-detector approach, first discussed by [6], uses "pixillated" proportional counters (gas electron multipliers) to record the avalanche of secondary electrons that result from gas-multiplication in a high field after drift into a region where this multiplication may take place. A third approach to such devices was suggested by Black[5] and exploits time of flight, and rotates the readout plane to be at right angles to the incident flux. This device sacrifices angular resolution when placed at the focus of a telescope but gains efficiency by providing a greater absorption depth.

Detecting the direction of the emitted photoelectron (relative to the direction of the incident flux) is not simple because the electrons, when they interact with matter, give up most of their energy at the *end* of their track, not the beginning. Of course, in the process of giving up energy to the local medium in which the initial photo-ionization took place, the electron changes its trajectory, thus losing the information as to the initial direction and hence polarization.

It is instructive to examine the image of a track: Figure 1.1 shows one obtained under relatively favorable conditions with an optical imaging chamber. The initial photoionization has taken place at the small concentration of light to the *left* in the figure. The size of the leftmost spot also indicates the short track of the Auger electron. As the primary photoelectron travels through the gas, it either changes direction through elastic scattering and/or both changes direction and loses energy through ionization. As these interactions occur, the path strays from the direction determined by the incident photon's polarization. Of course, the ionization process is energy dependent and most of the electron's energy is lost at the end, not the beginning, of its track. It should be clear from this discussion that, even under favorable conditions – where the range of the photoelectron is quite large compared to its interaction length – the ability to determine a precise angular distribution depends upon the capability and sophistication of the track-recognition software, not just the

Figure 1.1 The two-dimensional projection of a track produced when a 54 keV X-ray was absorbed in a mixture of argon (90%), CH_4 (5%), and trimethylamine (5%) at 2 atm. This particular track is $\simeq$14 mm in length.

spatial resolution of the detection system. The burden falls even more heavily on the software at lower energies, where the photoelectron track becomes very short and diffusion in the drifting photoelectron cloud conspires to mask the essential track information. Thus, the signal processing algorithms (rarely discussed) form an important part of the experiment, are a possible source of systematic effects, and may themselves reduce the efficiency for detecting polarized X-rays.

Polarimeters exploiting the photoelectric effect have been discussed in the literature, and two will be discussed in this conference. However, *no device of this type has ever been flown and those built have undergone relatively limited testing in the laboratory*. In some cases, performance claims depend more upon Monte-Carlo simulations than actual experiments. We eagerly await experimental verification of performance at lower energies, around 2–3 keV, where the overall sensitivity peaks.

1.3.2 Transmission filters

The potential advent of extremely large-area telescope missions, such as the International X-ray Observatory (IXO), may provide an opportunity to exploit the polarization dependence of narrow-band dichroic transmission filters, as discussed by G. Frazier elsewhere in this conference. The extremely narrow band (a consequent requirement for a detector of a few eV resolution), the low efficiency of the filter, and the association with a major observatory are all issues to be addressed. Regarding this last point, the history of X-ray polarimetry on major observatories has not been positive. The OSO-8 polarimeter received only a very limited amount of observing time as the result conflicting pointing requirements. The *Spectrum-X* (v1) polarimeter, one of at least two detectors at the focus of its telescope, was allocated only 11 days in the plan for the first year's observing. The polarimeter on the original *Einstein* observatory was "descoped". No polarimeter was selected to be part of either the *Chandra* or *XMM-Newton* missions, despite the important capabilities that each of these missions – especially *XMM-Newton* with its larger collecting area – might provide.

1.4 Concluding remarks: systematic effects

Only a few people in the world have any flight experience with X-ray polarimeters and it behoves one to take advantage of this experience. Precision X-ray polarimetry depends crucially on the elimination of potential systematic effects. This is especially true for polarimeters with modulation factors less than unity. Consider a polarimeter with a modulation factor of 40%, and a 5%-polarized source. In the absence of any background, this means one is dealing with a signal of only 2% modulation in the detector. To validate a detection means that systematic effects must be understood and calibrated well below the 1% level, a non-trivial task. If present, systematic effects alter the statistics discussed previously, further reducing sensitivity: it is harder to detect two signals at the same frequency than one!

To achieve high accuracy requires extremely careful calibration *with unpolarized beams*, as a function of energy, at the detector and at the system level! For example, suppose that the systematic error in the measured signal of an unpolarized source were 1%. Then for a modulation factor of 40%, the 3-sigma upper limit due to systematic effects alone would be 7.5% polarization. Thus, if a polarimeter is to measure few-percent-polarized sources with acceptable confidence, systematic effects in the modulated signal must be understood at $\leq$0.2%. Careful calibration – over the full energy range of performance – is essential.

Notes

1. http://projects.iasf-roma.inaf.it/xraypol/

References

[1] Angel, J.R.P., Novick, R., vanden Bout, P., & Wolff, R. (1969). *Phys. Rev. Lett.* **22**, 861.
[2] Austin, R.A. & Ramsey, B.D.A. (1992). *Proc SPIE* **1743**, 252.
[3] Austin, R.A. & Ramsey, B.D.A. (1993). *Optical Engineering* **32**, 1900.
[4] Austin, R.A., Minamitani, T., & Ramsey, B. D. (1993). *Proc. SPIE* **2010**, 118.
[5] Black, J. K. (2007). *Journal of Physics: Conference Series* **65**, 012005.
[6] Costa, E., Soffitta, P., Bellazzini, R., Brez, A., Lumb, N., & Spandre, G. (2001). *Nature* **411**, 662.
[7] Heitler, W. (1954). *The Quantum Theory of Radiation*, 3rd edn. New York, Dover Publications, Inc.
[8] James, R.W. (1965). *The Optical Principles of the Diffraction of X-rays*. Ithaca, NY, Cornell University Press.

[9] Kaaret, P.E. et al. (1994). *Proc. SPIE* **2010**, 22.
[10] Schnopper, H.W. & Kalata, K. (1969). *A.J.* **74**, 854.
[11] Novick, R., Weisskopf, M.C., Berthelsdorf, R., Linke, R., Wolff, R.S. (1972). *Ap.J.* **174**, L1.
[12] Weisskopf, M.C., Elsner, R.F., Kaspi, V.M., O'Dell, S.L., Pavlov, & G.G., Ramsey, B.D. (2009). In *Neutron Stars and Pulsars*, ed. W. Becker. Berlin, Heidelberg, Springer-Verlag.
[13] Silver, E.H., Weisskopf, M.C., Kestenbaum, H.L., Long, K.S., Novick, R., & Wolff, R.S. (1979). *Ap.J.* **232**, 248.
[14] Weisskopf, M.C., Berthelsdorf, R., Epstein, G., Linke, R., Mitchell, D., Novick, R., & Wolff, R.S. (1972). *Rev. Sci. Instr.* **43**, 967.
[15] Weisskopf, M.C., Cohen, C.G., Kestenbaum, H.L., Long, K.S., Novick, R., & Wolff, R.S. (1976). *Ap.J.* **208**, L125.
[16] Wolff, R.S., Angel, J.R.P., Novick, R., & vanden Bout, P. (1970). *ApJ* **60**, L21.

Part I

Polarimetry techniques

2

Scattering polarimetry in high-energy astronomy

M. L. McConnell

University of New Hampshire

At energies above a few keV, photon scattering provides an important means of measuring photon polarization. Here we review the fundamental principles of scattering polarimetry, present a summary of some of the more recent results, and review the prospects for new experimental results within the next few years.

2.1 Introduction

It has now been a little more than 100 years since the first reported laboratory measurements of γ-ray polarization based on the use of Compton scattering[2]. Although the first efforts to apply this technique in high-energy (X-ray and γ-ray) astronomy took place almost 40 years ago, this area of research is still in its infancy. This is a notoriously difficult area of research, compounded by the combination of low flux levels, high background rates and instrumental artifacts that can often mimic a polarization signature. Nonetheless, all of the recent polarization measurements have relied on this approach.

2.2 Experimental considerations

Scattering polarimetry relies on experimental methods that are based on the scattering of photons off electrons. The scattering of photons off single electrons is variably referred to as Compton scattering or, at lower energies, as Thomson scattering. Thomson scattering is the classical limit of Compton scattering in which there is no loss of energy to the electron. At lower energies, coherent scattering off

X-ray Polarimetry: A New Window in Astrophysics, eds. R. Bellazzini, E. Costa, G. Matt and G. Tagliaferri.
Published by Cambridge University Press.

the atomic electron cloud can also be important. This is known as Rayleigh scattering. Scattering polarimeter designs are based on the fact that photons scattered by any of these processes all tend to scatter at right angles to the incident photon's polarization vector.

A scatter polarimeter often (but not always) consists of two distinct detectors (or detector elements) to determine the energies of both the scattered photon and the scattered electron. The scattering detector provides the medium for the Compton interaction to take place, often composed of some low-Z material to minimize photoelectric interactions. The second detector, usually referred to as the absorber or calorimeter, absorbs the full energy of the scattered photon and, for this reason, is often composed of some high-Z material. The relative placement of the two detectors defines the scattering geometry and permits a measurement of the azimuthal scatter angle (η). A simplified arrangement is shown in Figure 2.1, where detector A represents a scattering element and detector B represents a calorimeter element. This arrangement shows the incident photon scattering at a scatter angle (θ) of 90°, a preferred geometry which serves to maximize the azimuthal modulation of polarized flux. The distribution of azimuthal scatter angles (η) carries a polarization signature from which the magnitude and direction of the polarization can be derived. An ideal event is one in which the photon scatters once in a scattering element and is subsequently absorbed in the calorimeter element, providing a measure of the full energy of the incident photon. Low-Z scintillator material (usually plastic) provides relatively poor energy resolution, so the energy resolution of

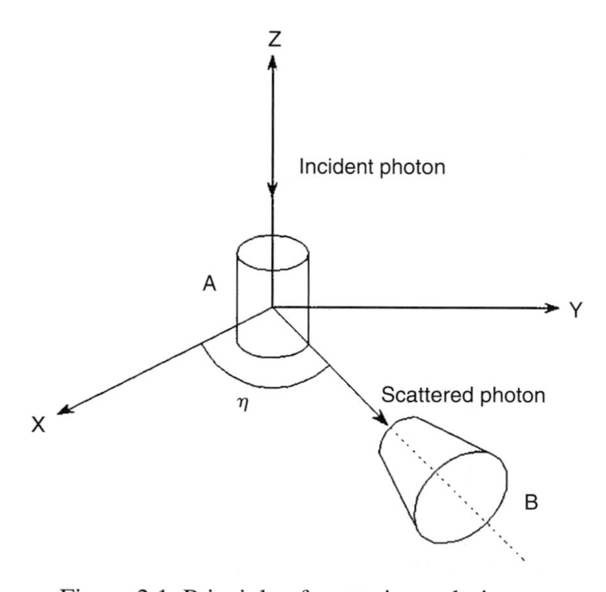

Figure 2.1 Principle of scattering polarimetry.

designs that incorporate low-Z detectors is often limited. Better energy resolution can be obtained, with some sacrifice in polarimeter performance, by using high-Z materials for both scattering and absorbing elements.

Experimentally, we can distinguish two different detection regimes: (1) passive scattering, in which the energy transferred to the electron is not measured; and (2) active scattering, in which the energy transferred to the electron is measured by some active scattering material. There is a significant advantage to active scattering, in that the instrumental background can be reduced by imposing a coincidence requirement. A very small coincidence window (to minimize accidental background events) can be achieved using fast detector materials (such as plastic scintillator). However, active scattering elements require that a sufficiently large amount of energy is lost to the scattered electron. At incident photon energies below 50 keV, measuring the small amount of energy lost to the electron (for preferred Compton scatter angles near 90°) becomes increasingly difficult. Low energy scattering polarimeters therefore often rely on passive scattering elements to reduce unnecessary complexity.

Finally, we note that instrumental systematics can be a severe problem in that they can often mimic the sort of polarization signal that is being measured. Although some of these systematic issues can be calibrated (such as variations in efficiencies between detector elements), the preferred approach to handling this problem is to rotate the instrument about its pointing axis.

2.3 Summary of recent results

Recent measurements have been restricted to bright sources: GRBs, solar flares and the Crab Nebula/Pulsar. A listing of recent polarization measurements, all based on Compton polarimeters, is given in Table 2.1. It is beyond the scope of this paper to review the full details of these results, but here we shall review some of the important experimental aspects of these measurements.

2.3.1 RHESSI

Although originally designed as a hard X-ray solar imager, the Ramaty High-Energy Solar Spectroscopic Imager (RHESSI), with its nine-element Ge detector array coupled to a rotating spacecraft platform, affords an opportunity for polarization measurements. There are actually two different (and complementary) techniques for measuring polarization with RHESSI. At higher energies (above ∼100–200 keV), scattering events between the Ge detectors within the spectrometer array can be used to measure polarization. Coincident events are identified in software. The lack of significant amounts of shielding surrounding the Ge array means that

Table 2.1 *Recent measurements*

Instrument	Energy Range	Source	Ref
RHESSI	0.2–1 MeV	GRB 021206	Coburn and Boggs [10]
RHESSI	0.2–1 MeV	Solar Flares	Boggs *et al.* [5]
RHESSI	100–350 keV	Solar Flares	Suarez-Garcia *et al.* [33]
CORONAS-F	20–100 keV	Solar Flares	Bogomolov *et al.* [7]
INTEGRAL/SPI	100–350 keV	GRB 041219a	Kalemci *et al.* [21]
INTEGRAL/SPI	0.1–1 MeV	GRB 041219a	McGlynn *et al.* [27]
INTEGRAL/SPI	0.1–1 MeV	GRB 061122	McGlynn *et al.* [28]
INTEGRAL/SPI	0.1–1 MeV	Crab	Dean *et al.* [11]
INTEGRAL/IBIS	200–800 keV	GRB 041219a	Götz *et al.* [16]
INTEGRAL/IBIS	200–800 keV	Crab	Forot *et al.* [13]

this mode is sensitive to photons arriving over a large area of the sky. At lower energies (below $\sim$100 keV) efforts to measure polarization rely on the front/rear segmentation of the Ge detectors and the presence of a small block of passive Be with the spectrometer array[25]. Collimation of the passive Be restricts low-energy measurements to a small FoV ($\sim 1^\circ$) centered on the Sun. In both cases, the rotation of the RHESSI spacecraft (required for hard X-ray imaging with RHESSI's rotation modulation collimators) greatly facilitates effective polarization measurements by reducing systematic uncertainties and providing a more uniform sampling in the azimuthal direction. All recent results from RHESSI were derived using the high-energy polarimetry mode. Efforts to use the low-energy polarimetry mode have so far been plagued with various instrumental issues. RHESSI was launched in 2002 and remains an active mission, so future measurements are still possible.

2.3.2 INTEGRAL

Launched in 2002, the INTEGRAL spacecraft carries four separate instruments, two of which are capable of polarization measurements – the SPectrometer on Integral (SPI) and the Imager on Board the Integral Spacecraft (IBIS). As was the case with RHESSI, the INTEGRAL design was not driven by polarimetry requirements, but it nonetheless retains significant capability for polarimetry. INTEGRAL differs from RHESSI in two important ways. First, the INTEGRAL spacecraft does not rotate, so instrumental systematics may be more problematic. Second, both SPI and IBIS are coded aperture imaging experiments, a feature that can be used to improve the signal-to-noise of weak sources. Because the detectors are heavily shielded, sources must lie within the imaging FoV for effective polarization measurements.

Using its array of 18 Ge detectors, polarimetry with the SPI instrument[36] is similar to the high-energy mode of RHESSI. Coincident events between adjacent Ge detectors define the azimuthal distribution of scattered photons. As with the RHESSI high-energy mode, the SPI data do not permit the identification of the first interaction site and coincident events in adjacent detectors (occurring within 350 ns) are identified in software. Polarimetry with IBIS[35] relies on a Compton scatter mode in which photons are scattered from an upper layer of CdTe detectors to a lower layer of CsI detectors (with the two layers separated by 12 cm). Compton scatter events are selected based on a single interaction in the CdTe layer (ISGRI) and a single interaction in the CsI layer (PICsIT), both occurring within a relatively wide hardware-based coincidence timing window (3.5 μs). In the case of intense GRBs, the throughput for IBIS events can be limited by telemetry bandwidth, so significant deadtime corrections are often required.

2.4 Summary of upcoming missions

In addition to both RHESSI and INTEGRAL, which are expected to remain active for at least a few more years, there are several ongoing efforts that are focused on the goal of obtaining high-energy polarization measurements. Here we summarize those missions (listed in Table 2.2) that have either recently been launched or that are currently scheduled for flight within the next few years.

PENGUIN-M. Recently launched (February, 2009) on the Coronas-Photon mission, the PENGUIN-M experiment will be able to measure 20–150 keV polarization from solar flares[12]. The design incorporates an array of three plastic scintillators surrounded by six NaI scintillators. It appears to be designed specifically for studying solar flare polarization.

GRAPE. The latest version of GRAPE (the Gamma RAy Polarimetry Experiment) is a modular design that includes an array of small plastic and CsI elements read out by a single multi-anode PMT[3]. A prototype module was flown on a high-altitude balloon in 2007. A large array of detector modules (of up to 16 modules) is currently being fabricated to be flown on a high-altitude balloon at the end of 2011[4]. It will be configured with a narrow FoV collimator to take data on the Crab and other bright point sources. For subsequent ultra-long duration balloon flights, the array will be expanded and configured without a collimator in order to measure polarization from γ-ray bursts (GRBs) and solar flares. A version of the GRAPE design has been proposed as part of the POET mission, whose primary goal is to measure GRB polarization[20; 26].

PHENEX. Although PHENEX (Polarimetry of High ENErgy X-rays) has evolved independently, the current design is very similar to that of GRAPE[23]. An array of five

Table 2.2 *Upcoming Missions*

Project	Energy Range	Target	First Flight(s)	Ref
PENGUIN-M	20–150 keV	flares	2009	[12]
GRAPE	50–500 keV	Crab/GRBs/flares	2007/2011	[3]
PHENEX	40–200 keV	Crab/Cyg X-1	2006/2009	[18]
NCT / GRIPS	0.2–1.0 MeV	Crab/flares	2009/2012	[9]
PoGOlite	25–80 keV	Crab/Cyg X-1	2010	[22]
GAP	50–500 keV	GRBs/flares	2011	[38]
Astro-H	50–500 keV	point sources	2013	[34]

polarimeter modules was flown on a high altitude balloon in 2006, which collected data on the Crab for about one hour. Polarization results were inconclusive[18]. A larger array is planned for flight in May of 2009. A modified version of the PHENEX design has been proposed as part of the POLARIS mission[19].

NCT. The Nuclear Compton Telescope (NCT) is a balloon-borne Compton telescope (0.2–10 MeV), based on the use of high-resolution Ge detectors, that is also capable of polarimetry[9]. This detector design is also incorporated into an imaging instrument that will be capable of arcsecond imaging polarimetry of solar flares (the Gamma-Ray Imager/Polarimeter For Solar Flares or GRIPS). The latest balloon flight of NCT was scheduled for May, 2009. The first flight of GRIPS is scheduled for 2012.

PoGOLite. PoGOLite, designed as a balloon experiment, is optimized for point-like sources in the 25–80 keV range[22]. It features a large active shielding arrangement, composed of plastic and BGO scintillator, that defines a FoV of $\sim 2°$. Both the scattering and calorimeter elements are plastic scintillator. The full instrument will consist of 271 detector elements. A “pathfinder” balloon flight (from Sweden to Canada) is scheduled for 2010 in a configuration with 61 elements[31].

GAP. GAP (GAmma-ray burst Polarimeter) is an experiment[38] that will be flown on the Japanese solar powered sail satellite, IKAROS. It is currently scheduled for launch in 2010 on a mission that will take it to Jupiter. The GAP instrument is a plastic/CsI polarimeter that can provide polarization measurements in the energy range 50–300 keV. GAP features a large-area plastic scintillator (14 cm in diameter) surrounded by an array of 12 CsI detectors. With a large FoV, it is expected to make measurements of about four GRBs per year with a MDP of 50%. The flight module has just recently been completed[39].

Astro-H. The Japanese Astro-H mission, scheduled for launch in 2013, will carry an array of small soft γ-ray Compton cameras (known as the Soft Gamma-ray Detector or SGD) that have been designed primarily as spectrometers, but will also be capable of polarization measurements[29; 34]. The scattering material is Si and the absorber material is CdTe. With a narrow FoV of 0.5–1.0°, the experiment will be most sensitive to polarization in the energy range 50–200 keV.

2.5 Potential new missions

In addition to these active missions, there are several projects, in various stages of development, seeking to develop experiments for future flight programs. POLAR is a modular polarimeter design, similar in many respects to GRAPE and PHENEX, except that is uses plastic scintillator elements as both scatterers and absorbers[32]. TSUBAME is a proposed Japanese university-built small satellite mission that is designed to do GRB polarimetry in the 30–100 keV energy band, using an arrangement of plastic and CsI detectors similar to that of GRAPE and PHENEX[1]. POLIX is another proposed small satellite mission (from India) that carries collimated scattering polarimeters for studying point sources of emission[30]. LaPOLCA represents a laboratory program to develop the technology for balloon-borne polarization measurements by combining a Laue lens imaging system with polarization-sensitive CZT detectors[8]. Polarization measurements have also been considered as a feature of the EXIST mission (one of the candidates for NASA's Black Hole Finder Probe) using a planar array of CZT detectors[14]. HX-POL is a design for a balloon-borne instrument utilizing Si and CZT detectors for polarization measurements[24; 15]. Both the Advanced Compton Telescope (ACT) concept[6] and the Gamma-Ray Investigations of Gamma-ray Bursts with Polarimetry and Spectroscopy (GRIPS) mission[17] represent mission concepts based on Compton imaging designs that would be capable of doing polarimetry.

2.6 Future prospects

In recent years, there has been a renewed interest in the field of high-energy astronomical polarimetry, largely resulting from several tantalizing new measurements. Experimentalists are now working hard to develop instrumentation that is specifically optimized for polarization studies. These new experiments promise to provide higher quality data that will hopefully move the field from an exploratory phase into a discovery phase. Scattering polarimeters, such as those described here, are expected to make significant contributions to this rapidly evolving field of study.

References

[1] Arimoto, M., et al. (2008). *AIP Conf. Proc.* **1000**, 607–610.
[2] Barkla, C. G. (1906). *Proc. Royal Soc. A* **77**, 247–255.
[3] Bloser, P. F., et al. (2009). *Nucl. Ins. Meth.* **A600**, 424–433.
[4] Bloser, P. F., et al. (2010). Ch 46, 304–321.
[5] Boggs, S. E., Coburn, W., & Kalemci E. (2006). *Ap. J.* **638**, 1129–1139.
[6] Boggs, S. E., et al. (2006). *Proc. SPIE* **6266**, 626624.
[7] Bogomolov, A. V., et al. (2003). *Sol. Sys. Res.* **37**, 112–120.
[8] Caroli, E., et al. (2010).
[9] Chang, H.-K., et al. (2007). *Adv. Sp. Res.* **40**, 1281–1287.

[10] Coburn, W. & Boggs, S. E. (2003). *Nature* **423**, 415–417.
[11] Dean, A. J., et al. (2008). *Science* **321**, 1183–1185.
[12] Dergachev, V. A., et al. (2009). *Bull. Russian Acad. Sci.: Physics* **73**, 419–421.
[13] Forot, F., et al. (2008). *Ap. J.* **688**, L29–L32.
[14] Garson, A. B., et al. (2010). These proceedings.
[15] Garson, A. B., et al. (2010). These proceedings.
[16] Götz, D., et al. (2009). *Ap. J.* **695**, L208–L212.
[17] Greiner, J., et al. (2010). Ch 48, 327–332
[18] Gunji, S., et al. (2007). *Proc. SPIE* **6686**, 668618.
[19] Hayashida, K., et al. (2010). Ch 49, 333–338.
[20] Hill, J., et al. (2008). *AIP Conf. Proc.* **1065**, 331–337.
[21] Kalemci, E., et al. (2007). *Ap. J. Supp.* **169**, 75–82.
[22] Kamae, T., et al. (2008). *Astroparticle Phys.* **30**, 72–84.
[23] Kishimoto, Y., et al. (2007). *IEEE Trans. Nucl. Sci.* **54**, 561–566.
[24] Krawczynski, H., et al. (2009). *IEEE NSS '08 Conf. Rec*, 111–117.
[25] McConnell, M. L., et al. (2002). *Solar Physics* **210**, 125–142.
[26] McConnell, M. L., et al. (2010). Ch 53, 355–358.
[27] McGlynn, S., et al. (2007). *Astron. Astrophys.* **466**, 895–904.
[28] McGlynn, S., et al. (2009). Submitted to *Astron. Astrophys.*, [arXiv:0903.5218].
[29] Odaka, H., et al. (2007). *Nucl. Ins. Meth.* **A579**, 878–885.
[30] Paul, B., et al. (2010). These proceedings.
[31] Pearce, M., et al. (2010). Ch 42, 291–298.
[32] Produit, N., et al. (2005). *Nucl. Ins. Meth.* **A550**, 616–625.
[33] Suarez-Garcia, E., et al. (2006). *Solar Physics* **239**, 149–172.
[34] Tajima, H., et al. (2010). Ch 40, 275–283.
[35] Ubertini, P. et al. (2003). *Astron. Astrophys.* **411**, L131–L139.
[36] Vedrenne, G., et al. (2003). *Astron. Astrophys.* **411**, L63–L70.
[37] Winkler, C., et al. (2003). *Astron. Astrophys.* **411**, L1–L6.
[38] Yonetoku, D., et al. (2006). *Proc. SPIE* **6266**, 62662C.
[39] Yonetoku, D., et al. (2010). Ch 50, 339–344.

3

Photoelectric polarimeters

R. Bellazzini & G. Spandre

INFN-Pisa, Italy

Polarimetry is widely considered a powerful observational technique in X-ray astronomy, useful to enhance our understanding of the emission mechanisms, geometry and magnetic field arrangement of many compact objects. However, the lack of suitable sensitive instrumentation in the X-ray energy band has been the limiting factor for its development in the last three decades. Up to now, polarization measurements have been made almost exclusively with Bragg crystal or Thomson scattering techniques and so far the only unambiguous detection of X-ray polarization has been obtained by the Weisskopf group in 1976 from observations of the Crab Nebula[1]. Only recently, with the development of a new class of high-sensitivity imaging detectors, the possibility to exploit the photoemission process to measure the photon polarization has become a reality. This paper will review the history of X-ray photoelectric polarimetry and discuss some innovative experimental techniques.

3.1 Introduction

X-ray astronomy deals with the most violent and compact spots in the Universe, such as the surfaces of pulsars, the close orbits around giant black holes, and the blast waves of supernova explosions. By making efficient use of the few photons emitted by disks around black holes and other objects, astronomers have successfully applied photometry, imaging and spectroscopy to these energetic and often variable sources. But polarimetry has been largely ignored at X-ray wavelengths because of the inefficiency of the existing instruments. Conventional X-ray polarimetry which relies on Bragg reflection or Thomson scattering actually suffer from low sensitivity

X-ray Polarimetry: A New Window in Astrophysics, eds. R. Bellazzini, E. Costa, G. Matt and G. Tagliaferri. Published by Cambridge University Press.

and inability to simultaneously combine imaging with polarization measurement. Being dominant in the low X-ray energy region (below 50 keV) and highly sensitive to polarization, the photoelectric effect can be, conversely, the basis of a new class of X-ray polarimeters. As a result of the photo-ionization, the electron is ejected, preferentially, in the direction of the electric field vector of the incident X-ray. A measurement of the photoelectron emission direction provides, then, a measure of the polarization state of the photon. Recently several new techniques have been developed which combine a good sensitivity of detection with unprecedented imaging capability. These techniques allow us to finely reconstruct the photoelectron track and derive angle and degree of polarization of the incoming radiation.

3.2 Polarization vs photoelectric effect

In 1926, by irradiating a cloud chamber with polarized X-rays, Auger[2] discovered, together with the effect of dielectronic recombination, the dependence of the emission direction of the photoelectrons on the linear polarization of the photons. The simplified analytical expression for the differential cross section distribution of the photoelectric effect, in the non-relativistic approximation, is[3]:

$$\frac{\partial\sigma}{\partial\Omega} = r_0^2 \frac{Z^5}{137^4}\left(\frac{mc^2}{h\nu}\right)^{\frac{7}{2}} \frac{4\sqrt{2}\sin^2(\theta)\cos^2(\varphi)}{(1-\beta\cos(\theta))^4}, \tag{3.1}$$

where θ is the polar angle between the direction of the incoming photon and the ejected electron (from K-shell), φ is the azimuth angle of the latter with respect to the X-ray polarization vector, Z is the atomic number of the absorption material, r_0 is the classical radius of the electron, and β its speed (in unit of c). The term in the denominator accounts for a slight bending forward of the distribution with increasing energy of the photoelectrons (Figure 3.1). At energies of few keV, photoelectrons are essentially emitted at 90° polar angle. In this plane, the direction of the photoelectron is 100% modulated by the photon polarization, following a $\cos^2(\varphi)$ angular dependence. In this sense the photoelectric effect is much more sensitive to photon polarization than the Compton effect, whose sensitivity to polarization decreases with the scattering angle. For this reason this process has been considered as the base principle of a good analyzer for a polarimeter, providing a strong polarization signature in the *few keV* energy range. In this energy band, which is extremely interesting for X-ray astronomers, other physical processes have very low cross section. In these last decades, scientists have conceived and developed a variety of photoelectric polarimeters, trying to overcome the two main limitations deriving from this technique:

- electrons propagate in matter less than photons at the energies of interest (short tracks);

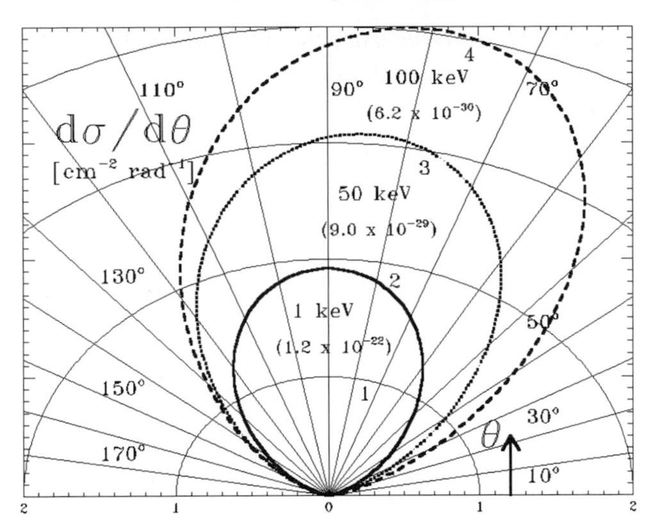

Figure 3.1 Dependence of polar angle of photoelectron in Ne.

- electrons scatter on atoms, randomizing their original direction while their energy decreases and during their motion toward the collecting electrode.

Indeed, the information about angle and degree of polarization of the incoming radiation resides in the first part of the photoelectron track so the main effort in the development of this class of polarimetry devices has been devoted to techniques allowing us to finely sample the electron path particularly in the vicinity of the absorption point.

3.3 Semiconductor-based polarimeters

In CCD-based, or 'pixel', polarimeters, the polarization information is obtained from the pixels crossed by the photoelectron that has been liberated by the absorption of the X-ray in the finely segmented silicon substrate. Because of anisotropy in the photoemission and the finite range of the photoelectron, an excess of events in contiguous pixels, aligned with the X-ray electric vector, are observed. Nevertheless, at energy below 10 keV where optics for X-ray astronomy are very effective and celestial sources more intense, the actual pixel size of a CCD is larger than the range of the photoelectron ($R_{Si}(\mu\text{m}) = (E/10\ \text{keV})^{1.7}$). At these energies single-pixel events outnumber two-pixels events (called *pairs*) while multi-pixel events are present only at higher energies as shown in [4]. Pair search, modulated with the polarization angle, is still possible but it is a border-line effect, extremely sensitive to systematics. The probability of firing two contiguous pixels might depend more on the probability of being absorbed in the frontier zone than on the polarization

Table 3.1 *Measurements of the modulation factor (M) for various CCD-based polarimeters*

CCD pixel size	M (%)	Energy (keV)	
12 μm × 12 μm	6	19	[5]
6.8 μm × 6.8 μm	2	15	[6]
4 μm × 9 μm	3	10	[7]

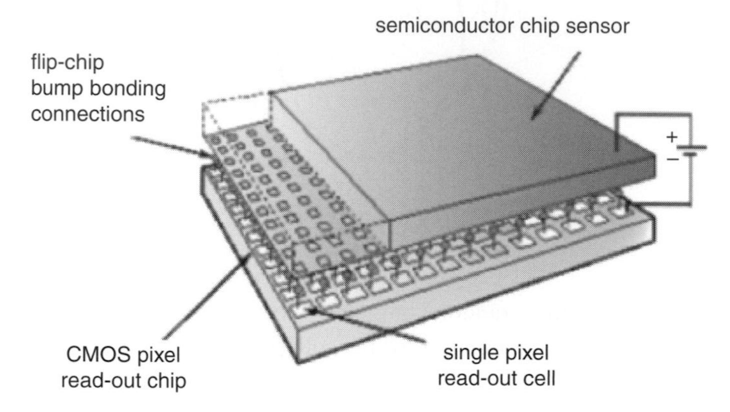

Figure 3.2 The Timepix detector.

itself. The control and calibration of these effects on flight depends on the point-spread function of the optics with respect to the CCD pixel size and on instabilities of the satellite pointing. The feasibility of this technique has been demonstrated by many authors and the literature reports values of the modulation factor (M) for different photon energies and CCD pixel sizes (Table 3.1).

Recently, the application of the hybrid photon-counting Timepix detector (Figure 3.2) to measure the linear polarization of hard X-ray radiation through the photoelectric effect, has been reported in literature[8]. Though the pixel pitch of the counting detector (55 μm) is larger than the pixel size of commercial CCDs, resulting in a significantly lower analyzing power, the typical sensor thickness of the device (300 μm of Si, CdTe or GaAs) is significantly larger than the active thickness of a CCD. This fact leads to a higher detection efficiency in the hard X-ray energy band. In the range 29–78 keV, where the energy of released photoelectrons is sufficient to cause significant track lengths in silicon and where the photo-effect still significantly contributes to absorption, the authors were able to measure an asymmetry between vertical and horizontal double-hit events in neighboring pixels of the Timepix detector, at room temperature. By using the spectroscopic capability of the device (time-over-threshold-mode), polarization asymmetries between 0.2%

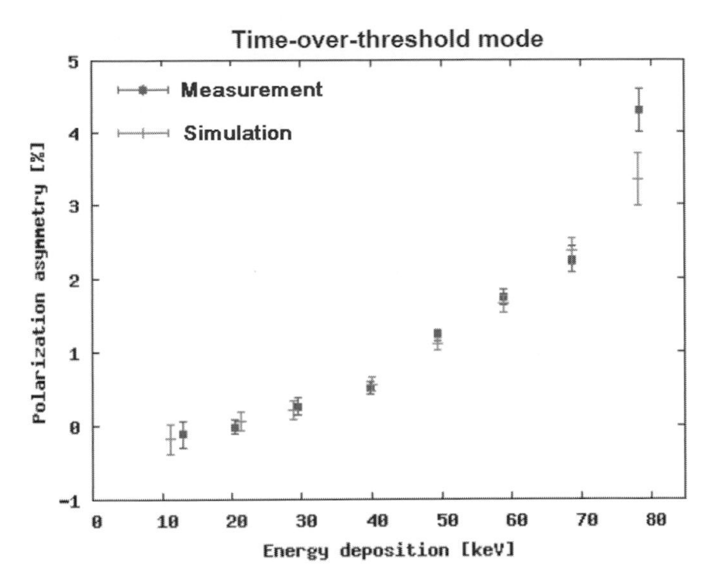

Figure 3.3 Measured polarization asymmetry as a function of the deposited energy.

at 29 keV and 3.4% at 78 keV were measured (Figure 3.3). Though the device is far from being optimized, nevertheless the high efficiency of the semiconductor sensor at high photon energies, the fast readout, and the space for improvements in the analyzing power, make investigations on the polarimetric performance of this detector very interesting. The hybrid silicon pixel device Timepix[9] has been developed at CERN by Medipix collaboration with support of the EUDET project.

3.4 Gas-based polarimeters

In a good photoelectric polarimeter the ionization pattern should not depend on the absorption point. This can be achieved if photoelectron track-length is sufficiently long and the charge collecting electrode of the imaging system is finely subdivided. This condition is achieved with the new class of micro-pattern gas detectors where the active medium, the gas mixture, allows the achievement of good detection efficiency in the soft X-ray energy band whilst still having sizable photoelectron track-length. The optimal polarimetric capability of these devices is due to the high resolution power of track imaging with which they are provided. Historically, the first scientist who studied X-ray polarization through the photoelectron track in a xenon-filled counter was P. W. Sanford, in 1970[10]. He related the rise-time of a signal of a proportional counter to the photoelectron emission angle with respect to the anode direction. More recently this technique has been proposed[11] for a balloon experiment of X-ray polarimetry in the 20–60 keV energy range. The

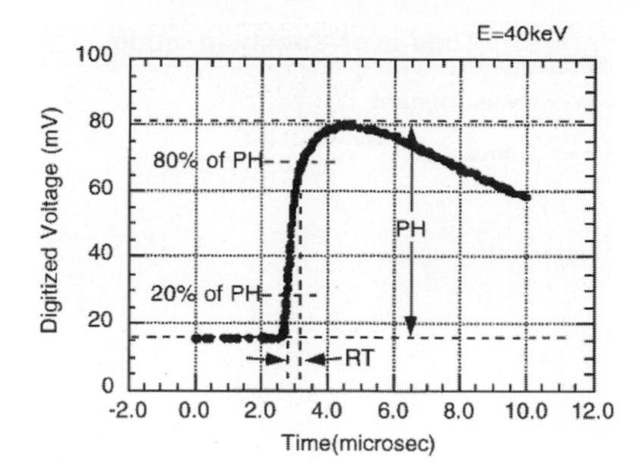

Figure 3.4 Digitized signal of the Xe gas proportional counter. Incident X-ray energy is 40 keV.

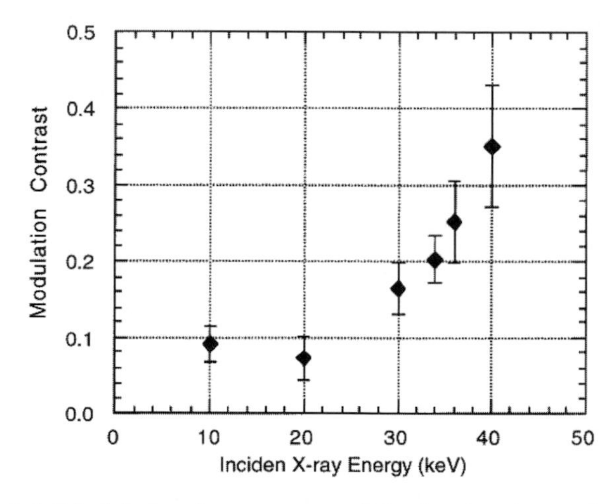

Figure 3.5 Modulation factor measured for the rise-time polarimeter.

authors have found that the rise-time of the signals, measured with a waveform digitizer (Figure 3.4), was significantly different if produced by two beams with orthogonal polarization states and that such a difference increased with energy[12]. With a xenon-CO_2-filled proportional counter, they measured modulation factors of $\sim$10% at 20 keV and 35% at 40 keV (Figure 3.5)

A different imaging technique has been followed by Austin et al., for polarimetry in the hard X-ray energy band from 40 keV to 100 keV[13; 14]. The instrument they have developed consists of an intensified, UV sensitive CCD camera coupled to a two-stage optical avalanche chamber. The track of each photoelectron ejected after the absorption of polarized radiation in the avalanche chamber is imaged onto a CCD camera (Figure 3.6). This technique allows the reconstruction of the

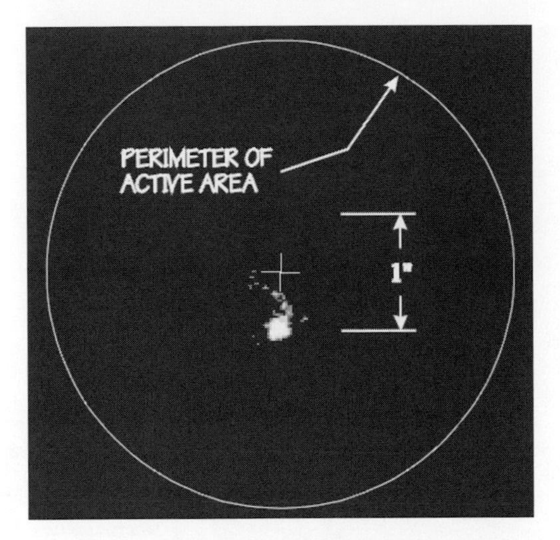

Figure 3.6 Image of a photoelectron track ejected following the absorption of a 54-keV photon.

emission direction of the photoelectron as well as the determination of the position of the absorbed X-ray within a region smaller than the size of the charge cloud. The detection medium is a mixture of 98% argon (converter) and 2% trimethylamine (wavelength shifter) at 2-atmospheres pressure. In the lab, with a small prototype polarimeter (Figure 3.7), the authors have measured a modulation factor of 30% with a 54-keV polarized X-ray beam. Another approach[15] uses an optical-imaging capillary gas proportional counter (CGPC) as a micro-segmented pixel detector to image the photoelectron track (Figure 3.8). The light signal from each capillary is read out using an image-intensified CCD camera with pixels of 24 μm $\times$ 24 μm, coupled to the CGPC through a lens optical system[16]. As the light emission region is confined in the capillary, the segmentation of the signals is very fine with little signal crosstalk. A sample image of a single photoelectron track produced by a 15 keV polarized photon is shown in Figure 3.9. The CGPC is filled with the Penning gas mixture Ar + 8% CH_4 + 2% TMA at 1 atmosphere. A $\cos^2$ fit to the angular distribution of the photoelectron tracks gives a modulation factor of 25.1% $\pm$ 2.0% for the 15 keV X-rays with a polarization degree of 99%.

3.4.1 GEM-based micro-pattern polarimeters

The idea to couple a gas electron multiplier (GEM) to amplify the ionization track of the photoelectron, with a finely segmented pixel readout system to image the track[17], has brought about the development of a class of non-dispersive devices with polarimetric sensitivity well above that achievable with traditional scattering

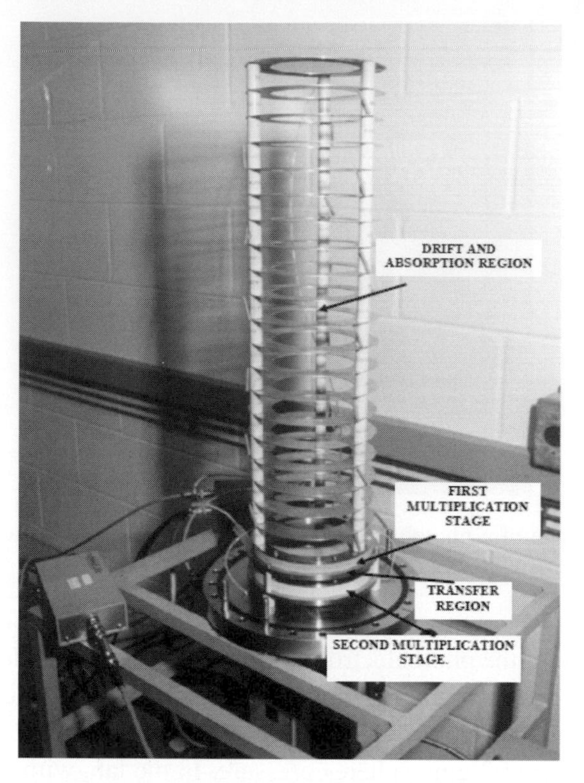

Figure 3.7 Photograph of the prototype of the photoimaging polarimeter. The pressure vessel has been removed to show interior details.

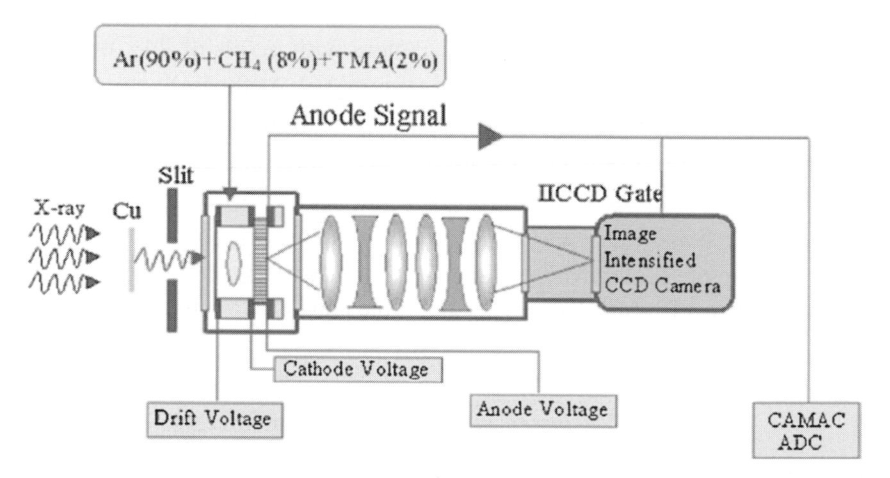

Figure 3.8 Schematic of the optical imaging CGPC setup. The CGPC is filled with the Penning gas mixture Ar + 8% CH_4 + 2% TMA at 1 atmosphere. The incoming polarized X-ray beam has an energy of 15 keV.

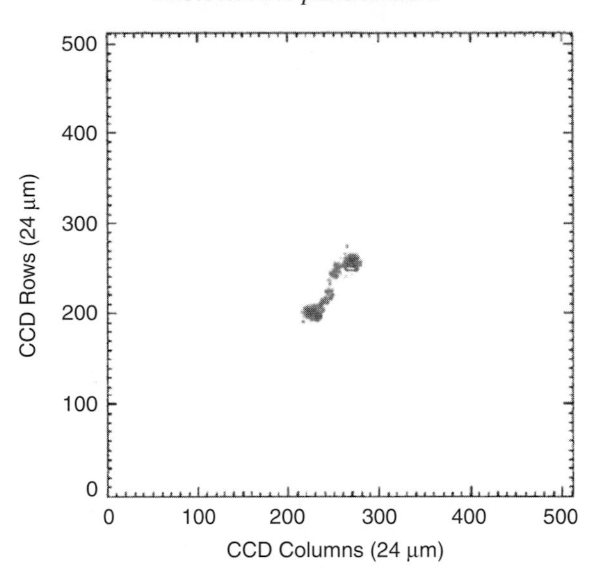

Figure 3.9 Image of a photoelectron track from a 15-keV polarized photon.

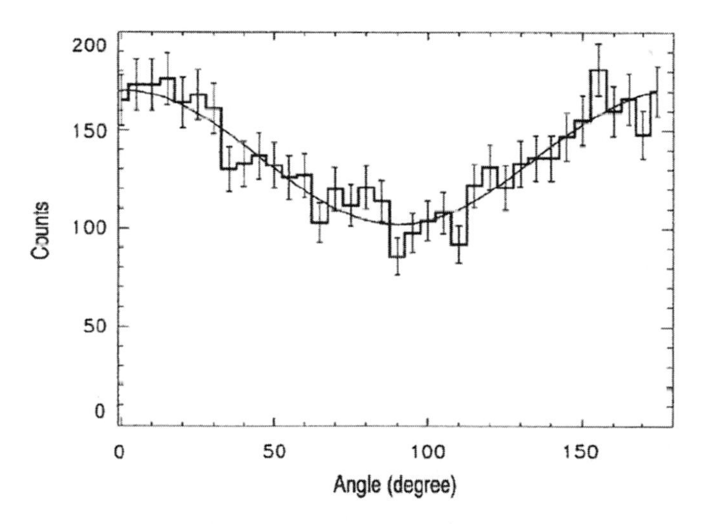

Figure 3.10 Angular distribution of the photoelectron tracks.

polarimeters. In a first pioneering work[18] the readout electrode of a micro-pattern gas detector (MPGD) was based on a multi-layer, fine-line PCB, patterned with 512 hexagonal pixels at 200 μm pitch. The signals were routed out from each pad to an external analysis chain, resident on an ASIC chip. Despite track-lengths of only few hundred microns, the photoelectron tracks were sufficiently sampled

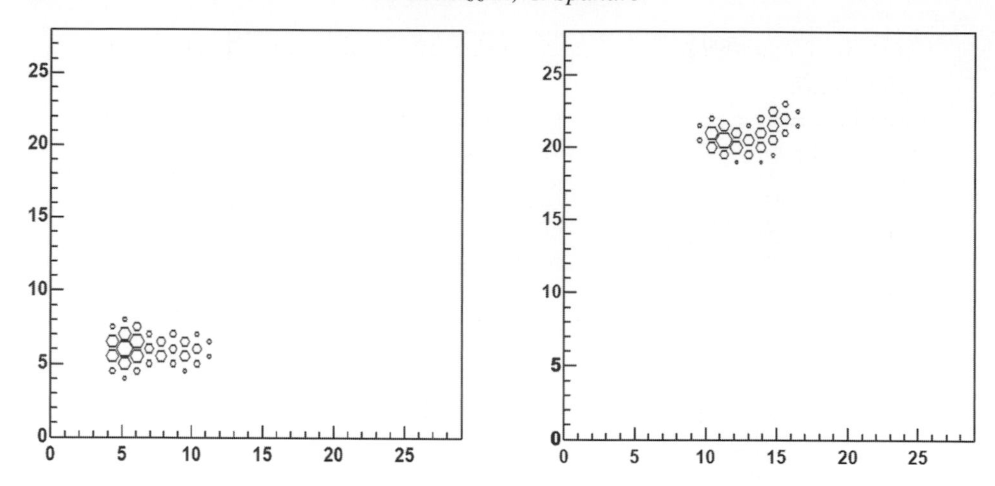

Figure 3.11 Photoelectron tracks from a 5.9-keV photon. Each hexagon's area is proportional to the collected charge on the corresponding pixel.

(Fig. 3.11) and a refined reconstruction algorithm allowed a modulation factor of $\simeq$40% to be obtained for 5.4 keV polarized photons using a neon-DME (80-20) gas mixture. To get rid of the constraints that limit the maximum number of independent electronics channels that can be brought to peripheral electronics, a custom CMOS readout plane fabricated in deep sub-micron VLSI technology, has been developed[19]. The chip is both the pixellated charge collecting electrode and the readout electronics. The pixels are realized in the CMOS top metal layer and are individually connected to a full electronics chain (pre-amplifier, shaping-amplifier, sample and hold, multiplexer) built in the underlying five layers of the chip. This readout approach has the advantage, compared to similar ones (CCD or TFT readout), of being fully asynchronous and externally triggerable. Furthermore, it supplies the complete analog information of the collected charge allowing imaging of the energy deposition process of the absorbed radiation. The whole device is extremely compact and a fast development of further ASIC generations of increased dimension with a progressively decreasing pixel size (see Table 3.2) was implemented (Figure 3.12). In the last generation more than 100 000 pixels have been integrated at 50 μm pitch in a 15 mm^2 active area. This ASIC, with more than 16.5 million transistors, has self-triggering capability and only the rectangular region where the track is located is read out, with a consequent reduction of the system dead time.

Figure 3.13 shows two examples of reconstructed tracks obtained with the generation II (left panel) and III (right panel) of the ASIC. In the two cases the same photon energy was used to irradiate the detectors (5.4 keV from Cr X-ray tube) but with different gas filling. Both tracks are reconstructed with a high degree of

Table 3.2 *Evolution of the readout of the MPGD from PCB prototype to CMOS VLSI*

Device	Pixels	Pitch (μm)	Size (mm)	Technology	Ref.
Prot.	512	120	3 (diam.)	multi-layer adv. PCB	[18]
Chip I	2101	80	4 (diam.)	CMOS 0.35 μm	[19]
Chip II	22080	80	11 μm × 11 μm	CMOS 0.35 μm	[20]
Chip II	105600	50	17 μm × 17 μm	CMOS 0.18 μm	[21]

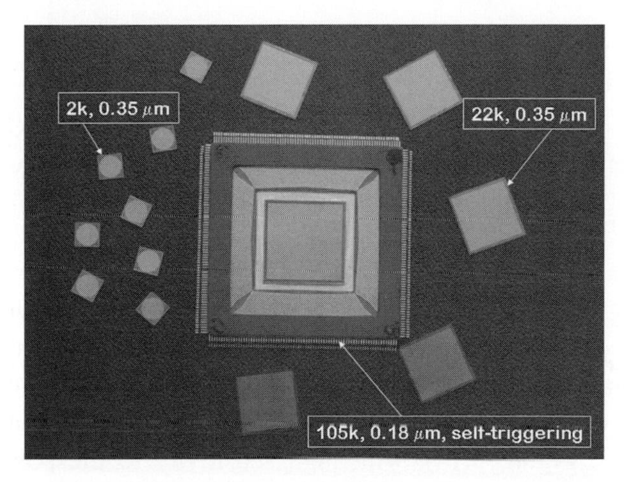

Figure 3.12 The three ASIC generations. The last version with 105k pixels is shown bonded to its ceramic package (304 pins).

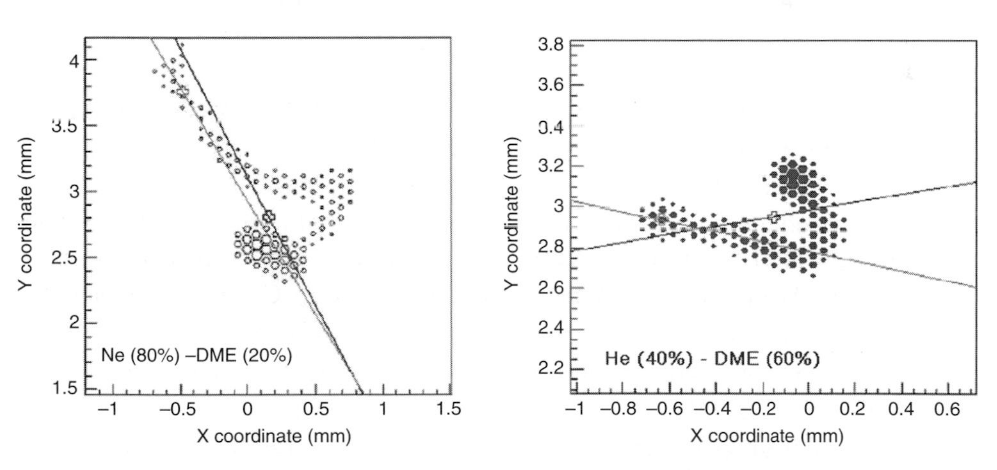

Figure 3.13 Real tracks obtained by irradiating the detector with X-rays from the Cr tube. Left panel refers to reconstruction obtained with ASIC II, right panel with ASIC III. Gas mixtures were different in the two cases.

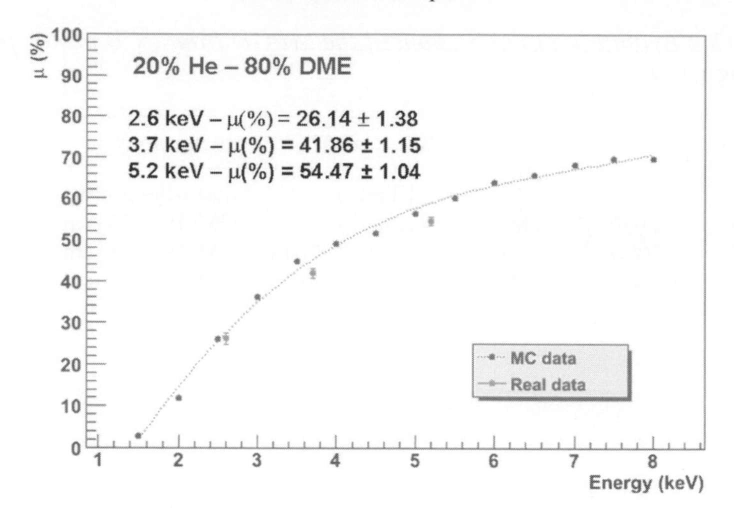

Figure 3.14 Measured modulation values at 2.6, 3.7, and 5.2 keV in He(20)-DME(80) gas mixture, superimposed on the Monte-Carlo simulation.

detail, allowing the Bragg peak and the absorption point to be very well distinguished. The cross in the figure refers to the evaluation of the photon conversion point and the line to the extrapolated emission direction of the photoelectron. Modulation factors ranging from 43.10% ± 1.18% for 3.7 keV to 54.47% ± 1.04% for 5.2 keV (see Figure 3.14), with high sensitivity to the polarization angle, have been obtained during a campaign of measurements performed using mosaic graphite and flat aluminum crystals to produce nearly completely polarized photons[22]. For comparison, the measured residual modulation obtained with totally unpolarized photons has resulted to be 0.18% ± 0.14% (Figure 3.15), consistent with zero.

The use of a different active-pixel anode plane coupled to a double-GEM, has been reported by Black et al.[23]. They have used an amorphous-silicon thin-film transistor (TFT) array similar to those utilized in flat-panel imagers. In this active-matrix-addressing scheme, each pixel is connected to the readout electronics through a TFT that acts as an analog switch. The charge is held on the pixel until the transistor is activated, allowing a multiplexed readout such that n^2 pixels are read out with $2n$ electronic channels. The active area of the detector is defined by the double-GEM filled with Ne-CO_2 (80-20) gas mixture. Photoelectron track images from 5.9- and 20-keV X-rays were recorded (Figure 3.16) and polarization measurements made at 4.5 keV. The results, shown in Figure 3.17, demonstrate the polarization sensitivity of the detector with the polarization angle. The three data sets, at different polarization angle, are consistent with an average modulation of 33% ± 3%.

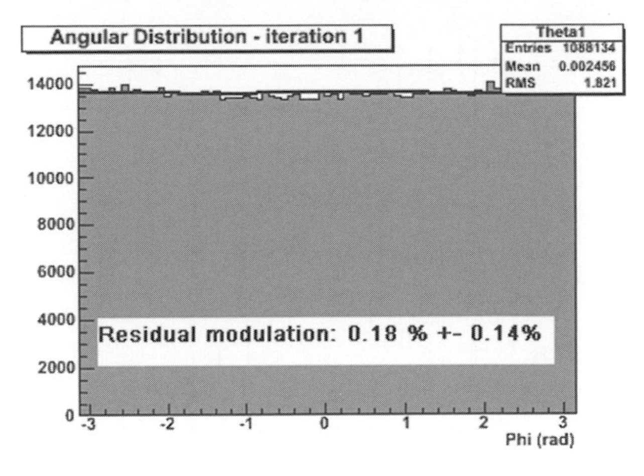

Figure 3.15 Residual modulation obtained with totally unpolarized photons from a ^{55}Fe X-ray source.

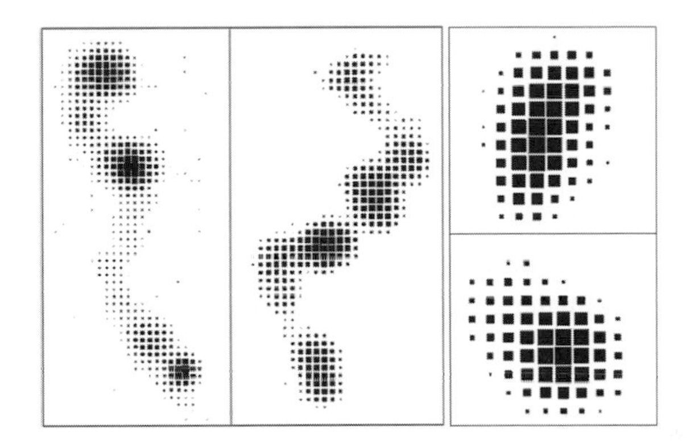

Figure 3.16 Track images from 20-keV X-rays (left) and 4.5-keV X-rays (right).

3.4.2 Micro-pattern TPC polarimeter

A novel method[24] to image the tracks of photoelectrons is the micro-pattern time projection chamber (TPC). Thanks to the TPC geometry, the diffusion of the primary ionization electrons, as they drift toward to the pixel read-out, is largely independent of the absorption depth. For this reason this technique offers the possibility of achieving large modulation factors simultaneously with high quantum efficiency. As shown in Figure 3.18 the TPC polarimeter is not a photon imager (i.e. it cannot localize the photon absorption point), but it uses a time projection technique to form the two-dimensional image of the photoelectron track from a one-dimensional strip readout. With a first TPC prototype having 1.8-cm absorption depth and Ne-DME (50-50) gas filling, a modulation factor of $45.0 \pm 0.6\%$ has been measured with 6.4-keV polarized photons. The residual modulation obtained with unpolarized

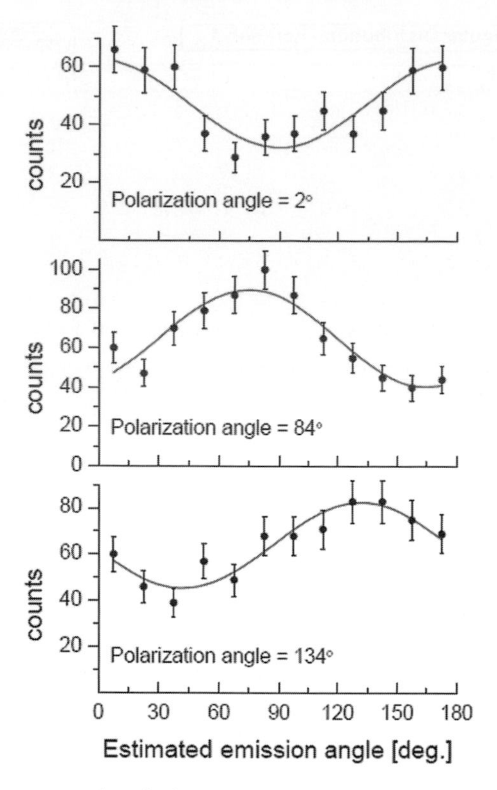

Figure 3.17 Reconstructed emission angles for three polarization angles. All data sets are consistent with the average modulation of 0.33 ± 0.03.

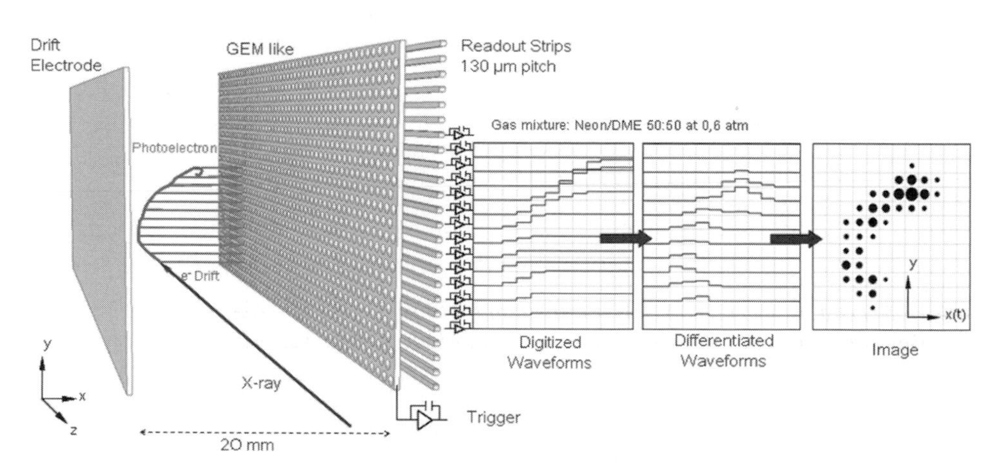

Figure 3.18 Photoelectron track imaging with a micro-pattern TPC. The digitized waveforms are represented as an image in which the areas of the circles are proportional to the charge deposited. The pixels are on a 130-μm spacing, while the waveforms are sampled with a 40-ns period.

5.9-keV X-rays, was $0.49\% \pm 0.54\%$. Efforts are currently underway to develop a next generation TPC polarimeter with thicker absorption depth and finer pixel pitch, to be tested over a wider range of X-ray energies.

3.5 Conclusions

All the polarization-detection techniques discussed in this review rely on the correlation between incident polarization and direction of photoelectron ejection and on the possibility offered by the new class of micro-pattern detectors to reconstruct in great detail the ionization track with high accuracy. Some of the presented technologies are already in a highly matured state of development. They offer a great advance in sensitivity, if compared with the state of the art of the conventional technique, and are planned to be used in near-future space missions.

References

[1] Weisskopf, M. C. et al. (1976). *The Astroph. J.* **208**, L125–L128.
[2] Auger, H. O. (1926). *Inst. Soc. Am. Trans.* **6**, 311.
[3] Heitler, W. (1970). *The Quantum Theory of Radiation*, London, Oxford University Press.
[4] Buschhorn, G. et al. (1994). *Nucl. Instr. and Meth. A* **346**, 578–588.
[5] Hayashida, K. et al. (1999). *Nucl. Instr. and Meth. A* **436**, 96–101.
[6] Schmidt, K. H. et al. (1999). *Proc. SPIE* **2808**, 230–241.
[7] Hill, J. E. et al. (1997). *Proc. SPIE* **3114**, 241–249.
[8] Michel, T. & Durst, J. (2008). *Nucl. Instr. and Meth. A* **594**, 188–195.
[9] Llopart, X. et al. (2007). *Nucl. Instr. and Meth. A* **581**, 485–494.
[10] Sanford, P.W. et al. (1970). *Non-Solar and Gamma Ray Astronomy*, Symposium No. 37, Dordrect-Holland, D. Reidel.
[11] Hayashida, K. et al. (2001). *ASP Conference Proceedings* **251**, 538–539.
[12] Hayashida, K. et al. (1999). *Nucl. Instr. and Meth. A* **421**, 241–248.
[13] Austin, R.A. et al. (1996). *The Astroph. J.* **208**, L125–L128.
[14] Austin, R.A. et al. (1993). *Proc. SPIE* **2010**, 118–125.
[15] Sakurai, H. et al. (2004). *Nucl. Instr. and Meth. A* **525**, 6–11.
[16] Sakurai, H. et al. (2002). *IEEE Trans. Nucl. Sci.* **49**, 1560.
[17] Costa, E. et al. (2001). *Nature* **411**, 662–665.
[18] Bellazzini, R. et al. (2002). *IEEE Trans. Nucl. Sci.* **49**, 1216–1220.
[19] Bellazzini, R. et al. (2004). *Nucl. Instr. and Meth. A* **535**, 477–484.
[20] Bellazzini, R. et al. (2006). *Nucl. Instr. and Meth. A* **560**, 425–434.
[21] Bellazzini, R. et al. (2006). *Nucl. Instr. and Meth. A* **566**, 552–562.
[22] Muleri, F. et al. (2006). *Nucl. Instr. and Meth. A* **584**, 149–159.
[23] Black, J. K. et al. (2003). *Nucl. Instr. and Meth. A* **513**, 639–643.
[24] Black, J. K. et al. (2007). *Nucl. Instr. and Meth. A* **581**, 755–760.

4

Bragg crystal polarimeters

E. Silver

Harvard-Smithsonian Center for Astrophysics

H. Schnopper

Deer Isle Observatory

The most successful measurements of cosmic X-ray polarization have been made with Bragg crystal polarimeters. We review the fundamental techniques of Bragg crystal polarimetry, describe how these were implemented on the *OSO-8* spacecraft and the SXRP polarimeteter intended for flight aboard the *Spectrum-X-Gamma* mission, and now, 35 years later, present an optimized design for a small satellite dedicated to polarimetric observations.

4.1 Introduction

A photon carries information about its direction, time of arrival, energy and polarization. Satellite missions such as *Uhuru*, *Ariel V*, *HEAO-1* and *ROSAT* located hundreds of sources in their all-sky surveys, and missions such as *HEAO-1*, *ASCA*, *Suzaku*, *RXTE*, *Chandra*, and *XMM/Newton* characterized source spectra and time variability. The X-ray community now needs polarization information to reveal the intrinsic small-scale geometry of astrophysical systems and to evaluate physical processes in regions near compact objects which cannot be resolved via imaging, spectroscopy or timing.

X-ray radiation is polarized when the production mechanism has an implicit directionality, such as when electrons interact with a magnetic field to produce cyclotron and synchrotron emission. In radio quasars, for example, the amplitude of the polarization is a diagnostic for the X-ray emission mechanism and the polarization direction associates the X-ray emission with particular regions identified in milli-arc-second radio images. Polarization also occurs when X-rays are scattered by electrons, a common process in the highly ionized environments of compact

X-ray Polarimetry: A New Window in Astrophysics, eds. R. Bellazzini, E. Costa, G. Matt and G. Tagliaferri.
Published by Cambridge University Press.

X-ray sources. Here the measurement of polarization and its dependence on energy provide information about source geometry and magnetic field on spatial scales comparable to a black hole event horizon. Eventually, X-ray polarization will likely show that in many cases, simple models of circular or spherical symmetry are naive.

To date, the Bragg crystal polarimeters flown aboard NASA's *OSO-8* spacecraft in 1975 provided the most successful measurements of cosmic X-ray polarization[1; 2; 3; 4]. In the late 1980s, a consortium led by Columbia University built the Stellar X-Ray Polarimeter (SXRP) consisting of a Bragg crystal and a Thomson scattering polarimeter[5]. SXRP was intended for flight aboard the Soviet *Spectrum Roentgen-Gamma* mission (*SRG*). With the break-up of the Soviet Union, it was never launched. Here, we review the fundamental techniques of Bragg crystal polarimetry and show how these were implemented in the *OSO-8* and SXRP instruments. Now, 35 years later, we describe APEX, an optimized polarimetric payload, designed as a Small Explorer for NASA.

4.2 Bragg crystal polarizers

A flat crystal oriented at 45° to a parallel beam of X-rays acts as a perfect polarization analyzer for those X-rays that satisfy the Bragg condition, $nhc/E = 2d \sin 45°$. As a polarimeter, the crystal panel takes advantage of the polarization dependence of Bragg reflection by preferentially reflecting photons whose electric vectors are perpendicular to the plane of incidence, which is the plane defined by the normal to the crystal at the intersection of the incoming and reflected rays. The diffracted X-rays may then be detected by a broad band energy dispersive instrument such as a proportional counter or a lithium-drifted silicon (Si(Li)) detector. This approach was first described by Schnopper and Kalata[6]. By rotating the crystal and detector combination with an angular frequency, ω around the line of sight to linearly polarized X-rays, the detected signal will exhibit a sinusoidal modulation at a frequency of 2ω.

In Figure 4.1, we show the modulation curve obtained at 2.6 keV when the *OSO-8* polarimeter viewed the Crab Nebula. *OSO-8* was a spinning spacecraft and its graphite polarimeters viewed the sky along the spin axis. The nominal rotation frequency, ω, of the *OSO-8* spacecraft was 6 rpm and the signature of polarization was observed at 2ω. The lower curve in Figure 4.1 is the average 2.6 keV background counting rate which showed no statistically significant modulation. The modulation data in the upper curve indicated that the polarization at 2.6 keV is $19.22\% \pm 0.92\%$. The phase of the modulation curve is a measure of the position angle of the electric vector, $155.770 \pm 1.3°$. This was the first and only high-precision X-ray polarization measurement obtained for any cosmic source and was in excellent agreement with optical polarization measurements of the Crab Nebula.

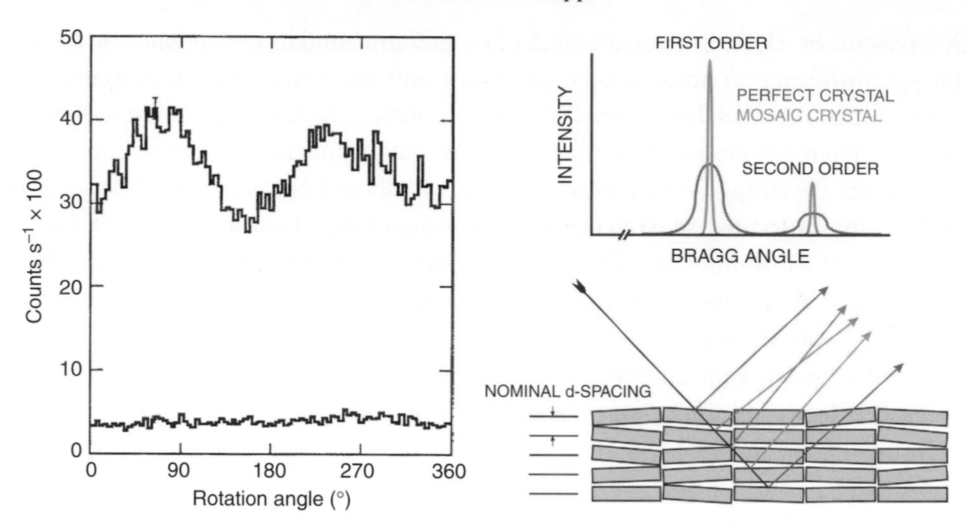

Figure 4.1 Left: The respective components to the polarization measurement of the Crab Nebula by *OSO-8*. Right: The increased bandwidth of the mosaic crystal produced by Gaussian distribution of perfect crystalets.

The rate at which continuum radiation is Bragg-reflected from a flat crystal is given by $S(E) = \int \frac{I(E)}{E} R(E) A_{proj} \epsilon(E) dE$ photons/s. Here, $I(E)$(keVcm^{-2}s^{-1} keV^{-1}) is the X-ray continuum intensity in an energy range dE, ϵ is the detector efficiency, $A_{proj}(E)$ is the projected area of a free-standing crystal or of a grazing incidence telescope that focuses X-rays on the crystal, and $R(E)$ is the crystal reflectivity. $R(E)$ is a sharply peaked function about the energy that satisfies the Bragg condition for the angle θ. The integral, $\int R(E,\theta)dE$, is defined as the effective width of the crystal, $\Delta E(\theta)$, and to an excellent approximation can be written as $\Delta E(\theta_B) = E_B \cot(\theta_B)\Delta\Theta$, where, $\Delta\Theta = \int R(E_b,\theta)d\theta$, is the integrated reflectivity over the rocking curve of the crystal. Since the X-ray continuum of cosmic X-ray sources varies slightly over the energy range reflected by the crystal, the predicted intensity is then $S(E_B) = I(E_B)A_{proj}\epsilon(E_B)\Delta\Theta$. The number of reflected photons is directly proportional to $\Delta\Theta$, which should be maximized to detect weak cosmic X-ray fluxes.

The peak reflection efficiency of nearly *perfect* crystals may be 50% or higher but only within a narrow energy bandwidth extending over a small fraction of an electron volt. To maximize the photon flux collected, the energy bandwidth can be broadened by utilizing *imperfect* crystals. An ideally imperfect crystal consists of a randomly oriented mosaic of small crystal domains. Each crystal domain has a regular atomic structure and may be thought of as a perfect crystalet. The crystalets are thin compared to the photoelectric absorption length in the crystal, allowing an X-ray to traverse many domains until it encounters one properly oriented for Bragg reflection. The random orientation of the small crystal domains greatly improves the possibility for simultaneous reflection of a range of photon wavelengths (see Figure 4.1). A detailed description of mosaic crystals and their use for astronomical applications can be found in the literature[7].

Both theoretical and experimental studies show that a graphite mosaic crystal is the best choice for a crystal polarization analyzer at energies above 1 keV. The 2-D crystal lattice spacing of 6.7Å in graphite corresponds to reflection by the (002) crystal planes and makes a flat graphite crystal panel sensitive at 2.6 keV in first order reflection and at 5.2 keV in second order. The energy bandwidth in first and second orders is primarily due to the mosaic spread of the specific graphite crystal chosen. The graphite on the *OSO-8* instrument had a mosaic spread of 0.8° which set the intrinsic bandpass at 0.04 keV and 0.08 keV in first and second order, respectively. The integrated reflectivities for *polarized* radiation were 1.2×10^{-3} radians and 5×10^{-4} radians, respectively. The mosaic spread for the graphite crystal used in SXRP was 0.5° and had the same integrated reflectivities as those for *OSO-8*.

To reduce the cosmic-ray-induced background signal associated with the detector on *OSO-8*, the crystal panel was curved to focus the diffracted X-rays into a detector with a small surface area. This is shown in Figure 4.2. A matrix of small, flat graphite

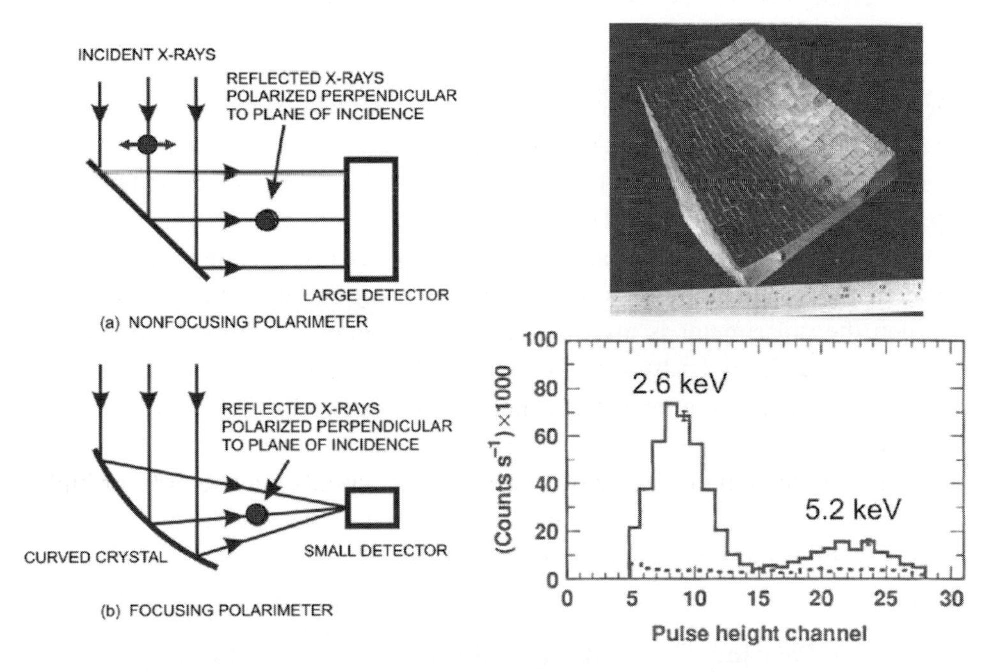

Figure 4.2 The focusing crystal polarimeter (left); One *OSO-8* graphite panel (top right); The first and second order proportional counter response from *OSO-8* (bottom right).

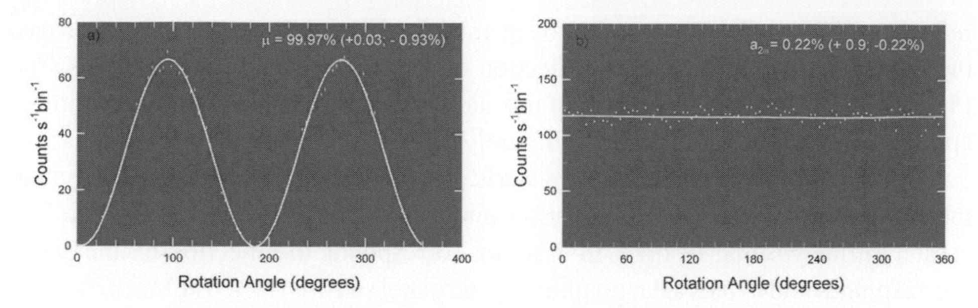

Figure 4.3 Left: Modulation curve for the SXRP using a 2.6 keV 100% polarized beam. The amplitude of the second harmonic (2ω) modulation is 99.97% (+0.03; −0.93%). Right: Modulation curve for the SXRP using an unpolarized beam. The amplitude of the 2ω modulation is 0.22% (+0.9; −0.22).

crystals 200–300 μm thick was mounted on the surface of a sector of a paraboloid of revolution (see Figure 4.2). This reduced the instrumental background counting rate by a factor of ~20. The curvature of the crystal panel on *OSO-8* encompassed a range of Bragg angles between 40° and 50° and increased the overall energy bandwidth of the crystal panel to 0.460 keV and 0.920 keV for first and second order Bragg reflections, respectively. In the SXRP a small flat crystal was located in the converging beam of the Soviet-Danish Roentgen Telescope (SODART) aboard the *SRG* satellite[5].

The X-ray detector must have sufficient energy resolution to separate the X-rays reflected in first and second order. This was easily accomplished in *OSO-8* and the *SRG* flight instrument. Figure 4.2 shows the pulse-height spectrum recorded during the 1976 observation of the Crab Nebula. The first (2.6 keV) and second (5.2 keV) order peaks are clearly shown. The dashed line represents the background counting rate measured while the instrument field of view was earth-occulted. The favorable signal-to-background ratio is the result of the focusing by the curved crystal panels.

When the polarimeter is used to measure a source of unknown polarization, the polarization is given by $P = M/\mu$, where M is the measured amplitude of modulation and μ is the instrument modulation factor. For *OSO-8*, the range of Bragg angles around 45°, combined with the angular extent in azimuth of the parabolic sector of 30°, allowed a small unmodulated signal to be transmitted continually. As a result, the modulation factor was 0.93. The modulation factor for SXRP is 0.99 as shown in Figure 4.3.

4.3 Polarimetric sensitivity and control of systematic effects

The sensitivity of a polarimeter is defined by the minimum detectable polarization (MDP) at the 3σ confidence level. For Poisson statistics, the MDP is related to the detected source counting rate S, length of the observation T, the modulation factor μ, and the background counting rate, B, according to MDP = $(4.29/\mu S)\sqrt{(S+B)/T}$.

Minimizing the MDP, which is obtained purely from counting statistics, is important, but it does not guarantee the ultimate sensitivity. Rather, the effects of the modulation factor and systematic uncertainties must be considered when building a polarimeter. For example, assume that two polarimeters can obtain a sufficient number of photons to make a statistically significant measurement at the 1% level. In other words, assume that both can obtain an MDP of 0.4%. To measure polarization of 1%, a crystal polarimeter with a modulation factor ~0.93 must measure a signal that is modulated at the 0.93% level. But a polarimeter with a modulation factor of ~0.40 must detect a modulation of 0.4%. Although the MDP is statistically possible, the systematic errors connected with the operation of the instrument could easily introduce a spurious modulation at this low level or even at a higher level. In this example, instrumental effects must be measured and understood at a level $\leq$0.4%.

The crystal polarimeter has a higher modulation factor than any other polarimeter design. Rotating the crystal/detector pair not only generates the 2ω signature of polarization but also serves to minimize spurious modulation. With rotation, individual detector sensitivities must be constant to a precision of <1% during only one rotation period of the instrument. Fourier analysis of the data stream as a function of rotation angle also can identify spurious modulations at other frequencies. *OSO-8* was not calibrated before launch, but the SXRP crystal polarimeter response to an unpolarized source of X-rays was measured in the lab before flight. Figure 4.3 shows that the response to unpolarized radiation was below the 0.22% level. Understanding and/or eliminating the systematic effects that may cause spurious modulation is essential to being able to measure polarization at the $\leq$1% level.

4.4 The Astrophysical Polarimetric EXplorer(APEX): an optimized Bragg crystal polarimeter

Whereas the *OSO-8* polarimeters were small instruments, APEX has a 30-fold increase in total collecting area, a factor 14 times better background rejection, and the ability to make long observations of classes of extragalactic as well as Galactic sources. Measurement redundancy, with seven independent graphite crystal panel/detector pairs plus satellite rotation, makes it possible to reduce systematic uncertainties to the level of < 0.3% which allows measurements of 1% amplitude of polarization to be obtained for bright Galactic sources.

Each polarimeter is identically configured and consists of a parabolic petal with a set of curved pyrolitic graphite Bragg crystals mounted to its surface that focus

X-ray photons into a Si(Li) detector (see Figure 4.4). There are no deployed portions of the experiment. The Bragg angles range between 39° and 53° and the azimuthal extent of each parabolic sector is 28° resulting in a modulation factor of 92.5%. Small graphite crystals curved in two dimensions to match the shape of the paraboloid of revolution are used instead of the small flat crystals employed for *OSO-8*. These improve the focusing properties of the panel and fewer crystals are

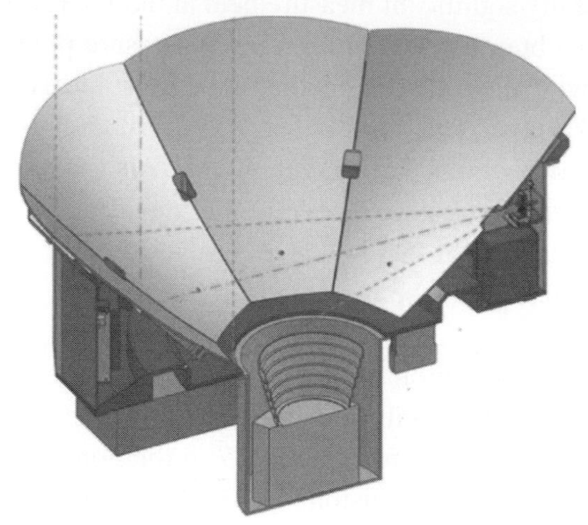

Table 1 Polarimeter Properties

APEX	1st Order (2.6 keV)	2nd Order (5.2 keV)
Projected Area (cm^2)	8600	8600
Effective Area (cm^2)	750	213
Reflectivity	0.091	0.025
$\Delta\Theta$ (rad) (unpolarized)	6×10^{-4}	1.7×10^{-4}
Detector Efficiency	0.96	0.99
Detector Energy Resolution (keV)	0.150	0.190
Detector Background (cts/cm²/s/keV)	0.003	0.003
Detector area (cm^2)	1.77	1.77

Figure 4.4 A cutaway view of APEX showing three Bragg panels. Sample trajectories for three X-rays are shown reflected from one panel and focused into a Si(Li) detector.

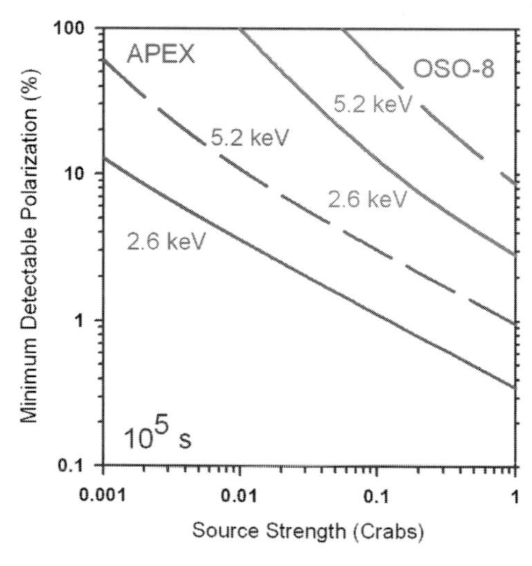

Table 2 APEX Instrument Capabilities

Polarization Energies	2.6 and 5.2 keV
Time Resolution	300 μsec
MDP (3σ)	
Stellar Mass Black Holes e.g. Cygnus X-1	<1% at 2.6 keV and 1% at 5.2 keV in 200 ks
Accreting Pulsars e.g. Cen X-3	~1% in 50 ks per 20% phase bin at 2.6 keV and 5.2 keV
Blazars, Quasars, Seyferts e.g. MK421, 3C273, N4151	2-4% in 300 ks at 2.6 keV

Figure 4.5 Left: The MDP for APEX and *OSO-8*. Right: APEX capabilities.

required to cover the parabolic surface. The radius of curvature for each complete crystal panel is 975 mm and fits easily within the Pegasus shroud. The seven APEX polarimeters provide a total projected area of 8600 cm^2. (The total projected area of the two *OSO-8* crystal panels was 285 cm^2.) At the focus of each crystal panel is a Si(Li) detector. Each is located where two crystal panels on the opposite side meet. The Si(Li) detectors for APEX offer superior energy resolution to the proportional counters of *OSO-8* and SXRP. This, combined with their smaller size, reduces the APEX background by a factor of $\sim$14 compared to *OSO-8* and is necessary towards achieving high sensitivity for faint extragalactic sources. The properties of the instrument are summarized in Table 1. The MDP obtainable with APEX is plotted in Figure 4.5 as a function of source strength in units of the Crab Nebula. The values for OSO-8 are shown for comparison. A summary of its polarimetric capabilities is listed in Table 2.

References

[1] Novick, R. et al, (1977). In *Proc.of the Twentieth COSPAR, Tel Aviv, Israel, June 1977.*

[2] Weisskopf, M.C., et al. (1978). *Ap J*, **220**, 117.

[3] Silver, E., Weisskopf, M., Kestenbaum, H., Long, K., Novick, R., and Wolff, R. (1978). *Ap J*, **225**, 221.

[4] Silver, E., Weisskopf, M., Kestenbaum, H., Long, K., Novick, R., and Wolff, R. (1979). *Ap J*, **232**, 248.

[5] Kaaret, P. et al. (1990). *IAU Colloquium No. 123: "Observatories in Earth Orbit and Beyond"*.

[6] Schnopper, H. W. & Kalata, K. (1969). *AJ*, **74**, 854.

[7] Angel, J.R.P. & Weisskopf, M.C. (1970). *The Astronomical Journal*, **75, N 3**, 231.

5

X-ray polarimetry with the photon-counting pixel detector Timepix

T. Michel, J. Durst & F. Bayer
ECAP, University of Erlangen-Nürnberg

J. Jakubek
IEAP, Czech Technical University in Prague

We investigated the capability of the hybrid photon counting pixel detector Timepix to measure the degree of linear X-ray polarization between 27 and 84 keV. Due to its ability to measure energy deposition or the detection time in each pixel, both photoelectric effect and Compton scattering in the sensor can be exploited. The analyzing power exploiting photoelectric effect was found to be small compared to X-ray-sensitive CCDs due to the larger pixel pitch (55 μm) of the Timepix. We were able to measure a polarization asymmetry of $(0.96 \pm 0.02)\%$ between vertical and horizontal double-hit events in neighbouring pixels. The polarization asymmetry was measured with dependence on the energy deposition in the sensor. Asymmetries range between 0.2% at 29 keV and 3.4% at 78 keV. In order to exploit the polarization signature of Compton scattering in the sensor, the time-to-shutter mode of the Timepix was used. We measured a large modulation factor of about 68.1% in good agreement with simulations.

5.1 Introduction

It has already been demonstrated that X-ray-sensitive CCDs can be used to measure the degree of linear polarization of X-rays[1] using the effect that photoelectrons are emitted with a nonisotropic angular distribution with respect to the orientation of the electric field vector of impinging photons. Up to the last year hybrid semiconductor pixel detectors like the Timepix-detector have never been used for X-ray polarimetry. The main reason for this is that the pixel pitch is large compared to CCDs which results in a much smaller analyzing power. The Timepix is also able to measure the detection times of events in the time-to-shutter mode. Thus, it should

X-ray Polarimetry: A New Window in Astrophysics, eds. R. Bellazzini, E. Costa, G. Matt and G. Tagliaferri. Published by Cambridge University Press.

be possible to identify Compton-scattering events in the pixel matrix. The aim of the described experiments was to demonstrate the capability of Timepix to perform polarimetry using photoelectric effect and also Compton scattering.

5.2 Experimental setup, simulation and measurements

The Timepix detector[2] has been developed by the international Medipix collaboration with support of the EUDET project to be used in a TPC or as an X-ray-imaging detector. The ASIC comprises 256×256 electronic cells at a pitch of $55\,\mu m$ with preamplifier, discriminator and counter, and can be bump-bonded to a semiconductor sensor layer. In our experiments we used a $300\ \mu m$ thick silicon sensor. Each pixel can operate in one of three modes. In the counting mode the counter counts the number of events in each pixel having deposited an amount of energy larger than the equivalent discriminator threshold. In the time-over-threshold mode the counter counts the number of pulses of a clock signal during the time for which the discriminator input pulse is above threshold. This time is proportional to the energy deposition in the pixel. In the time-to-shutter mode the counter starts counting clock pulses upon detection of an X-ray photon. Counting of clock pulses is stopped in all triggered pixels when the shutter signal occurs. We produced linearly polarized photons between 27 keV and 84 keV by Compton scattering an X-ray tube spectrum off a polymethylacrylate target. We used an X-ray tube with tungsten anode at 100 kV acceleration voltage. The emitted X-ray spectrum was filtered with a 1 mm thick iron plate. The radiation was collimated to a beam of 2 mm diameter. The target had dimensions of $14 \times 3 \times 5\ mm^3$ and was placed 8 cm in front of the Timepix. The detector could be rotated around the normal of the sensor surface. It was positioned in such a way that mainly photons which have been Compton scattered by $\theta \approx 90°$ in the target were accepted. The scattered X-rays, incident on the detector, are linearly polarized to a high degree. We used the Monte-Carlo simulation package ROSI to simulate setup and detector.

Our measurements concerning photoelectric effect are described in more detail in [3]. Due to the fact that the photoelectrons are emitted with a certain distribution preferentially in the plane of linear polarization, we expect more coincident triggering of adjacent pixels in one row than triggering of two adjacent pixels in one column, if the polarization plane is oriented along the rows. We define an asymmetry as $A(\alpha) := (N_r(\alpha) - N_c(\alpha))/(N_r(\alpha) + N_c(\alpha))$, where N_c (N_r) is the number of double hit events in the column (row). A is proportional to the degree of linear polarization. The frame time was set to 60 ms in order to have a tolerable occupancy of about 190 triggered pixels in each image. The threshold was set to 3.5 keV in order to be sensitive to small amounts of energy deposited in one pixel at the beginning of the track of the photoelectron.

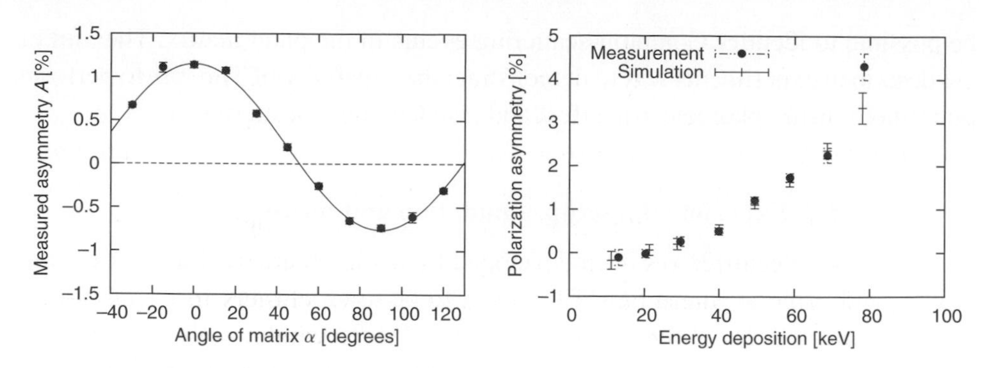

Figure 5.1 Measured asymmetry $A(\alpha)$ (left-hand side) and measured/simulated polarization asymmetry A_{pol} in dependence on energy deposition (right-hand side).

The asymmetry A was measured for different angles α between the plane of linear polarization and the rows of the pixel matrix. The angle α was varied by rotating the detector in steps of 15° from −30° to 120°. The numbers of double hits measured without target are subtracted from the corresponding numbers measured with the target. The measured asymmetry as a function of α is shown in Figure 5.1. The function $A(\alpha) = A_{pol} \cdot cos(2\alpha) + A_{app}$ was fitted to the data to extract polarization asymmetry A_{pol} and apparative asymmetry A_{app}. We found a polarization asymmetry of $A_{pol} = (0.957 \pm 0.019)\%$ and an apparative asymmetry of $A_{app} = (0.194 \pm 0.013)\%$. The simulation resulted in a polarization asymmetry of $A_{pol} = (0.888 \pm 0.042)\%$ which agrees with the measured value taking into account uncertainties in the simulation, e.g., changes of the emission spectrum of the X-ray tube during the long measurements, the estimated thickness of the silver-containing glue and the estimated diameter of the bump-bonds. The Timepix was operated in the time-over-threshold mode to measure the energy dependence of A_{pol}. We obtained a relative energy resolution of 7% at 59.5 keV. The energy depositions of the two adjacent pixels in each double hit event were added. The number of double hits (N_c and N_r) was counted in energy bins with a width of 10 keV. Measurements at the angles $\alpha = 0°$ and $\alpha = 90°$ have been performed and subtracted from each other to remove apparative asymmetries. Figure 5.1 shows the measured and simulated polarization asymmetry in dependence on energy deposition. The agreement over a wide energy range is excellent showing that the behaviour of the Timepix is well understood. The simulated polarization asymmetry increases from 0.2% at an energy deposition of 29 keV to 3.4% at 78 keV. This behaviour is expected because of the increasing track lengths of the photoelectrons. The average apparative asymmetry found in the time-over-threshold measurement is $A_{app} = (0.149 \pm 0.066)\%$.

Our measurements concerning Compton scattering are described in more detail in [4]. The time-to-shutter mode was used to identify coincidences of Compton scattering events in the sensor. We applied a clock frequency of 28.9 MHz to the detector to measure the detection times. The frame time was set very short to 311 μs so that the mean occupancy was 5.44 pixels. The coincidently triggered pixels in coincidence windows are identified in each frame. Next, the spatial distribution of coincidently triggered pixels is analyzed and clusters of adjacent pixels are identified. The coordinates of the cluster are calculated as the average x- and y-positions of all involved pixel centres. These averaged coordinates are used to calculate the angle β of the line connecting the centres of the two clusters in respect to the direction of the rows of the matrix. Only coincidence windows with exactly two separated triggered clusters are considered in the data analysis. In order to reduce the impact of random coincidences, which are characterized by larger distances between the clusters, the distance between two clusters has to be smaller than 3 mm to be identified as a Compton scattering event. We also specified a minimum distance of $r_{min} = 0.441$ mm to obtain an angular resolution of 10°. We restricted the analysis to the central area with a diameter of 254 pixels to avoid corner effects. Due to the pixelation of the sensor matrix, the geometrical acceptance of the angle bins in β is not uniform. We determined the relative acceptance of each angle bin by a simulation with unpolarized radiation impinging on the detector. These values are used to correct the number of detected events in the angle bins of the measurement and the simulation for polarized radiation by -7% to $+14\%$. The number of events with two clusters, having an angle β between 0° and 180°, was determined with a binning of 10°. Target and nontarget measurements were subtracted from each other. The measured angular distribution was corrected for random coincidences. Figure 5.2 shows the resulting modulation curve in comparison to simulation results.

For the orientation angle of the plane of linear polarization we obtained $\phi_0 = 7.0° \pm 3.7°$ which is consistent with our expectation of $\phi_0 = 0°$ taking into account

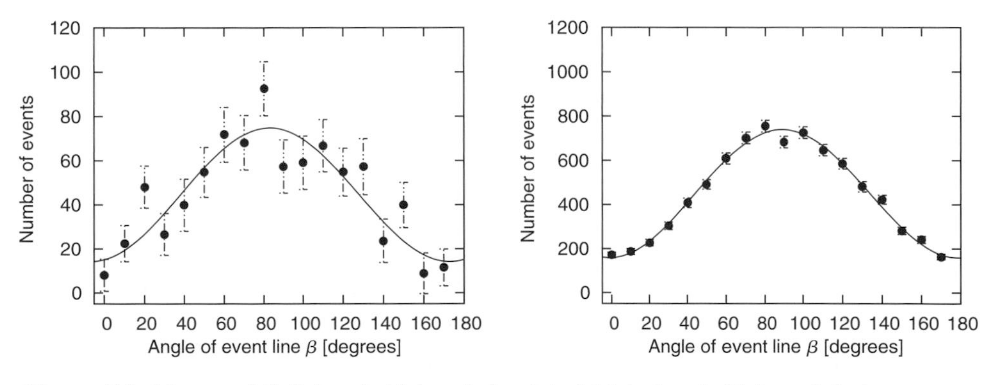

Figure 5.2 Measured (left-hand side) and simulated (right-hand side) modulation curve.

also inaccuracies in the alignment of our setup. The measured modulation factor is $\mu^{meas} = (68.1 \pm 16.4)\%$. With the estimated degree of polarization of the detected Compton scattering events of $P = 97.1\%$, we obtain a modulation factor of $\mu^{sim} = \mu^{sim}_{100} \cdot P = (62.8 \pm 5.0)\%$.

The following considerations only give a feeling of the best performance that could be achieved in future. The given values for the MDP thus have to be seen as absolute minimum values. We ignore the influence of background radiation, the problem of powering and cooling a large array of such detectors and the challenge of reducing and transmitting the large amount of data. Therefore these considerations investigate more the potential of using a highly segmented silicon sensor array rather than using the Timepix detector itself. With these assumptions the MDP is given by $MDP_{99} = 4.29/(\mu_{100} \cdot \sqrt{\epsilon \cdot S \cdot T})$[5]. Efficiency and modulation factor, μ_{100}, were determined for different silicon sensor thicknesses and for irradiation of a Timepix with X-ray photons in the energy range between 40 keV and 100 keV from the Crab Nebula. S was calculated for an array of 5000 Timepix detectors with a sensor area of one square metre. We assumed that a coded mask array reduces the incoming flux by 50%. The MDP is 2.9% for 300 μm sensor-thickness, 1.15% for a thickness of 1000 μm and 0.81% for a 2-mm thick silicon sensor. The modulation factor seems to be lower for a thick silicon sensor, which may be due to the increased acceptance in the scattering angle θ, especially at short distances between the clusters. We find a better polarimetric performance for thicker sensor layers. Figure 5.3 shows the MDP in 10 keV broad energy bands for a silicon sensor thickness of 2 mm.

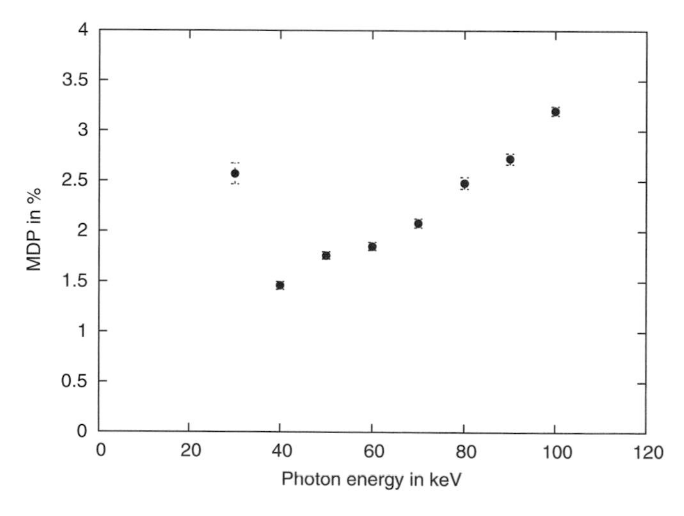

Figure 5.3 MDP in energy bands of 10 keV width in the Crab spectrum (observation time of 10^6 seconds, effective detector area of 5000 cm^2, 2-mm thick silicon sensor).

5.3 Conclusion

We have successfully demonstrated that the Timepix detector with silicon sensor can be used as X-ray polarimeter between about 27 and 84 keV. Imaging polarimetry is already possible using the photoelectric effect, but with a very small analyzing power. This technology cannot compete with polarimeters based on micro-pattern gas detectors, like the one described in [6], but addresses higher photon energies. One drawback of the Timepix is that it is not possible to determine both the energy deposition and the detection time simultaneously. Therefore it is not possible to exploit Compton scattering to determine the degree of linear polarization as function of photon energy. Additionally, the information on the point of impact of the photon is not directly accessible. A large modulation factor was found when using Compton scattering. Further investigations are needed to determine optimal pixel pitch, sensor thickness and geometry of the detector arrangement and to optimize the ASIC to see if an application in X-ray astronomy is possible.

References

[1] Tsunemi, H., Hayashida, K., Tamura, K., Nomoto, S., Wada, M., Hirano, A. & Miyata, E. (1992). *Nucl. Instr. Meth.* **A321**, 629–631.
[2] Llopart, X., Ballabriga, R., Campbell, M., Tlustos, L. & Wong, W. (2007). *Nucl. Instr. Meth.* **A581**, 485–494.
[3] Michel, T. & Durst, J. (2008). *Nucl. Instr. Meth.* **A594**, 188–195.
[4] Michel, T., Durst, J. & Jakubek, J. (2009). *Nucl. Instr. Meth.* **A603**, 384–392.
[5] Weisskopf, M.C., Elsner, R.F., Hanna, D., Kaspi, V.M., O'Dell, S.L., Pavlov, G.G. & Ramsey, B.D. `http://arxiv.org/abs/astro-ph/0611483`
[6] Bellazzini, R., et al. (2007). *Nucl. Instr. Meth.* **A579**, 853–858.

6

High-energy polarized photon interactions with matter: simulations with Geant4

F. Longo

University of Trieste and INFN, sezione di Trieste, Italy

G. O. Depaola & M. L. Iparraguire

Universitad Nacional de Cordoba, Argentina

A detailed simulation of the interactions of polarized photons is required to design new gamma-ray telescopes. Two new classes have been designed and implemented to describe the interactions of polarized photons in the pair-production regime.

6.1 Introduction

A detailed simulation of the geometry and of the involved physics processes is required from the very early stages of a new instrument development. Indeed both the development and the optimization of new-instrument concepts, performed with detailed comparisons between different configurations, as well as the detailed characterization of instrument responses and the verification of the scientific objectives of a new instrument, require the same detailed simulation.

Geant4 is an object-oriented toolkit for the simulation of high-energy physics detectors that is now widely used in nuclear physics, medical physics, astrophysics, space applications, radiation background studies, etc.[1]. Geant4 is supported by a large international collaboration with the participation of various institutes around the world. It is also an experiment in the application of rigorous software engineering methodologies.

In particular the physics is open to the user, who has the possibility to select among different models of the same physics process or to extend existing models to cover new requirements.

X-ray Polarimetry: A New Window in Astrophysics, eds. R. Bellazzini, E. Costa, G. Matt and G. Tagliaferri. Published by Cambridge University Press.

6.2 Polarized photon interactions in Geant4

The Compton and Rayleigh processes are affected by the polarization of the incoming radiation, and even an unpolarized beam acquires a certain degree of polarization after a Compton or Rayleigh event. So a description of these processes in which the polarization is present is particularly relevant even though the incoming radiation is not polarized.

The basic formulae to sample the azimuthal angle and the polarization vector of the scattered photon were obtained by Depaola[3] following the procedure suggested by Heitler[2], who described the Compton process in the case of an incident linearly polarized radiation. The usual description of polarization phenomena requires a careful consideration of all the angles involved and the general features of polarization sensitive processes are not readily apparent from the complicated equations that result. The results obtained in the analysis of the linear polarization case can also be used to describe the unpolarized case.

The cross section for the Compton scattering is:

$$d\sigma = \frac{1}{4} r_0^2 d\Omega \frac{k^2}{k_0^2} \left[\frac{k_0}{k} + \frac{k}{k_0} - 2 + 4\cos^2 \Theta \right]$$

where k_0 is the energy of incident photon, k is the energy of the scattered photon, Θ is the angle between the two polarization vectors, and r_0 is the electron classical radius. It gives the absolute value of the probability that a primary photon of energy k_0, while passing through an absorber whose thickness is such that the absorber contains one electron per square centimeter, will suffer a particular collision, with a free electron at rest, from which the scattered photon emerges with energy k, within solid angle $d\Omega$ and is polarized with the electric vector that makes an angle Θ with the electric vector of the incident vector.

To test the capability of our method to give the polarization and depolarization effects of the scattered radiation for the unpolarized and polarized incident radiation respectively we performed a detailed simulation and compared the theoretical predictions with the results we obtained[3]. The simulations were done using the Geant4 tool kit introducing classes for the polarized Rayleigh and Compton process. In the unpolarized simulation, the direction of the polarization vector of the incident radiation was taken at random. The method developed by Depaola[3], starting from the Klein-Nishina cross section, is able to reproduce the polarization status of the scattered radiation, independently from the polarization of the incoming beam. The polarization should always be taken into account in the description of Compton scattering, because even an incoming unpolarized radiation acquires a certain degree of polarization after a scattering process, and then its azimuthal distribution can no longer be taken as isotropic in a second scattering process. Our

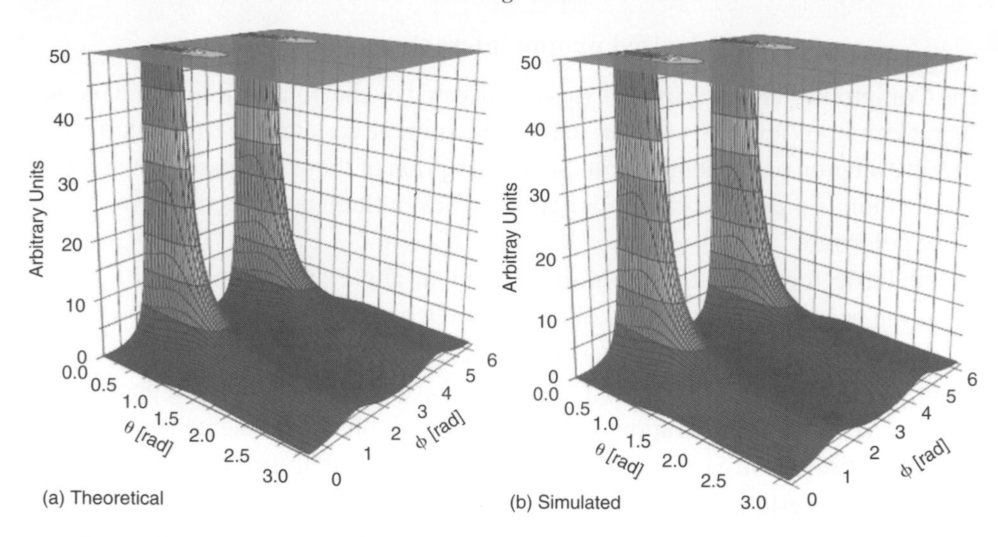

Figure 6.1 Intensity ratio between the cross section for events scattered in the plane perpendicular to the polarization vector and the cross section for events scattered in the parallel plane for a 100% linearly polarized 1-MeV photon[3].

code is able to reproduce this effect in a correct and precise manner and it is suitable to be used in simulation programs of detectors in the Compton regime.

6.3 High-energy polarized photon interactions

6.3.1 Pair production in the electric field of the nucleus

The knowledge of the polarization direction of incoming high-energy gamma-rays is very important in astrophysics since linearly polarized rays constitute a distinctive byproduct of matter accretion onto black holes or neutron stars. At present, the Large Area Telescope (LAT) on board the Fermi Gamma-ray Space Telescope[4] shows an improved angular and energy resolution in comparison to EGRET on board the Compton Gamma Ray Observatory[5]. This project is designed to measure gamma-rays in the energy range from 20 MeV to more than 300 GeV where the interaction with matter at this energy range is dominated almost exclusively by pair production. In a previous work[6] the cross section of pair production was extensively studied looking for what characteristics of this process can be useful to determine the polarization of gamma-rays. It was found that the polarization information lies in the azimuthal distribution of the particle moments. The differential cross section for high-energy linearly polarized gamma-rays was found by Depaola[7]. Using a Monte-Carlo procedure to integrate this cross section over polar angles and energy one obtains the spatial azimuthal distribution. The entire surface of this distribution

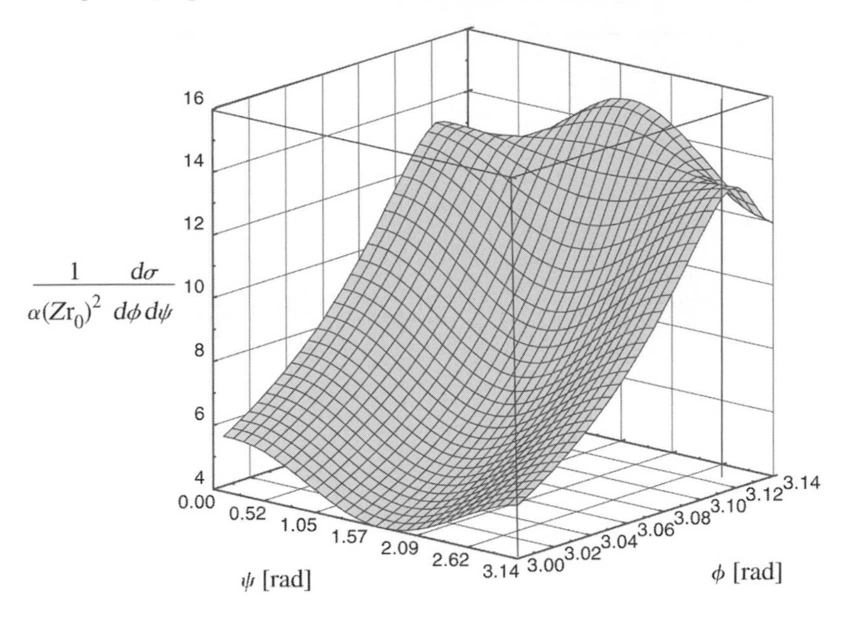

Figure 6.2 Spatial azimuthal distribution of a pair created by a 100% linearly polarized 100 MeV photon[7].

was evaluated with the superposition of two fitting functions, easier to calculate in a Monte-Carlo code, smoothly joining in a particular angular region[7].

The Monte-Carlo method, originally developed in Geant3, was reimplemented in Geant4 to sample the azimuthal distribution for linearly polarized gamma-rays. When integrating the azimuthal distribution over the polarization dependence to obtain the distribution for nonpolarized gamma-rays, it was found that the coplanars approximation used by some codes is correct only for high energies (>200 MeV). From the results obtained in the Monte-Carlo simulations to test the capability of the instruments similar to Fermi/LAT to detect gamma-ray polarization, we conclude that it will be very difficult for these missions to detect it.

6.3.2 Pair production in the electric field of the electrons

The triple production shows very interesting characteristics for detecting gamma-ray polarization of high-energy beams. One can consider that the azimuthal distribution of the recoil electron does not depend on the photon energy. The double-differential cross section was calculated by Depaola and Iparraguire[8] through numerical integration of the cross section obtained using only Feynman diagrams of the process obtained by Borsellino[9]. This calculation was compared with the asymptotic expression obtained by Boldyshev et al.[10].

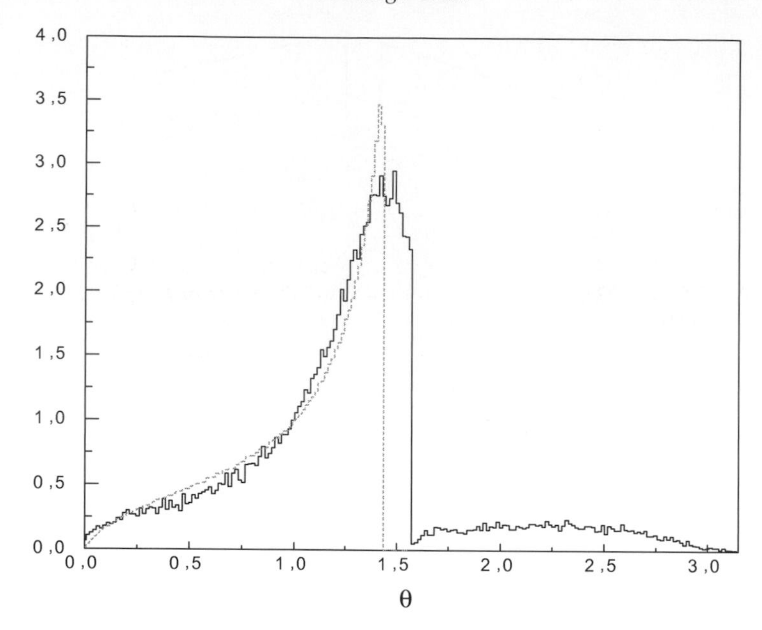

Figure 6.3 Polar angle distribution for the recoil electron. Continuous line: preliminary simulation; dashed line: theoretical expectations[8].

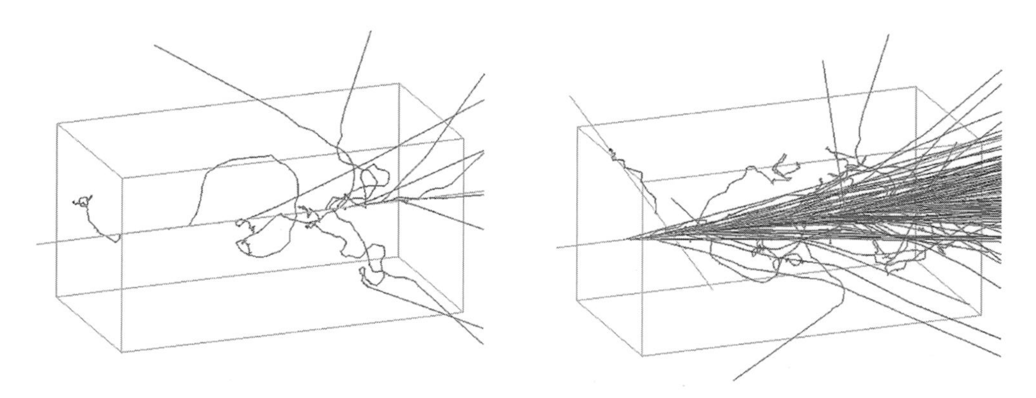

Figure 6.4 Example of pair production simulation with Geant4. Right: only recoil electron. Left: including the created pair.

A new algorithm was implemented in[8] with a very high efficiency and it is suitable to be included in any of the most popular Monte-Carlo tool kits for simulating the radiation transport. A preliminary simulation result, using a very simple model (a gas cube) shows that the method used to determine the particle tracks in gamma-ray telescopes do not produce any distinguishable polarization modulation. However the method used to analyze the photoelectrons in a gas photoelectric detector shows a difference between the polarized and nonpolarized case.

These results are also indicative of the potential to determine the high-energy gamma-ray polarization by pair production in the electric field of the electrons.

References

[1] Agostinelli, S., et al. (2003), *Nuclear Instruments and Methods in Physics Research Section A*, **506**, 250.
[2] Heitler, W. (1954), *The Quantum Theory of Radiation*. Oxford: Clarendon Press.
[3] Depaola, G.O. (2003), *Nuclear Instruments and Methods in Physics Research Section A*, **512**, 619.
[4] Atwood, W.B., et al. (2009), *The Astrophysical Journal*, **697**, 1071.
[5] Kanbach, G. et al (1988), *Space Science Reviews*, **49**, 69.
[6] Depaola, G.O., Kozameh, C.N., Tiglio, M.H., (1999), *Astroparticle Physics*, **10**, 175.
[7] Depaola, G.O. (2000), *Nuclear Instruments and Methods in Physics Research Section A*, **452**, 298.
[8] Depaola, G.O. & Iparraguire, M.L. (2009), submitted to *Nuclear Instruments and Methods in Physics Research Section A*.
[9] Borsellino, A. (1947), *Nuovo Cimento*, **4**, 112.
[10] Boldyshev, et al., (1994), *Phys. Part. Nucl.*, **25**, 3

7

The GPD as a polarimeter: theory and facts

P. Soffitta, E. Costa, S. Fabiani, F. Lazzarotto, F. Muleri & A. Rubini

IASF-Roma/INAF, Italy

R. Bellazzini, A. Brez, M. Minuti, M. Pinchera & G. Spandre

INFN-Pisa, Italy

The advent of a new generation of X-ray polarimeters based on the photoelectric effect poses the problem of their calibration. We devised and built a calibration facility aimed at the study of the performances of photoelectric X-ray polarimeters such as the Gas Pixel Detector (GPD). The calibration facility exploits the 45° Bragg diffraction from crystals of both X-ray lines characteristic of X-ray tubes and from continuum. A set of linear and rotary stages allows the GPD to be calibrated on its whole surface. We successfully tested the GPD filled with a mixture of He-DME 30-70 at one atmosphere. We measured the modulation factor at 2.69 keV and 4.51 keV. We also studied the homogeneity of the modulation factor, of the angular phase and of the position reconstruction capability on the surface of the GPD.

7.1 Introduction

Since early 2000[1; 2] we have been developing true 2-D X-ray polarimeters based on the photoelectric effect. The Gas Pixel Detector (GPD), an evolution based on the use of an ASIC CMOS readout chip [3; 4], has been calibrated at energies greater then 5 keV with a Thomson based X-ray polarizer. However, the sensitivity of the GPD at the focus of conventional X-ray optics peaks below 5 keV. In this regime the photoelectric effect severely competes with Thomson scattering. For this reason we developed a facility[5] aimed at the production of polarized X-rays in the few keV band. We exploited the 45° diffraction of X-ray lines or continuum produced by relatively low-power (50 W) or very low-power (200 mW) X-ray tubes. Close to this reflection angle the X-ray lines are highly polarized. Continuum is also highly polarized provided that the acceptance angle around 45° is small enough. To keep it small,

X-ray Polarimetry: A New Window in Astrophysics, eds. R. Bellazzini, E. Costa, G. Matt and G. Tagliaferri. Published by Cambridge University Press.

below 1°, we inserted two capillary plates with $L/D = 40$ at the input of the polarizer and $L/D = 100$ at the output of the polarizer. L is the length of the channel and D is diameter of the channel (usually 10 μm). We found that this combination maximizes both the flux and polarization. The use of capillary plates is mandatory to eliminate the scattered X-rays and the possible fluorescence produced by the crystal. The GPD is placed in a fully motorized positioning system composed either of linear or rotary stages. The control software is integrated with the positioning software, so the polarimeter can obtain data and can be oriented virtually in any position. The alignment of the crystal with respect to the capillary plates is performed by using a dedicated commercial Si-PiN detector and maximizing the counting rate while tilting the crystal itself. When using the continuum, by the width of the spectrum we can estimate the amount of polarization. The detector we tested was filled with a mixture 30-70 He-DME, with an absorption gap of 1 cm and a Be window of 50 μm. The ASIC CMOS readout chip has 105 600 pixel with a pitch of 50 μm in a hexagonal pattern.

7.2 Measurement of the modulation factor at two energies

By using the diffracted continuum, and the diffracted X-ray line at 3.7 keV we already demonstrated the low-energy polarization sensitivity of the GPD[6]. Here we measure the modulation factor of the GPD at 2.69 keV (Lα) and 4.51 keV (Kα) by using respectively a rhodium tube with a Ge (111) crystal and a titanium tube with a CaF_2 (220) crystal. We usually set the HV just below the next order. In Figure 7.1 we present the modulation curves obtained. X-rays, in both configurations, were collimated with the usual 40 and 100 L/D. The modulation factor found was 33% at 2.69 keV (see Figure 7.1(a)) and 42% at 4.51 keV (see Figure 7.1(b)). The modulation at 2.69 keV is particular encouraging since we reached the Monte-Carlo simulation expectations at low energy[6].

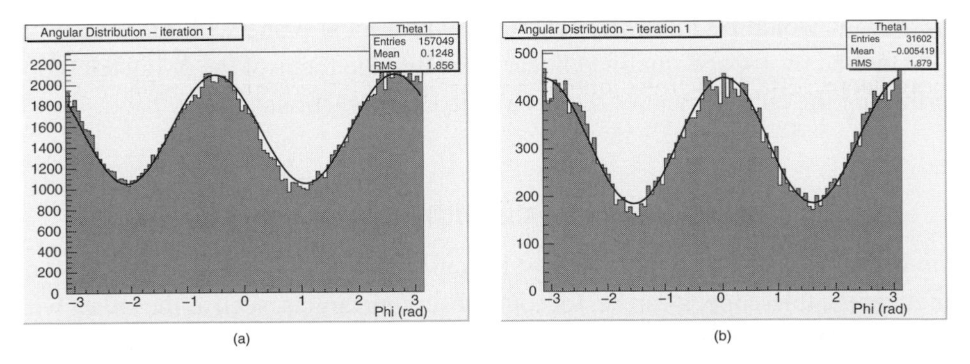

Figure 7.1 (a) Modulation factor at 2.69 keV (33.20% ± 0.27%). (b) Modulation factor at 4.51 keV (42.0% ± 0.6%).

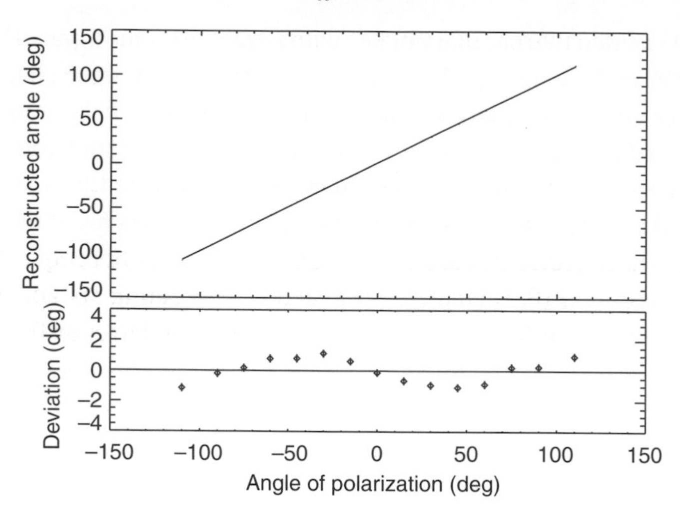

Figure 7.2 Sensitivity to polarization angle of the GPD.

7.3 Sensitivity to polarization angle

The sensitivity of the polarimeter to measure the polarization angle has been studied[7], rotating the GPD with respect to the polarized X-ray beam produced by the titanium X-ray tube. We used the usual combination of capillary plates, and we collected about 10^5 counts for each position angle. The results are shown in Figure 7.2 where the reconstructed polarization angle is plotted against the mechanical angle. The data are fitted with a straight line with angular coefficient consistent with unity (being the parameter of the linear fit: $y = m * X + q$, with $m = 0.9969 \pm 0.0011$, $q = 2.087° \pm 0.068°$). The residuals show systematic deviation of at most $1°$, a small value which can be easily calibrated since they have a well-defined sinusoidal behavior. This measure gave us confidence that the GPD does not need rotation, reducing very much the complexity and the reliability of a focal plane assembly. Also it would be relatively easy to devise a focal plane with more than one GPD moved by a space-qualified linear stage into the focus of the X-ray telescope, permitting the enlargement of the polarimetric observational energy band.

7.4 Characterization of the GPD on its detecting surface

The active area of the polarimeter is 1.5 cm $\times$ 1.5 cm. Depending on the focal length of the telescope, it can be, for some observing targets, such as the pulsar wind nebulae (PWN) or the supernovae remnants (SNR) without a plerion completely covered by the source during observation. Also possible pointing errors can be present during a typical observation. Pointing errors can be either absolute or as a consequence of drifting. It is therefore mandatory to measure the characteristics of the GPD on the detecting surface. The most important parameters to measure are the modulation factor, the angular phase and the position reconstruction. We exploited the usual diffracted photons at 4.51 keV from the titanium X-ray tube, with diaphragm of 0.5 mm diameter at the X-ray output to have a small spot. We set the detector in 25 different positions organized in a 5×5 matrix of points separated by 2.25 mm. We collected about 30 000 events in two hours of acquisition for each position. We used the (0,0) position as reference, collecting six acquisitions each one every 10 hours to look for possible drift. We did not detect drift in the relevant parameters.

7.4.1 Modulation factor and angular phase on the detector surface

We measured the modulation factor for all the 25 positions on the surface. The results are shown in the Figure 7.3(a). The measurement of the modulation factor is constant suggesting good stability of polarization measurement. We also measured the angular phase in the same 25 different positions and we plotted the results in Figure 7.3(b). Each point in the abscissa represents a measure of the angular phase at a given position on the detector surface. The angular phase is constant within 3σ. The GPD therefore appears to be very homogeneous as far as modulation factor and angular phase are concerned.

7.4.2 Position resolution

The GPD has a very good position resolution. This derives from the use of algorithms that permit the calculation, for each track, of the impact point distinct from the

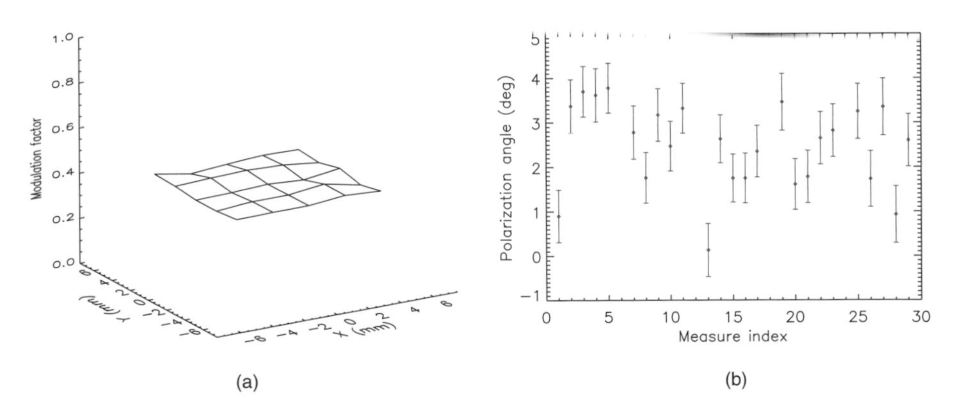

Figure 7.3 (a) Modulation factor (4.51 keV) on the detector surface. (b) Polarization angle (4.51 keV) on the detector surface.

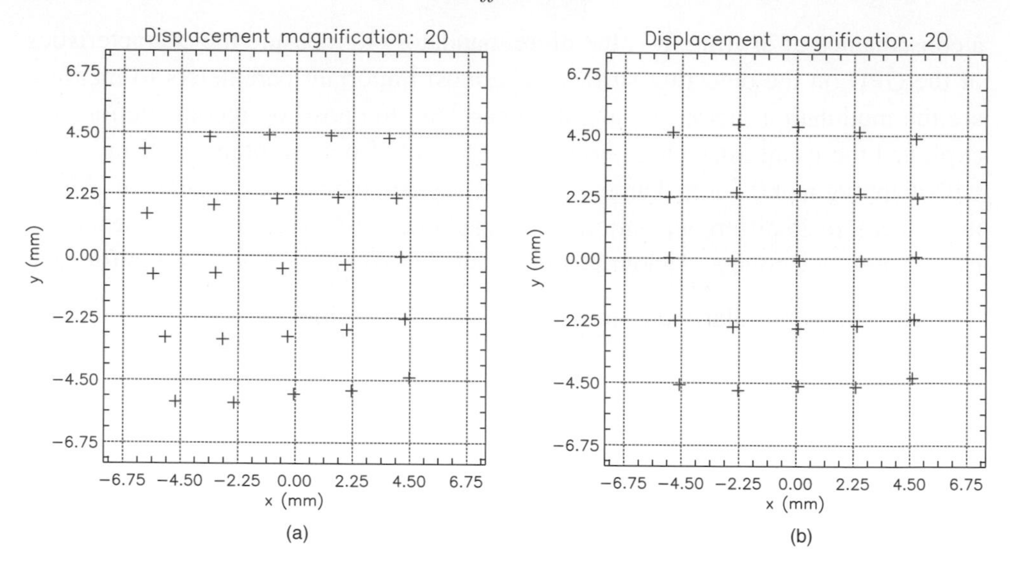

Figure 7.4 (a) Raw mean positions of the collimated X-ray beam on the detector surfaces in the 25 position matrix. The displacement with respect to the nominal position are magnified a factor of 20. (b) The same mean positions of Figure 7.4 but after the software compensation of the rotation of the linear stages. The displacements with respect to the nominal position are magnified a factor of 20.

barycenter. We measured the mean position obtained by the matrix of 25 positions. We compared the reconstructed mean positions with the nominal positions in the detector reference. It is important to measure at what level the position reconstruction is homogeneous to study how much the electric field is parallel to itself within the GPD. If there was a focusing effect or a de-focusing effect this would give a definite pattern in the 2-D plot of the mean of the impact points. Moreover, we cannot exclude the possibility that the motorized linear positioning system introduces rotation with respect to the GPD axis and, actually, this is the case (see Figure 7.4(a)). By applying a linear fitting procedure, we compensated for the rotation of linear motorized stages with respect to the GPD axis (we assume that the axis of the GPD is perfectly perpendicular). This rotation is anyway very small being 19.1 arcseconds in Y and only 12.8 arcseconds in X. The displacements shown are, actually, magnified by a factor of 20. In Figure 7.4(b) we plotted the mean positions after applying the compensation (i.e. shift and rotation). The displacements are still magnified of a factor of 20. The quadratic residuals in X and Y show that the maximum displacement is 25 μm. This value is very small, being only half of the pitch. The electric field is, therefore, homogeneous inside the detector, the electrodes being parallel to each other.

7.5 Conclusions

We devised and built a facility for the calibration and the characterization of the GPD. We used the facility to explore the properties of the GPD on its active surface and to measure its sensitivity to polarization angle rotation. We found good homogeneity of the modulation factor and of the angular phase. Also we obtained an almost perfect measure of the polarization angle rotation. We also detected good modulation at 2.69 keV and 4.51 keV. The GPD is found to be suitable to be used as spectro-imaging-polarimeter in upcoming X-ray missions.

References

[1] Costa, E., Soffitta, P., Bellazzini, R., Brez, A., Lumb, N., & Spandre, G. (2001). *Nature*, **411**, 662.

[2] Bellazzini, R., Baldini, L., Brez, A., Costa, E., Latronico, L., Omodei, N., Soffitta, P., & Spandre, G. (2003). *Nucl. Instr. and Meth. A*, **510**, 176–184.

[3] Bellazzini, R., Spandre, G., Minuti, M., Baldini, L., Brez, A., Cavalca, F., Latronico, L., Omodei, N., Massai, M.M., Sgroó, C., Costa, E., Soffitta, P., Krummenacher, F., & de Oliveira, R. (2006). *Nucl. Instr. and Meth. A*, **566**, 552–562.

[4] Bellazzini, R., Spandre G., Minuti, M., Baldini, Brez, A., Latronico, L., Omodei, N., Razzano, M., Massai M.M., Pesce-Rollins, M., Sgró, C., Costa, E., Soffitta, P., Sipila, H., & Lempinen E. (2004). *Nucl. Instr. and Meth. A*, **579**, 853–858.

[5] Muleri, F., Soffitta, P., Bellazzini, R., Brez, A., Pinchera, M., Rubini, A., & Spandre G. (2008). *Society of Photo-Optical Instrumentation Engineers (SPIE) Conference Series*, **7011** doi: 10.1117/12.789605.

[6] Muleri, F., Soffitta, P., Baldini, L., Bellazzini, R., Bregeon, J., Brez, A., Costa, E., Frutti, M., Latronico, L., Minuti, M., Negri, M.B., Omodei, N., Pesce-Rollins, M., Pinchera, M., Razzano, M., Rubini, A., Sgró, C., & Spandre, G. (2008). *Nucl. Instr. and Meth. A*, **584**, 149–159.

[7] Muleri, F. (2009). 'Expectation and Perspectives of X-ray Photoelectric Polarimetry' *PhD Thesis*, Tor Vergata University, Rome, Italy.

8

Ideal gas electron multipliers (GEMs) for X-ray polarimeters

T. Tamagawa, A. Hayato, T. Iwahashi, S. Konami & A. Fumi
RIKEN

on behalf of the RIKEN X-ray Polarimeter team

We have developed gas electron multipliers (GEMs) for space science applications, in particular for X-ray polarimeters. We have employed a laser etching technique instead of the standard wet etching for the GEM production. Our GEMs showed no gain increase after applying high voltage and kept the gain for more than two weeks at a level of 2% (RMS). We show the gain properties and the results of some acceleration tests to mimic a two-years low-Earth-orbit operation in this paper.

8.1 Introduction

The GEM is one of the recently developed micro-pattern gas detectors[1]. A dense pattern of through-holes is drilled in an insulator substrate, which is typically polyimide, sandwiched by thin copper foils. The surface and cross-section micrographs of a GEM are shown in Figure 8.1. When high voltage is applied to the copper electrodes in an appropriate gas, the GEM works as an electron multiplier. GEMs are used in many fields such as high energy and nuclear physics, X-ray imaging, etc. In astrophysics, photoelectric X-ray polarimeters, in which the GEM is a key device to multiply an electron cloud whilst retaining its shape, are the most interesting application[2].

We have produced GEMs since 2002 for X-ray polarimeters[3; 4; 5]. The standard method to produce GEMs is a wet etching technique, while our method is laser etching, which has many advantages. Cylindrical holes are easily formed with the laser etching. The capability to drill cylindrical holes helps in forming finer-pitch holes on a thicker substrate. The finest pitches we have produced were 50 and 80 μm on 50 and 100 μm thick substrates respectively.

X-ray Polarimetry: A New Window in Astrophysics, eds. R. Bellazzini, E. Costa, G. Matt and G. Tagliaferri. Published by Cambridge University Press.

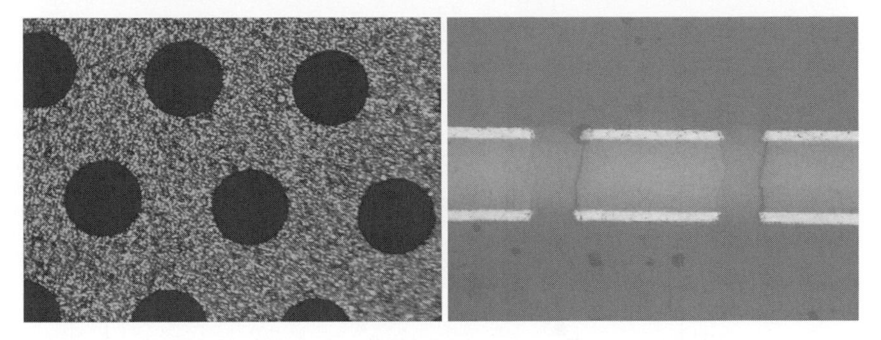

Figure 8.1 The micrographs of a GEM processed with the laser etching. The hole pitch of the GEM is 140 μm and the hole diameter is 70 μm. The thickness of the substrate is 100 μm.

8.2 Gain properties

8.2.1 Gain curves

The thickness of a standard GEM is 50 μm, while we have successfully produced the GEM with a 100 μm thick substrate. Figure 8.2 shows the effective gain of a 100 μm thick GEM in an Ar/CO_2 (70%/30%) gas mixture. A gain of 10^4 was achieved at an applied voltage of 700 V per GEM. For comparison, a double 50 μm thick GEM stack is superposed on the same figure. The effective gain of the 100 μm thick GEM was about 60 times higher than that of the double 50 μm GEM stack, when we operated the double GEM at the same combined voltage as the 100 μm thick GEM. Obtaining an appropriate gain at lower applied voltage decreases the risk of discharges.

8.2.2 Gain stability

It is known that charging-up inside the GEM holes increases gain as a function of time after applying high voltage. Figure 8.3 shows the gain variation of a GEM manufactured with laser etching. RMS of the gain variation was less than 4%. For comparison, the gain curve of a GEM produced with wet etching is shown in the figure. Although both GEMs had identical geometry and were tested with the same setup, the gain of the wet etched GEM was gradually increased.

The gain stability is quite important for operating a GEM in low-earth orbit (LEO). The satellites pass through the south Atlantic anomaly (SAA) every 90 minutes. During the passage of the SAA, high voltage applied to the GEM should be turned off to avoid unexpected discharges, and the high voltage will be ramped up again after the SAA passage. It is necessary to keep the gain stable at least for 90 minutes. Our GEMs satisfied the requirement.

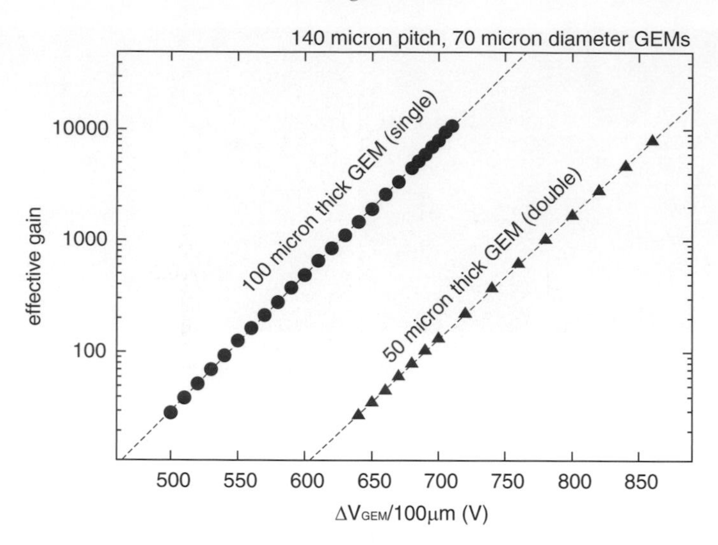

Figure 8.2 Gain curves of our laser etched GEMs. The gain curves of a single layer of 100 μm thick GEM and double 50 μm thick GEMs are shown.

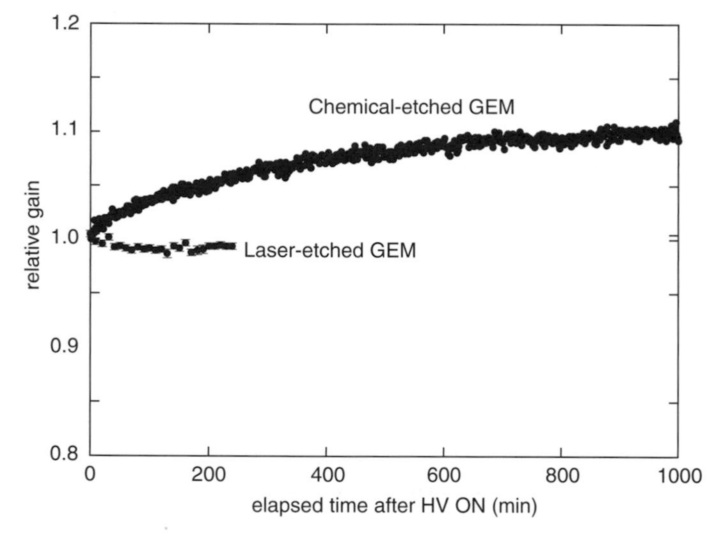

Figure 8.3 The gain stability of laser- and wet-etched GEMs. The laser-etched GEM shows no gain increase as a function of elapsed time after turning on high voltage.

8.2.3 Gain uniformity

Gain uniformity is one of the important factors to track photoelectrons precisely. We measured the gain uniformity of our GEMs with a collimated X-ray beam with a spot diameter of 200 μm. The beam spot was moved across the GEM in steps of

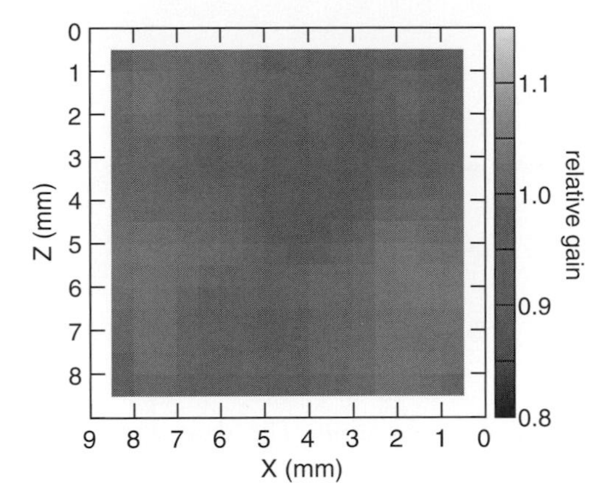

Figure 8.4 The gain uniformity of the 50 μm pitch GEM.

500 μm. Figure 8.4 shows the distribution of relative gain normalized by the mean value of measured gain. RMS of the distribution was 3%. The gain uniformity of the other GEMs were almost the same level as shown in Figure 8.4 (RMS = 3–6%).

8.3 Space environment tests

8.3.1 Accelerated test of high voltage on and off

Although the gain of our GEMs is stable after turning on high voltage, the high voltage is frequently ramped up and down when a satellite passes the SAA. The total number of the ramp up and down is about 6000 times for two years mission life. To mimic the SAA passage, we measured the gain whilst turning on and off high voltage for two weeks. The gain was corrected with measured pressure and temperature. The correction method was mentioned in [5]. No damage was observed during the test, and the gain stability was 4% (RMS) (Figure 8.5).

8.3.2 Heavy ion irradiation

Many charged particles are present in space; some of them are trapped in the magnetic field of the Earth, and the others are randomly incident galactic cosmic rays. The main component of cosmic rays is the proton. However, the energy deposit of protons is at the same level as X-rays. The most dangerous element is iron, which has significant flux in space and large energy loss in the detector ($dE/dx \propto Z^2$). Since heavier elements than iron have a few orders of magnitude lower flux, we can ignore the impact of those elements. The estimated mean flux of the iron ions at the LEO with an altitude of 600 km and an inclination angle of 30° is 2.0×10^3 cts yr^{-1}cm^{-2}[6].

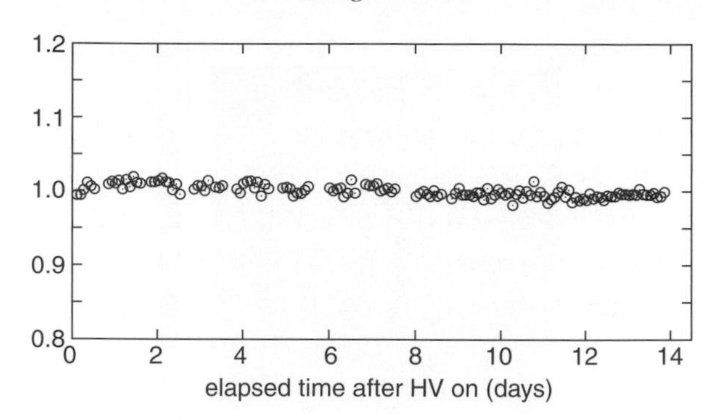

Figure 8.5 Two weeks' gain variation. The data were taken during intervals between high voltage ramp up and down. A total of 6500 on/off cycles were performed during the test. The gain was corrected with measured temperature and pressure, and normalized by the gain of the first measurement ($t = 0$).

We have irradiated the GEMs with fully stripped Fe ions with an energy of 500 MeV/n. The Fe beam was provided by the synchrotron heavy ion accelerator (HIMAC) located in Chiba, Japan[7]. The beam flux was about 430 cts spill^{-1}cm^{-2}, where the spill consists of a 1.7 s flat top and a 1.6 s off beam. Total amount of irradiation corresponded to about 40 years operation in the LEO. During the test, we monitored the gain of the GEMs with 5.9 keV X-rays from an ^{55}Fe radioactive source, with the result that no damage and no significant variation of the gain were observed.

8.3.3 *Proton irradiation*

Inside the SAA, many trapped protons penetrate through the GEM. Since high voltage will be turned off during the SAA, we are not concerned about damage caused by electric discharges. However, the passage of a significant flux of protons may damage the insulator layer of GEMs. The mean flux of trapped protons at the LEO with the same orbital parameters mentioned above is 1.7×10^9 cts yr^{-1}cm^{-2}.

We irradiated a GEM with a proton beam with an energy of 160 MeV at HIMAC. The beam flux was about 5×10^7 cts spill^{-1}cm^{-1}, and the total exposure time was about 80 minutes, representing 40 years' operation in the LEO. We measured leakage current between the GEM copper electrodes before and after the proton irradiation, with the result that no degradation was observed.

8.4 Summary and perspectives

We have developed GEMs with laser etching for X-ray polarimeters. Our GEMs have very stable gain as a function of time after applying high voltage. The stability is a huge advantage in operating the polarimeter in LEO. From our space environment tests by irradiating with protons and Fe ions, no significant damage was observed. Our GEMs are almost ready to fly.

Our GEMs are used for the X-ray polarimeters onboard Gravity and Extreme Magnetism SMEX (GEMS) led by NASA/GSFC. On the other hand, imaging X-ray polarimeters, in which our GEMs are combined with the ASIC developed by INFN/Pisa, are being considered for a future Italian or Japanese X-ray polarimeter mission, and for the International X-ray Observatory (IXO).

References

[1] Sauli, F. (1997). *Nucl. Instr. and Meth.* **A386**, 531.
[2] Costa, E., et al. (1997). *Nature* **411**, 662.
[3] Tamagawa, T., et al. (2006). *Nucl. Instr. and Meth.* **A560**, 418.
[4] Hayato, A., et al. (2006). *Proc. SPIE* **6266**, 117H.
[5] Tamagawa, T., et al. (2009). *Nucl. Instr. and Meth.* accepted, doi:10.1016/j.nima.2009.07.014.
[6] Iwahashi, T., et al. (2008). *IEEE NSS Conference Record 2008*, 1959.
[7] HIMAC `http://www.nirs.go.jp/`

9

Broad-band soft X-ray polarimetry

H. L. Marshall, R. Heilmann, N. Schulz & K. Murphy

MIT Kavli Institute

We developed an instrument design capable of measuring linear X-ray polarization over a broad band using conventional spectroscopic optics. A set of multilayer-coated flats reflects the dispersed X-rays to the instrument detectors. The intensity variation with position angle is measured to determine three Stokes parameters: I, Q, and U – all as a function of energy. By laterally grading the multilayer optics and matching the dispersion of the gratings, one may take advantage of high multilayer reflectivities and achieve modulation factors >50% over the entire 0.2–0.8 keV band. This instrument could be used in a small orbiting mission or scaled up for the International X-ray Observatory. Laboratory work has begun that would demonstrate the capabilities of key components.

9.1 Introduction

The soft X-ray band (0.1–1.0 keV) should prove to be a fruitful region to explore for polarized emission. One concept, the Polarimeter for Low Energy X-ray Astrophysical Sources (PLEXAS), proposed the use of multilayer-coated mirrors tuned to 0.25 keV as Bragg reflectors[9]. As in similar Bragg reflection systems, the PLEXAS design had a narrow bandpass, reducing its attractiveness for astrophysical observations because one expects polarization to be energy dependent, so a wide bandpass is desired.

Marshall[8] described a method to overcome this limitation by using transmission gratings to disperse in the incoming X-rays. Following up on this approach, Marshall[7] suggested an arrangement that can be used in missions ranging from a small explorer to the International X-ray Observatory (IXO).

X-ray Polarimetry: A New Window in Astrophysics, eds. R. Bellazzini, E. Costa, G. Matt and G. Tagliaferri. Published by Cambridge University Press.

9.2 Polarimetry with the Extreme Ultraviolet Explorer

For the record, it is worth noting that the first attempt to measure the polarization above 0.1 keV of an extragalactic source used the Extreme Ultraviolet Explorer (EUVE). However, the telescope was not designed for polarimetry. The expected modulation factor was small, 2.7%, so the experiment failed due to systematic errors that could not be eliminated with existing calibration data[10]. Even using an in-flight null calibrator (a white dwarf) was insufficient. The program described here grew out of this attempt, so one might say that it was not a complete loss.

9.3 Science goals

Here we describe two potential scientific studies to be performed with an X-ray polarimetry mission with sensitivity in the 0.1–1.0 keV band.

Probing the relativistic jets in BL Lac objects – Blazars are all believed to contain parsec-scale jets with $\beta \equiv v/c$ approaching 0.995. The jet and shock models make different predictions regarding the directionality of the magnetic field at X-ray energies: for knots in a laminar jet flow it should lie nearly parallel to the jet axis[5], while for shocks it should lie perpendicular[6]. McNamara *et al.* (these proceedings, Chapter 21) suggest that X-ray polarization data could be used to deduce the primary emission mechanism at the base, discriminating between synchrotron, self-Compton (SSC), and external Compton models. Their SSC models predict polarizations between 20% and 80%, depending on the uniformity of seed photons and the inclination angle[11]. The X-ray spectra are usually very steep so that a small instrument operating below 1 keV can be quite effective.

Polarization in disks of active galactic nuclei and X-ray binaries – Schnittmann & Krolik (these proceedings, Chapter 15) show that the variation of polarization with energy could be used as a probe of the black hole spin and that the polarization position angle would rotate through 90 deg between 1 and 2 keV in some cases, arguing that X-ray polarization measurements are needed both below and above 2 keV[12]. As Blandford *et al.* (2002) noted, "to understand the inner disk we need ultraviolet and X-ray polarimetry"[1].

9.4 Basis of a soft X-ray polarimeter

The approach to this polarimeter design was inspired by a new blazed transmission-grating design, called the critical-angle transmission (CAT) grating[4], and the corresponding application to the Con-X mission in the design of a transmission-grating spectrometer[3]. This type of grating can provide very high efficiency in first order in the soft X-ray band. For a spectrometer, one places detectors on the Rowland torus, which is slightly ahead of the telescope's imaging surface.

A dispersive spectrometer becomes a polarimeter by placing multilayer-coated flats on the Rowland circle that redirect and polarize the spectra. The flats are tilted about the spectral dispersion axis by an angle θ. For this design[7], the graze angle θ was chosen to be 40°. The detectors are then oriented toward the mirrors at an angle $90 - 2\theta = 10°$ to the plane perpendicular to the optical axis of the telescope. Figure 9.1 shows how the optics might look. In this case, the entrance aperture is divided into eight sectors, with gratings aligned to each other in each sector and along the average radial direction to the optical axis. This approach generates seven

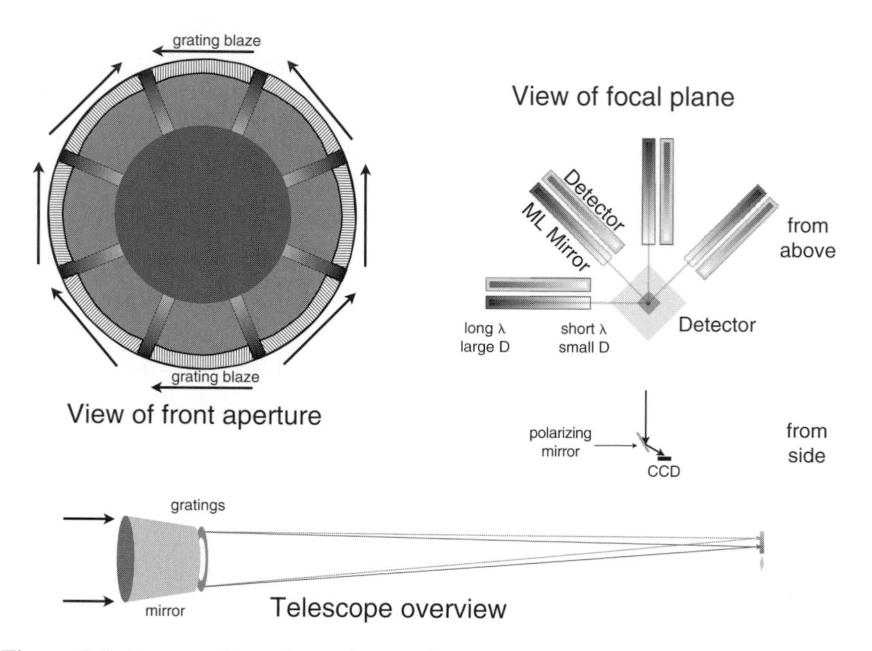

Figure 9.1 *Bottom:* Overview of a small telescope designed for soft X-ray spectropolarimetry, based on a design suggested by Marshall[7]. A small set of nested mirror shells focusses X-rays through gratings that disperse to an array of detectors about 2 m from the entrance aperture. *Top left:* View from the front aperture. Gratings are placed behind the mirror with the grating bars oriented along the average radius to the mirror axis. The gratings are blazed in the directions shown. *Top right:* The view from above the detectors and polarizers. Spectra from the gratings are incident on multilayer-coated flats that are tilted along the dispersion axis which contains the zeroth order (dot in center). The angle of the tilt is the same for all mirrors and always redirects the X-rays to the adjacent detector in a clockwise direction. The multilayer coating spacing, D, increases linearly outward from zeroth order, just as the wavelength increases, in order to match the first order wavelength to the peak of the multilayer reflectivity. The detectors could be aligned to the appropriate Rowland torus (as shown in the side view) or be tilted to face the corresponding mirror. Measuring the intensity at a given wavelength as a function of clocking angle then provides $I(\lambda)$, $Q(\lambda)$, and $U(\lambda)$.

spectra that are reflected by four polarizing flats to four detectors. See Marshall [7] for details.

The fundamental requirement of this approach is to vary the multilayer spacing, d, on the polarizing flats laterally, along the dispersion, in order to provide optimal reflection at graze angle θ. The multilayer period varies linearly with x, providing high reflectivity in a narrow bandpass at large graze angles[7]. At Brewster's angle, $\theta = 45°$, reflectivity is minimized when the E-vector is in the plane containing the incident ray and the surface normal (p-polarized) and maximized when the E-vector is in the surface plane (s-polarized). The polarization position angle (PA) is the average orientation in sky coordinates of the E-vector for the incoming X-rays. Sampling at least three PAs is required in order to measure three Stokes parameters (I, Q, U) uniquely, so one would require at least three separate detector systems with accompanying multilayer-coated flats. For this study, four detector-flat combinations are assumed, as shown in Figure 9.1.

Critical angle transmission (CAT) gratings have excellent prospects for high efficiency in first order, up to 50% over a wide band[4]. The CAT gratings are free-standing (so no support membrane is needed) and have higher intrinsic efficiency than the gratings used on the *Chandra* program, making them excellent candidates for use in X-ray spectrometers as well as a possible X-ray polarimeter. The effective area of small mission for unpolarized light, and the minimum detectable polarization across the bandpass of the instrument, for two possible observations, are shown in Figure 9.2.

A CAT grating spectrometer has been proposed as one of two approaches to use on the IXO, so there is a development path for this type of soft X-ray polarimeter. Sampling at least three PAs is required in order to measure three Stokes parameters (I, Q, U) uniquely, so one would require at least three separate detector systems with accompanying multilayer-coated flats, as shown in Figure 9.3.

9.5 A soft X-ray polarimeter prototyping facility

We have recently recommissioned the X-ray grating evaluation facility (X-GEF), a 17-m beamline that was developed for testing transmission gratings fabricated at MIT for the *Chandra* project[2]. With MKI technology development funding, we will soon adapt the existing source to produce polarized X-rays at the O-Kα line (0.525 keV). A new chamber will house a polarized-source multilayer (PSML) mirror, provided by Reflective X-ray Optics (RXO). The source will be mounted in the chamber at 90° to the existing beamline. A rotatable flange will connect the PSML chamber output to the vacuum 17-m facility so that the polarization vector may be rotated with respect to the grating and its dispersion direction. This prototyping effort will help us demonstrate the viability of soft X-ray polarimetry

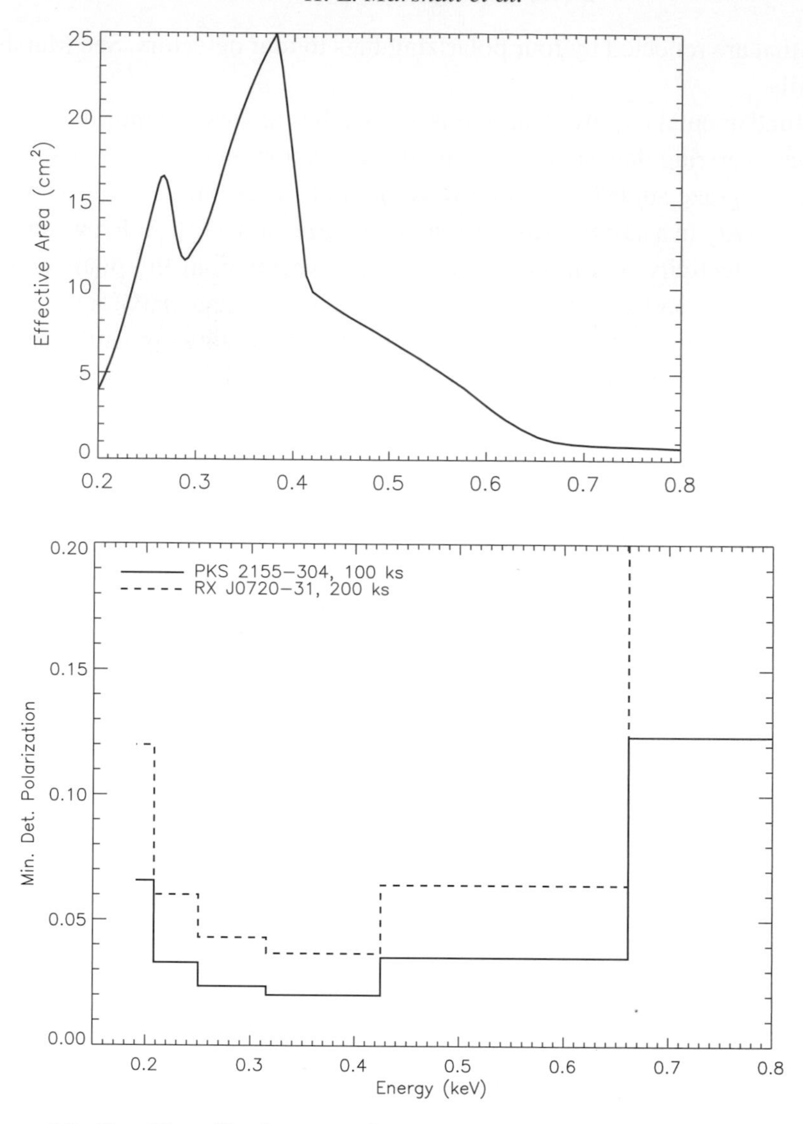

Figure 9.2 *Top:* The effective area of a small mission for unpolarized light. The modulation factor is about 0.90 all across the bandpass. The geometric area of the broad-band mirror is assumed to be 200 cm^2. For details, see Marshall[7]. *Bottom:* Minimum detectable polarization as a function of energy across the bandpass of the instrument for two different possible observations. The solid line shows how we could detect linear polarization at a level of 15–20% across the entire energy band from 0.2 to 0.8 keV for PKS 2155-304 in 100 ks. For RX J0720-31, spectroscopy allows one to obtain the polarization below, in, and above absorption features.

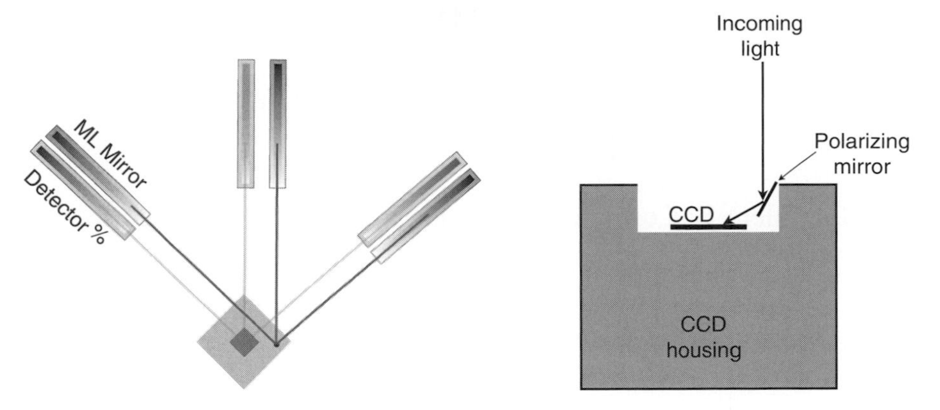

Figure 9.3 *Left:* Top view of a focal plane layout that could be used for IXO, in the manner suggested by Marshall[7]. When used as a spectrometer (gray lines), the zeroth order of the spectrometer is centered in the dark square, representing an IXO wide-field imager. When used as a polarimeter (black lines), the zeroth order is placed at the location of the black dot so that the dispersed spectrum first intercepts the laterally graded multilayer mirror that is angled at 30° to the incoming X-rays. The lines leading to zeroth order are all the same length, showing that specific wavelengths appear at different distances from the near end of the mirror but are at the same distances when used as a spectrometer. *Right:* Side view of the CCD housing where the dispersion is perpendicular to the plane of the drawing and the multilayer mirror is oriented 30° to the incoming, dispersed X-rays.

and show how a broad-band version can be developed. A proposal to expand the facility to four or five more energies between 0.25 and 0.8 keV is under consideration by NASA.

References

[1] Blandford, R., et al. (2002). In *Astrophysical Spectropolarimetry*, ed. J. Trujillo Bueno, F. Moreno-Insertis, and F. Sánchez.
[2] Dewey, D., et al. (1994). *SPIE* **2280**, 257–271.
[3] Flanagan, K., et al. (2007). *SPIE* **6688**, 66880Y.
[4] Heilmann, R. K., et al. (2008). *Opt. Express* **16**, 8658–8669.
[5] Marscher, A. P. (1980). *Ap. J.* **235**, 386–391.
[6] Marscher, A. P. & Gear, W. K. (1985). *ApJ* **298**, 114–127.
[7] Marshall, H. L. (2008). *SPIE* **7011**, 701129.
[8] Marshall, H. L. (2007). *SPIE* **6688**, 66880Z.
[9] Marshall, H. L., et al. (2003). *SPIE* **4843**, 360–371.
[10] Marshall, H. L., et al. (2001). *Ap. J.* **549**, 938–947.
[11] McNamara, A. L., et al. (2009). *MNRAS* **395**, 1507–1514.
[12] Schnittman, J. D. & Krolik, J. H. (2009), [arXiv:0902.3982].

10

Feasibility of X-ray photoelectric polarimeters with large field of view

F. Muleri, E. Costa, S. Di Cosimo, S. Fabiani, F. Lazzarotto, A. Rubini & P. Soffitta

INAF/IASF-Rome

R. Bellazzini, A. Brez, M. Minuti, M. Pinchera & G. Spandre

INFN-Pisa

The construction of X-ray polarimeters with large field of view could have a paramount scientific importance. They would unveil polarimetry of fast transient sources, GRBs and bursts from magnetars, to cite only the most important examples. This paper investigates the possibility of preserving the sensitivity of photoelectric polarimeters in case of beams inclined at angles $>10°$. The behaviour with both polarized and unpolarized radiation is studied through a consistent analytical treatment and measurements. We discovered deep changes in the intrinsic response of the instruments, implying a significant reduction of the amplitude of the response for polarized radiation and a strong spurious signal even in case of unpolarized photons. However these effects could be at least in principle corrected by the methods and the algorithms presented, if the direction of inclination and the energy of absorbed photons are known.

10.1 The photoelectric differential cross section

The emission directions of photoelectrons carry a significant memory of the polarization of the absorbed photons. Indeed the probability that a photoelectron is emitted in a certain direction is modulated as a $\cos^2 \phi$ function, where ϕ is the azimuthal angle between the direction of emission and polarization. In the working condition of current polarimeters, inner spherical symmetric orbitals (e.g. 1s and 2s) give the largest contribution to photoelectric absorption. Neglecting outer

X-ray Polarimetry: A New Window in Astrophysics, eds. R. Bellazzini, E. Costa, G. Matt and G. Tagliaferri. Published by Cambridge University Press.

shells, the differential cross section of the photoelectric effect for the 1s orbital is:

$$\frac{d\sigma}{d\Omega} \propto \frac{\sin^2\theta\cos^2\phi}{(1+\beta\cos\theta)^4}, \tag{10.1}$$

where θ is the latitudinal angle between the direction of the photon and that of the photoelectron and β is the velocity of the emitted electron in units of the speed of light.

The angular distribution $\mathcal{D}(\phi,\theta)$ of emitted photoelectrons is proportional to the differential cross section and then:

$$\mathcal{D}^{pol}(\phi,\theta) = \frac{\sin^2\theta\cos^2\phi}{(1+\beta\cos\theta)^4}\sin\theta \quad \text{for polarized photons,} \tag{10.2}$$

$$\mathcal{D}^{un}(\phi,\theta) = \frac{\sin^2\theta}{(1+\beta\cos\theta)^4}\sin\theta \quad \text{for unpolarized photons,} \tag{10.3}$$

where the term $\sin\theta$ derives from $d\Omega = \sin\theta d\theta d\phi$. The term $\cos^2\phi$ expresses the azimuthal dependency on the polarization of absorbed photons. The latitudinal distribution is more complex and includes the relativistic correction $1/(1+\beta\cos\theta)^4$. If this term is neglected, $\mathcal{D} \propto \sin^3\theta$ and then the photoelectrons are emitted preferentially into the plane perpendicular to the direction of incidence, $\theta = \pi/2$. In this case a complete symmetry between the emission above and below this plane is present, namely the directions $\theta = \bar{\theta}$ are equivalent to those $\theta = -\bar{\theta}$. However, the presence of the relativistic correction causes an effect of "forward folding", making more probable the emission in the semi-space opposite to the incident direction of photons (see Figure 10.1(a)).

10.2 The modulation curve for on-axis photons

Photoelectric polarimeters[1; 2; 3; 4] are able to resolve the tracks of photoelectrons in a gas cell. The polarization is derived by reconstructing the initial direction of photoelectrons. Collecting a sufficient number of photons (several tens of thousands), the direction and the degree of polarization are derived by the $\cos^2$ modulation of the azimuthal response of the instrument.

Current instruments can measure only the projections of the photoelectron tracks on a plane. Then the expected modulation curve is derived by summing all photoelectrons emitted in the same azimuthal angular bin, regardless of their latitudinal distribution (see Figure 10.1(a)). This can be performed by integrating the angular

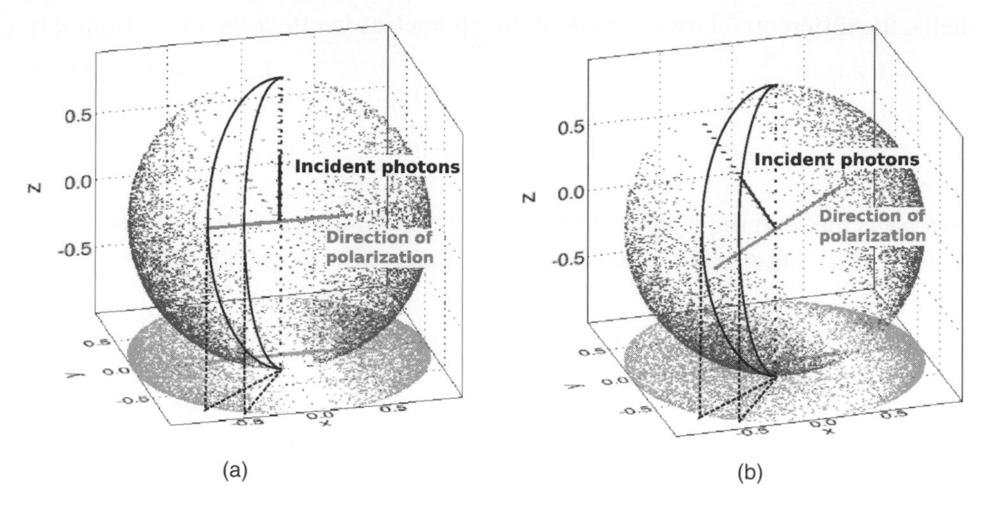

Figure 10.1 Emission directions of photoelectrons, including the relativistic correction. The incident direction of (polarized) photons is the black solid line, while the direction of polarization is grey. The photons are absorbed in the point (0,0,0), and the emission directions are represented with the black points on the sphere of unitary radius. The grey points in the xy plane characterize the projections of the tracks on the detector. Photons are incident orthogonally to the detector in (a). Note that the forward folding makes more probable the emission in the semi-space $z < 0$. The beam is inclined 30° in (b). The energy of photoelectrons is 50 keV ($\beta \approx 0.41$) to show the systematic effects discussed below more clearly.

distribution of photoelectrons $\mathcal{D}(\phi,\theta)$ over the angle θ:

$$^{pol}\mathcal{M}_0(\phi) = \frac{d\sigma}{d\phi} \propto \int_0^\pi \mathcal{D}^{pol}(\phi,\theta)d\theta \propto \cos^2\phi, \tag{10.4}$$

$$^{un}\mathcal{M}_0(\phi) = \frac{d\sigma}{d\phi} \propto \int_0^\pi \mathcal{D}^{un}(\phi,\theta)d\theta \propto 1. \tag{10.5}$$

The function $\mathcal{M}_0(\phi)$ hereafter will be called *intrinsic modulation* to distinguish it from that actually measured by the instruments $\mathcal{M}(\phi)$. The latter must include an unmodulated contribution which takes into account the errors in the measurement of the initial direction of emission:

$$\mathcal{M}(\phi) = M \cdot \mathcal{M}_0(\phi) + C \quad \text{with} \quad \begin{cases} ^{pol}M_0(\phi) & = & \cos^2\phi \\ ^{un}M_0(\phi) & = & 1 \end{cases} \tag{10.6}$$

where M and C are constants. Equation 10.6 is the function generally used to fit the modulation of the histogram of the photoelectrons' direction.

10.3 The case of inclined photons

The modulation curve for inclined beams can be calculated with the same procedure as in the case of orthogonal photons. However, care must be used to express the function $\mathcal{D}(\phi,\theta)$ in the right frame of reference. In the following we will assume that the direction of polarization is along the x axis and the inclination of an angle δ is around y (see Figure 10.1(b)). In this case the intrinsic modulation for polarized and unpolarized radiation is[5]:

$$\begin{aligned} {}^{pol}\mathcal{M}_0(\phi,\delta,\beta) = &\left[-\frac{3}{2}\pi\beta\sin\delta\cos^2\delta\right]\cos^3\phi + \left[\frac{4}{3}\cos^2\delta\right]\cos^2\phi + \\ &+ \left[\frac{\pi}{2}\beta\left(2\sin\delta - 3\sin^3\delta\right)\right]\cos\phi + \left[\frac{2}{3}\sin^2\delta\right], \end{aligned} \tag{10.7}$$

$$\begin{aligned} {}^{un}\mathcal{M}_0(\phi,\delta,\beta) = &\left[\frac{3}{2}\pi\beta\sin^3\delta\right]\cos^3\phi + \left[-\frac{4}{3}\sin^2\delta\right]\cos^2\phi + \\ &+ \left[-\frac{\pi}{2}\beta\left(\sin\delta + 3\sin^3\delta\right)\right]\cos\phi + \left[2 - \frac{2}{3}\cos^2\delta\right], \end{aligned} \tag{10.8}$$

where the relativistic correction was approximated at its first order to perform the integration, $(1+\beta\cos\theta)^{-4} \approx 1 - 4\beta\cos\theta$.

Despite the presence of several terms, the functions ${}^{pol}\mathcal{M}_0(\phi,\delta,\beta)$ and ${}^{un}\mathcal{M}_0(\phi,\delta,\beta)$ depend only on the parameters β and δ, namely on the energy and inclination of the absorbed photons. When $\delta = 0$ both functions are proportional to those obtained when photons are incident perpendicularly to the detector. A constant of proportionality is required since the intrinsic modulation calculated above is not normalized to unity.

There are two kinds of systematic effects on the modulation curve, related to the inclination or the forward folding. The former causes a decrease of the modulation factor due to the presence of an unmodulated intrinsic term proportional to $\sin^2\delta$. Moreover it introduces a spurious modulation for unpolarized photons, whose amplitude is proportional to $\sin^2\delta$. Instead the relativistic correction makes the modulation curve no more periodic over 180° because terms which are proportional to $\cos\phi$ and $\cos^3\phi$ are present if $\beta \neq 0$.

10.4 Measurements

The systematic effects presented above were measured with the Gas Pixel Detector[3] filled with a mixture 30% helium and 70% DME. Both completely polarized and unpolarized beams were produced at the X-ray facility built at INAF/IASF of Rome[6]. We exploited Bragg diffraction at 45° on a fluorite crystal to produce

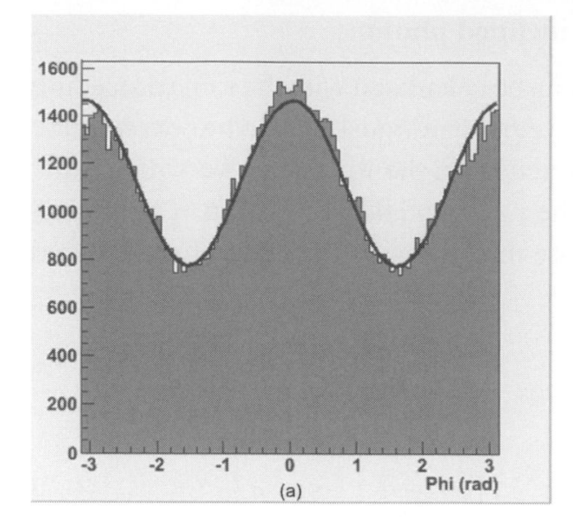

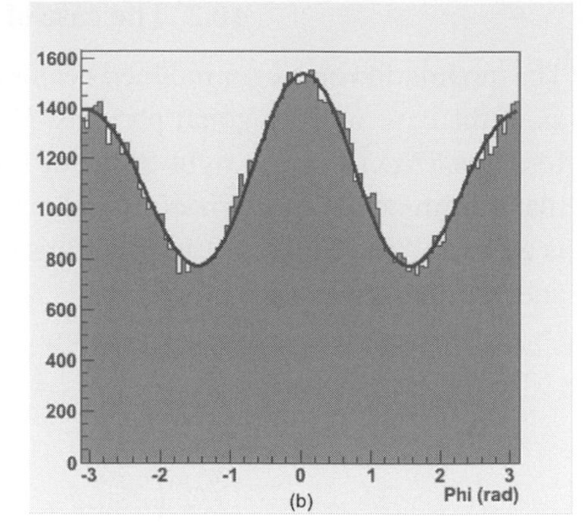

Figure 10.2 Modulation curve for polarized photons at 4.5 keV and inclined at 30°. The curve is fitted with the usual $\cos^2$ in (a) and the $\mathcal{M}$ function (b).

polarized photons at 4.5 keV, while a ^{55}Fe radioactive source was used as source of unpolarized radiation.

An example of the modulation curve obtained for inclined and polarized photons is reported in Figure 10.2, where the inclination is 30°. The fit with a $\cos^2$ function is unsatisfactory ($\chi_r^2 = 2.18$, see Figure 10.2(a)) since the two peaks corresponding to the direction of polarization are asymmetric, as expected. Moreover the amplitude of the modulation is strongly affected by the inclination. The modulation factor as a function of the inclination is reported in Figure 10.3 (dotted-line) and it is reduced by a factor of ~50% at $\delta = 40°$.

Instead the modulation is properly fitted with the intrinsic modulation derived in Section 10.3 (see Figure 10.2(b), $\chi_r^2 = 1.38$). We used the function:

$$\mathcal{M}(\phi,\bar{\delta},\bar{\beta}) = M \cdot \left\{ \frac{1}{\frac{4}{3} - \frac{2}{3}\sin^2\bar{\delta} + \frac{\pi\bar{\beta}}{2}\left|\sin\bar{\delta}\right|} \left[{}^{pol}\mathcal{M}_0(\phi,\bar{\delta},\bar{\beta}) \right] \right\} + C, \tag{10.9}$$

where M and C are constants and the factor $\left(\frac{4}{3} - \frac{2}{3}\sin^2\bar{\delta} + \frac{\pi\bar{\beta}}{2}\left|\sin\bar{\delta}\right|\right)^{-1}$ normalizes the intrinsic modulation to unity. We assumed that the inclination and the energy of the incident photons are known and then the values of δ and β are fixed at the expected ones, $\bar{\delta}$ and $\bar{\beta}$ respectively. The modulation factor, calculated with the usual formula $\mu = M/(M + 2C)$, is reported in Figure 10.3 (dashed line). In this case the modulation factor is almost constant, since the presence of an intrinsic unmodulated term is properly accounted for.

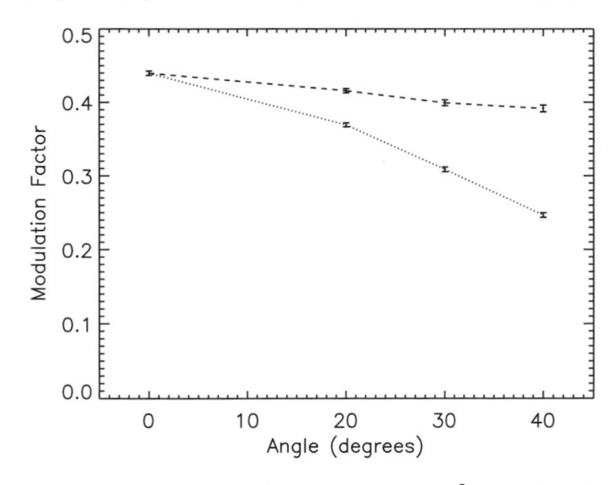

Figure 10.3 Modulation factor obtained with the $\cos^2$ function (dotted line) and the proper intrinsic modulation $\mathcal{M}$ (dashed line) as a function of the inclination angle.

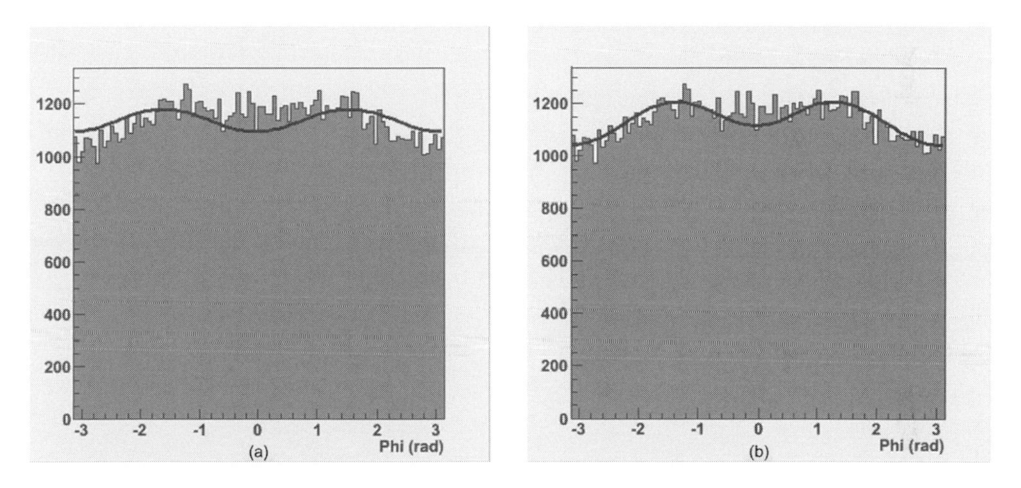

Figure 10.4 Modulation curve measured for unpolarized photons at 5.9 keV inclined at 30°, fitted with the usual $\cos^2$ (a) and the $\mathcal{M}$ functions (b).

The modulation curve in the case of a collimated ^{55}Fe source inclined at $\delta = 30°$ is reported in Figure 10.4. A fit with a $\cos^2$ function (see Figure 10.4(a)) returns a spurious modulation of 3.6%.

10.5 Conclusions

We investigated the response of photoelectric polarimeters in case of beams which are inclined with respect to the optical axis of the instrument. Significant systematic effects emerge if the inclination δ is larger than $\sim 10°$. However, they could be removed at least in principle if δ and the energy of absorbed photons are known.

References

[1] Costa, E. et al., (2001). *Nature* **411**, 662.
[2] Bellazzini, R. et al. (2006). *Nucl. Instr. and Meth. A* **566**, 552.
[3] Bellazzini, R. et al. (2007). *Nucl. Instr. and Meth. A* **579**, 853.
[4] Black, J.K. et al. (2007). *Nucl. Instr. and Meth. A* **581**, 755.
[5] Muleri, F. et al. (2009). In preparation.
[6] Muleri, F. et al. (2008). *Proc. of SPIE* **7011**, 701127-1.

11

Angular resolution of a photoelectric polarimeter

F. Lazzarotto,[1] S. Fabiani, E. Costa, F. Muleri,
P. Soffitta, S. Di Cosimo, G. Di Persio & A. Rubini
IASF - INAF Roma

R. Bellazzini, A. Brez & G. Spandre
INFN Pisa

V. Cotroneo, A. Moretti, G. Pareschi & G. Tagliaferri
INAF Brera

The INFN and INAF Italian research institutes developed a space-borne X-ray polarimetry experiment based on an X-ray telescope, focussing the radiation on a Gas Pixel Detector (GPD). The instrument obtains the polarization angle of the absorbed photons from the direction of emission of the photoelectrons as visualized in the GPD. Here we will show how we compute the angular resolution of such an instrument.

11.1 Introduction

X-ray telescopes are based on the grazing angle principle. The radiation is reflected with small incidence angles on the surfaces of hyperboloid and paraboloid mirrors and then is focused. The GPD is a gas detector which is able to image the photoelectron tracks. The polarization is measured using the dependence of the photoelectric cross section on the photon polarization direction[1; 3]. The photoelectron is emitted with more probability in the direction of the electric field of the photon. The track created by the photoelectron path is drifted and amplified by the Gas Electron Multiplier (GEM) and collected on a finely sub-divided pixel detector. Using different mixtures of gas it is possible to properly select the energy band of the instrument in the range of about 1–30 keV. This GPD has the capability to preserve the imaging while reaching a good sensitivity in polarization as well as in spectroscopic and timing measurements.

X-ray Polarimetry: A New Window in Astrophysics, eds. R. Bellazzini, E. Costa, G. Matt and G. Tagliaferri.
Published by Cambridge University Press.

Table 11.1 *GPD characteristics.*

Characteristic	Value	Unit
Optics energy band	0.1–10	keV
GPD energy band	1–30	keV
GPD area	15×15	mm^2
GPD height	10	mm
GPD transistors	16.5×10^6	n.
GPD pixels	105 600	n.
GPD pixel matrix	300×352	n.

11.2 Resolution calculation and related simulation software

We studied a system composed of an X-ray telescope and the GPD. We considered only the on-axis radiation. In this case an ideally perfect optical system can focus the radiation exactly in a single point on the detector, assuming that it has:

- Perfect quality reflective surfaces;
- Perfectly coaxial alignment of the mirrors;
- A detector with negligible thickness.

For the GPD the thickness of the gas cell is not negligible: 1 cm. To express the real behavior of radiation intensity distribution, the point-spread function (PSF) is obtained taking into account:

- Blurring introduced by imperfections of the optics[2];
- Blurring due to the approximations of the photon-track reconstruction algorithm;
- Blurring due to the radiation absorption in the gas.

We developed a simulation software based on Monte-Carlo techniques to study the angular resolution of the instrument (see Figure 11.2[4]). At this level the intrinsic resolution of the detector is neglected, the simulation program takes as input:

- The surface density of the incident radiation (photons cm^{-2});
- The geometry and the effective area of the optical system;
- The geometrical and physical characteristics of the gas detector.

The program calculates the absorption point of the photons in the gas cell taking into account the effects of the optical aberrations and gas blurring. In output it produces graphics and statistics on the photon detection positions around the focal plane.

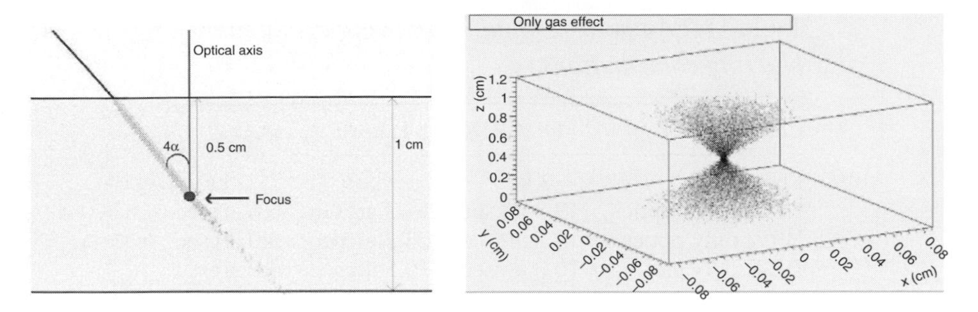

Figure 11.1 (a) Photon path in the GPD. (b) Distribution of the absorbtion points in the gas cell of the GPD causing the gas blurring.

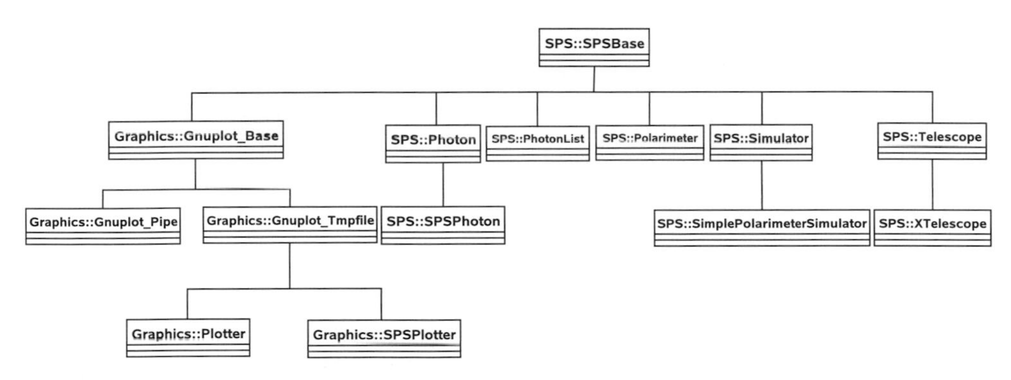

Figure 11.2 Simulation software class tree.

11.3 Conclusion

We report in Table 11.2 the angular resolution results expressed as the HPW (*Half Power Width*) of the radiation intensity on the detector plane for a simulation with a gas mixture composed 70% of DME and 30% of He. In Figure 11.3 the related error circles show that small missions as POLARIX and HXMT can be used to achieve the first results for angular resolved X-ray polarimetry. For instance it will be possible to measure the polarization of the main regions of extended sources such as the pulsar wind nebulae. Advanced missions such as IXO will be able to investigate the detailed properties of such sources or to reach the resolution needed to resolve the knots in AGN jets.

Notes

1. email: francesco.lazzarotto@iasf-roma.inaf.it; url: `http://bigfoot.iasf-roma.inaf.it/~agile/Polar/SPSdoc/index.html`

Table 11.2 *Angular resolution, showing the different blurring contributions.*

Characteristic	POLARIX	HXMT	IXO
Energy	3 keV	3 keV	1.5 keV
HPW gas + optics	19.3 arcsec	34.7 arcsec	6.6 arcsec
HPW only optics	14.7 arcsec	23.2 arcsec	5.0 arcsec
HPW only gas	10.0 arcsec	19.5 arcsec	3.0 arcsec

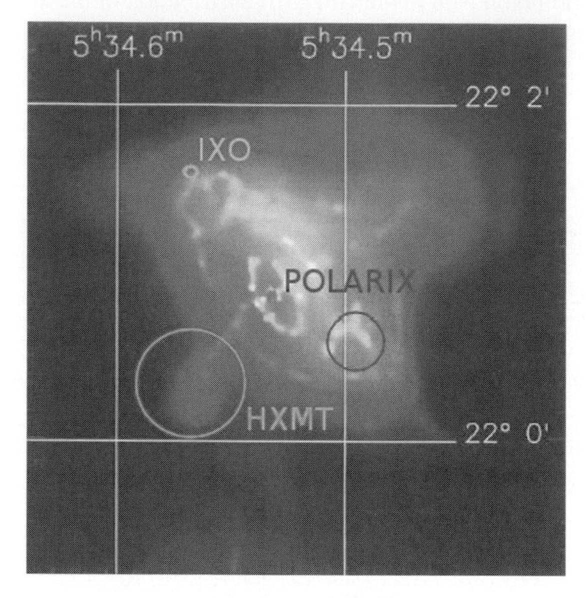

Figure 11.3 Qualitative representation of the resolution results on the image of the Crab PWN.

References

[1] Costa et al., (2001). "An efficient photoelectric X-ray polarimeter for the study of black holes and neutron stars", *Nature* **411**, 662–665.

[2] Citterio O. et al., (1993). "X-Ray optics for the JET-X experiment aboard the SPECTRUM-X Satellite", *SPIE* Vol. 2279.

[3] Bellazzini R. et al., (2007). "A sealed Gas Pixel Detector for X-ray astronomy", NIMPA 579 (853).

[4] Fabiani et al., (2008). "The Study of PWNe with a photoelectric polarimeter", PoS(CRAB2008)027.

12

Development of a Thomson X-ray polarimeter

P. V. Rishin, B. Paul & R. Duraichelvan

Raman Research Institute, Sadashivanagar, India

J. Marykutty & D. Jincy

School of Pure and Applied Physics, Mahatma Gandhi University, India

R. Cowsik

McDonnell Center for the Space Sciences, Department of Physics, Washington University, St. Louis

We describe the current status of the design and development of a Thomson X-ray polarimeter suitable for a small satellite mission. Currently we are considering two detector geometries, one using rectangular detectors placed on four sides of a scattering element and the other using a single cylindrical detector with the scattering element at the center. The rectangular detector configuration has been fabricated and tested. The cylindrical detector is currently under fabrication. In order to compensate any pointing offset of the satellite, a collimator with a flat-topped response has been developed that provides a constant effective area over an angular range. We have also developed a double crystal monochromator/polarizer for the purpose of test and calibration of the polarimeter. Preliminary test results from the developmental activities are presented here.

12.1 Introduction

A Thomsom X-ray polarimeter experiment has been proposed for a small satellite mission of the Indian Space Research Organization (ISRO). Currently, a laboratory model has been developed. This experiment will be suitable for X-ray polarization measurement of hard X-ray sources like accretion powered pulsars, black hole candidates in low-hard states etc.[1; 2]. For about 50 brightest X-ray sources a minimum detectable polarization of 2–3% will be achieved with the final configuration.

X-ray Polarimetry: A New Window in Astrophysics, eds. R. Bellazzini, E. Costa, G. Matt and G. Tagliaferri. Published by Cambridge University Press.

Two configurations are considered based on the geometry of the detector element: (1) rectangular detectors and (2) cylindrical detectors. The X-ray polarization will be measured by spinning the platform around the viewing axis. In both the cases, energy range covered will be 5–30 keV. Here we describe the design and current state of development of the Thomson X-ray polarimeter and also the development of a test and calibration facility.

12.2 Thomson X-ray polarimeter design

Rectangular detectors: in Figure 12.1 we have shown the configuration of the Thomson polarimeter with rectangular detectors. Here the detector elements are flat multi-wire proportional counters placed on four sides of a 900-cm^2 disk-like scattering element made of Be or Li. Each detector also has a photon collection area of 900 cm^2.

Cylindrical detectors: the second configuration uses cylindrical detectors as shown in Figure 12.2. In this configuration, we have a 360° coverage of the scattered photons which is an advantage compared to the rectangular detectors. The experiment will have two such detectors, each with a photon collection area of

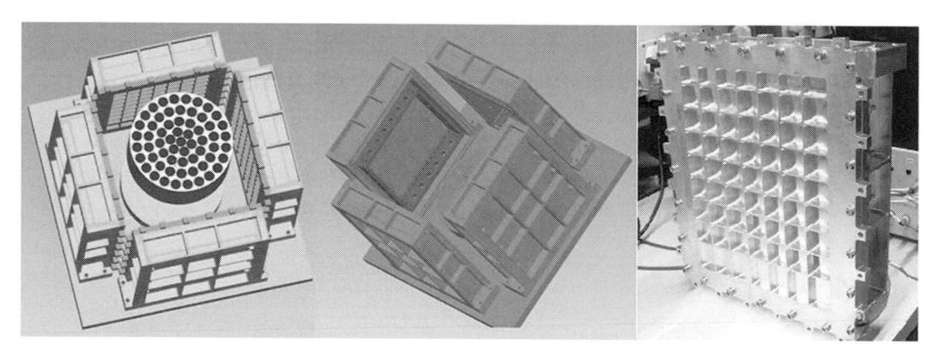

Figure 12.1 Polarimeter configuration with rectangular detectors.

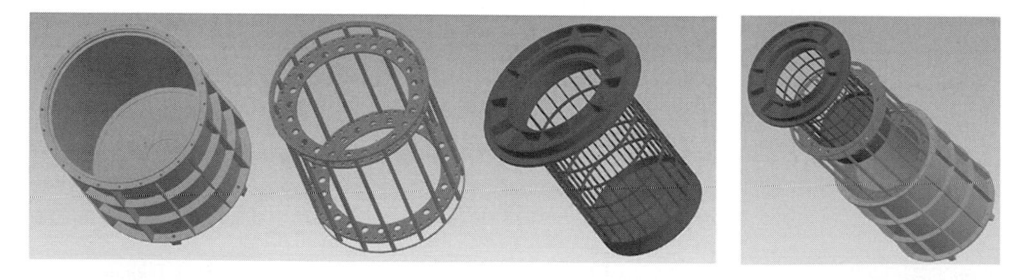

Figure 12.2 Cylindrical detector fabrication design and the assembly.

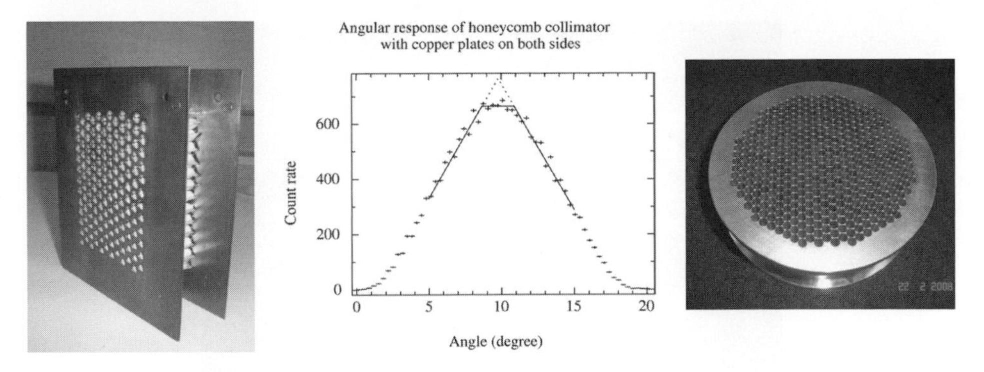

Figure 12.3 Collimators with honeycomb based and solid Al based construction.

about 350 cm^2. In a cylindrical detector, we expect to have uniform gain and quantum efficiency in all directions which will help to reduce systematic errors while searching for faint signals.

Collimator: in order to compensate for inaccuracy in satellite pointing and to attain constant effective area, a collimator with a flat-topped response is required. For the Thomson polarimeter, we have tested two collimator constructions which give flat-topped responses. The first construction uses aluminium honeycomb, sandwiched between copper plates with circular holes as shown in Figure 12.3. The holes on the front plate have slightly smaller diameter than the holes on the back plate, which leads to a flat-topped response. Since aluminium does not provide enough X-ray absorption above 15 keV, a silver coating of ~10-micron thickness is provided on the aluminium surface to increase absorption. This construction has the advantage of small weight per unit area coverage. The second construction uses solid aluminium block with machined tapered holes to acheive a flat-topped response. For the flight unit, a collimator with a flat-topped response of 0.5° will be developed.

Test electronics: the different electronic units required for pulse processing and data acquisition from the detectors have been developed. This includes (1) HV distribution and charge-sensitive preamplifiers, (2) anti-coincidence counters and (3) coincidence counters. We have also developed a general purpose fixed-threshold pulse counter system capable of acquiring four 8-bit channels simultaneously. An FPGA-based multi-channel analyzer with ADC input and configurable pulse selection logic (coincidence, anti-coincidence, etc.) is under development.

12.3 Test and calibration system

In order to test and calibrate the polarimeter, the following systems are at various stages of development:

Figure 12.4 The rotary stage with the detectors mounted.

(1) A double-crystal monochromator/polarizer working in the energy range 2–60 keV to facilitate calibration of detectors over the entire energy band of interest. The total energy range will be covered using three pairs of crystals with $d = 1.15$, 1.9 and 3.1 Å. The first crystal plate with $d = 3.1$ Å has been fabricated, with the crystals parallelized to arc-second level. The crystal plate has been tested in the energy range of 7–11 keV and has FWHM of the beam $\sim$2% in this range.
(2) A polarized source based on Thomson scattering of X-rays from a low Z scatterer.
(3) A polarized source based on filtering out the highest energy (last 10%) photons from an electron beam X-ray generator.
(4) A rotary stage to simulate the spinning satellite platform during polarization measurements (shown in Figure 12.4)

12.4 Results

A laboratory model of the Thomson X-ray polarimeter with a Be scatterer and rectangular detectors has been completed and tested. Figure 12.5 shows the test results. In the 9–23 keV range, we obtained a modulation factor of up to 35%. The modulation factor is improved to about 45% when only one third of the detector is exposed, thereby reducing the opening angle. The experiment has been proposed for a small satellite mission of the Indian Space Research Organization (ISRO).

Acknowledgements

We thank members of the RAL and MES of Raman Research Institute for their excellent support in development of the instrument.

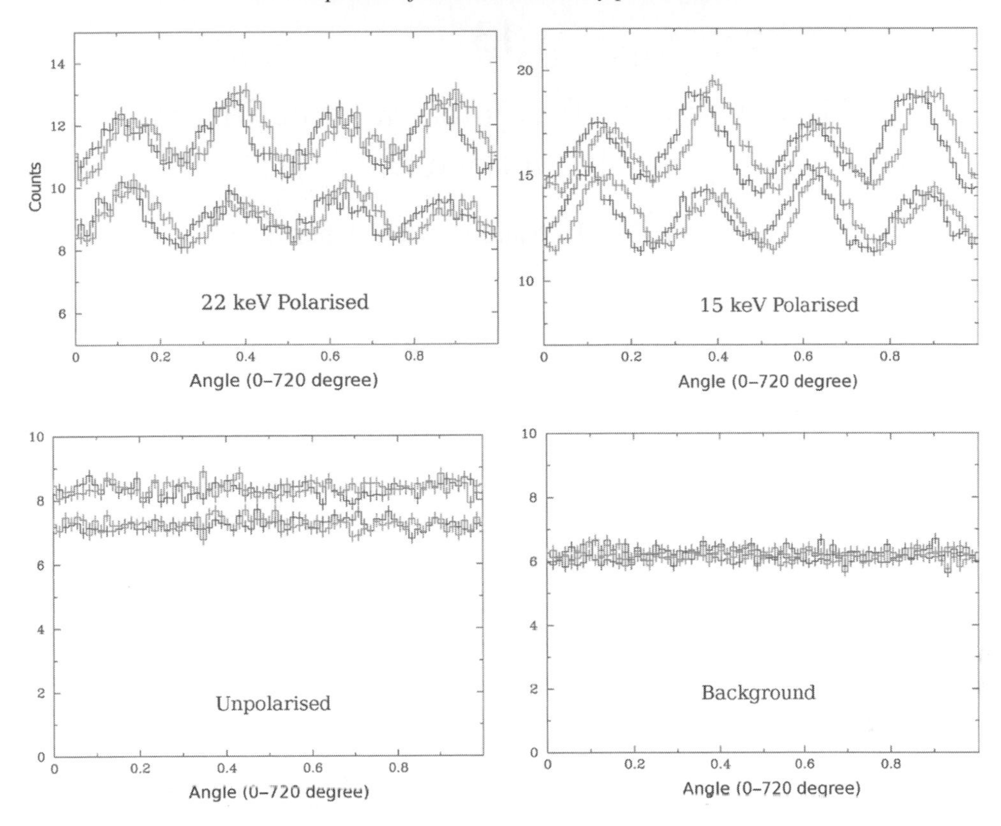

Figure 12.5 Modulation obtained at different energies.

References

[1] Weisskopf, M. C. et al. (2006). [astro-ph/0611483].
[2] Kallman, T. (2004). *Advances in Space Research* **34**, 2673.

13

Hard X-ray / soft gamma-ray polarimetry using a Laue lens

N. Barrière, L. Natalucci & P. Ubertini

INAF - IASF Roma

Hard X-ray / soft gamma-ray polarimetric analysis can be performed efficiently by the study of Compton scattering anisotropy in a detector composed of fine pixels. But in the energy range above 100 keV where source fluxes are extremely weak and instrumental background very strong, such delicate measurement is actually very difficult to perform. The Laue lens is an emerging technology based on diffraction in crystals allowing the concentration of soft gamma-rays. This kind of optics can be applied to realize an efficient high-sensitivity and high-angular-resolution telescope, though at the cost of a field of view reduced to a few arcmin. A 20-m focal-length telescope concept focusing in the 100–600 keV energy range is taken as example here to show that recent progresses in the domain of high-reflectivity crystals can lead to very appealing performance. The Laue lens being fully transparent to polarization, this kind of telescope would be well suited to perform polarimetric studies since the ideal focal plane is a stack of finely pixelated planar detectors – in order to reconstruct the point spread function – which is also ideal to perform Compton tracking of events.

13.1 Introduction

A Laue lens concentrates gamma-rays using Bragg diffraction in the volume of a large number of crystals arranged in concentric rings and accurately orientated in order to diffract radiation coming from infinity towards a common focal point (e.g. [1; 2]). This principle is applicable from $\sim$100 keV up to 1.5 MeV, but with a unique lens it is difficult to cover efficiently a continuous energy band of more than about half an order of magnitude width. Thanks to the decoupling between

X-ray Polarimetry: A New Window in Astrophysics, eds. R. Bellazzini, E. Costa, G. Matt and G. Tagliaferri.
Published by Cambridge University Press.

collecting area and sensitive area, a Laue lens produces a dramatic increase of the signal to background ratio. This is the key to achieve the long-awaited sensitivity leap of one to two orders of magnitude with respect to past and current instruments (IBIS and SPI onboard INTEGRAL, Comptel onboard CGRO, BAT onboard SWIFT).

Despite the Laue lens being a concentrator (i.e. not a direct imaging system) it allows the reconstruction of images with an angular resolution of $\sim$1 arcmin in a field of view of $\sim$10 arcmin. Moreover, combined with a focal plane capable of tracking the various interactions of an event, the telescope becomes sensitive to polarization: the lens being fully transparent to polarization[6; 4] it makes possible its anaysis in the focal plane. Using the non-uniformity of the azimuthal angle of Compton diffusion, it becomes possible to perform a statistical study of the signal in order to measure its polarization angle and fraction.

A telescope featuring a Laue lens would be perfectly adapted to furthering our understanding of high-energy processes occurring in a variety of violent events. High-sensitivity investigations of point sources such as compact objects, pulsars and active galactic nuclei should bring important insights into the still poorly understood emission mechanisms. For this purpose the polarization-detection capabilities inherent to a Laue-lens telescope in conjunction with its high angular resolution would be extremely helpful, for instance, to distinguish between various models of pulsars' magnetospheres. Also the link between jet ejection and accretion in black hole and neutron star systems could be clarified by observation of the spectral properties and polarization of the transition-state emission.

13.2 Proposed telescope

The main driver of the studied concept is to perform high-sensitivity hard X-ray pointed observations in the energy range 100–600 keV. The Laue lens, which has a focal distance of 20 m, can be carried on extensible booms allowing a medium-sized mission to carry the telescope.

13.2.1 Laue lens

The Laue lens designed for this study is composed of 20 000 crystal tiles of 15 mm $\times$ 15 mm in a complex assembly of CaF_2, Cu, Ge, Ag, V and Mo. The required mosaicity for the crystal is 2 arcmin, which is not too tight a constraint for their production (growth and cut) and mounting. Figure 13.1 shows the effective area of the lens and the repartition of the various crystals over the lens radii. Crystals development for Laue lenses is currently being addressed in various institutes in Europe, and gives very positive results[5; 3]. Crystals total mass equals 150 kg,

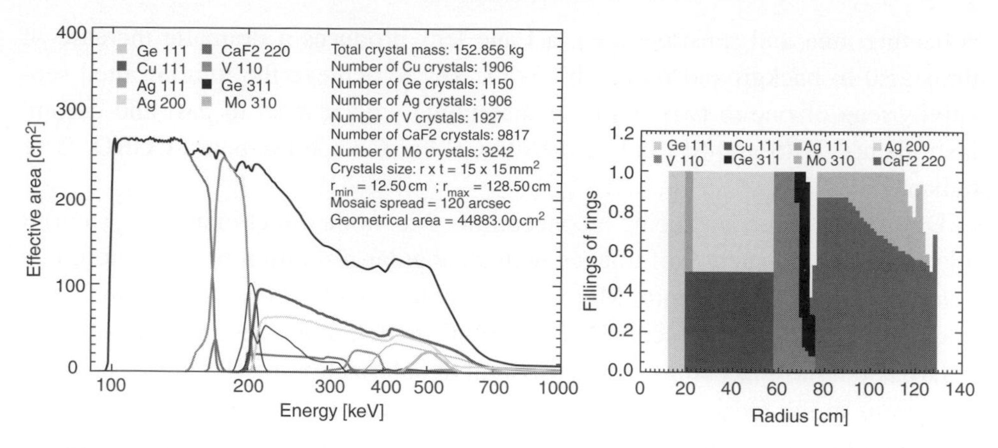

Figure 13.1 *Left.* Effective area of the lens. *Right.* Repartition of the various crystals and reflections used to reach the effective area.

and based on MAX[7] and GRI[8] mission studies, its structure can be estimated to ∼70 kg/m^2, which makes a total weight of 460 kg for the lens.

13.2.2 Focal plane instrument

The advantages of having a Compton camera at the focus of the lens are multiple. The inherent fine pixelization allows the reconstruction of the point spread function (PSF) via the localization of the first interactions, which is fundamental to take full benefit of the lens. Since the signal is confined in a defined area, simultaneous monitoring of the background is possible using the non-lighted detector areas allowing efficient evaluation of its time variability. Background rejection can be further enhanced if Compton kinematics is applied to the reconstruction of events. At the cost of a non-negligible fraction of the signal, it is possible to constrain the original direction of the events within an annulus, and in many cases, to establish whether the photon energy was fully recorded or not[11]. This makes a powerful tool to efficiently reject background by discriminating photons whose 'event circle' does not intercept the lens direction.

Performing a statistical number of Compton event reconstructions directly allows the measurement of the signal polarization angle and fraction. However, it is necessary that the first layer is made out of silicon (Si) to apply Compton kinematics down to 100 keV. At 100 keV Compton interaction is, by a factor of ∼10, the most probable interaction in Si, but the drawback is that it has a low interaction cross section: mean free path at 100 keV in Si is already 2.5 cm. Consequently Si layers have to be complemented by additional layers of a much more absorbent material of which CdTe (or CdZnTe) is a prime choice[9; 10].

A focal plane of 8 cm × 8 cm gives a field of view of 10 arcmin FWHM (limited by the spread of the signal over the focal plane that induces a sensitivity loss). For point sources, the reconstruction of the PSF allows the localization to less than 1 arcmin within the field of view. In case of an extended source, combined observations in a dithering pattern still allow advantage to be taken of the imaging capabilities.

13.3 Conclusion

According to preliminary studies, continuum sensitivity better than 2×10^{-7} ph/s/cm^2/keV (3σ, 10^5 s, $\Delta E/E = E/2$, point source on-axis) is attainable using the Laue lens described in this paper. This sensitivity could allow polarimetric studies of bright sources with an unprecedented angular resolution (especially interesting for the Crab nebula). An appealing option to complete the Laue lens telescope would be to perform all-sky monitoring covering half the sky continuously over the entire mission lifetime. Thanks to a dedicated design it is conceivable that the instrument at the focus of the lens could act simultaneously as focal-plane and large-field-of-view Compton camera.

Acknowledgement

NB and LN are grateful to ASI for the support of Laue lens studies through grant I/088/06/0.

References

[1] Lund, N. (1992). *Exp. Astron.* **2**, 259–273.
[2] Halloin, H. & Bastie, P. (2005). *Exp. Astron.* **20**, 151–170.
[3] Barrière, N. et al., (2009). *J. Crystal Growth*, in press.
[4] Curado da Silva, R. M. et al. (2008). *J. Applied Physics* **104(8)**, 084903.
[5] Ferrari, C. et al., (2008). *Proc. SPIE* **7077**, 70770O.
[6] Barrière, N. et al. (2009). *Proc. Science*, CRAB2008_019.
[7] Duchon, P. (2005). *Exp. Astron.* **20**, 483–495.
[8] Knödlseder, J. et al. (2009). *Exp. Astron.* **23**, 1, 121–138.
[9] Takahashi, T. et al. (2006). *Proc. SPIE* **6266**, 62660D.
[10] Ubertini, P. et al. (2003). *A&A* **411**, L131.
[11] Zoglauer, A. et al. (2007). *Proc. IEEE NSS/MIC* **6**, 4436–4441.

Part II

Polarized emission in X-ray sources

14

Probing strong gravity effects with X-ray polarimetry

M. Dovčiak

Astronomical Institute ASCR, Prague

Light coming from the innermost regions of active galactic nuclei or galactic black hole systems is heavily influenced by the strong gravity of their central compact body. High velocities of the emitting matter also modify the properties of the light received by the detector. In this contribution the influence of the combined special and general relativistic effects on the observed intensity and polarization is summarized. The application on the systems with a geometrically thin and optically thick Keplerian accretion disc is shown.

14.1 Introduction

The light emitted in the vicinity of the compact object has different properties when absorbed by a detector at infinity. First of all the photons are usually emitted by fast-moving matter – orbiting, falling into or being ejected from the central body with very high speeds. Thus the effects of Einstein's special relativity, the Doppler shift and aberration, change the photon energy and the direction of its polarization. The beaming effect in the direction of emitting matter motion may be quite significant as well. It is worth mentioning that although these effects are those of special relativity, the high velocities causing them are due to large gravity of the central object. Therefore we need to use general relativity in order to evaluate them properly.

However, general relativity has even more direct impact on the properties of these photons. They move from a strong gravity region, often dragged along by the rotating space-time if the compact body has large angular momentum. Again, the photon energy and direction of its polarization is affected due to the gravitational

X-ray Polarimetry: A New Window in Astrophysics, eds. R. Bellazzini, E. Costa, G. Matt and G. Tagliaferri. Published by Cambridge University Press.

redshift and by rotation of the parallelly transported polarization vector. Moreover, the strong gravitation bends the photon trajectories, diluting or concentrating them, which either diminishes or amplifies the observed intensity. We refer to these as light bending and lensing effects.

The emission is often nonisotropic and the intensity and polarization may depend on the local emission angle. In this case the aberration and light bending may play a very important role in yet another way.

14.2 The observed radiation

To visualise the relativistic effects at work, we have chosen a geometry of an accretion disc in the equatorial plane of either the nonrotating Schwarzschild or the extremely rotating Kerr black holes. We assume the Keplerian accretion disc which is geometrically thin and optically thick. The properties of the observed radiation coming from such a disc are described by Stokes parameters evaluated at infinity:

$$I_{\mathrm{obs}} = \int_{\Sigma} \mathrm{d}S \; G I_{\mathrm{loc}} \,, \tag{14.1}$$

$$Q_{\mathrm{obs}} = \int_{\Sigma} \mathrm{d}S \; G I_{\mathrm{loc}} P_{\mathrm{loc}} \cos 2(\chi_{\mathrm{loc}} + \psi) \,, \tag{14.2}$$

$$U_{\mathrm{obs}} = \int_{\Sigma} \mathrm{d}S \; G I_{\mathrm{loc}} P_{\mathrm{loc}} \sin 2(\chi_{\mathrm{loc}} + \psi) \,. \tag{14.3}$$

For simplicity, we considered that the local disc emission is not circularly polarized, i.e. both the local and the observed fourth Stokes parameter V is zero. The quantities I_{loc}, P_{loc} and χ_{loc} denote the local intensity, the local polarization degree and the local polarization angle, respectively. The relativistic effects are hidden in the transfer function G and in the change of the polarization angle ψ. The integration over the disc can be performed either directly in Boyer-Lindquist coordinates of the disc r, φ or in the image of the disc on the observer's sky (detector plane) with the impact parameters α, β as coordinates (see Figure 14.1). It is convenient to use the former case when interested in the local emission of the disc whereas the latter one is useful for describing the observed radiation. In the first case the transfer function can be expressed as $G = g^{\Gamma-1}\mu_{\mathrm{e}}\ell$, in the second case it is $G = g^{\Gamma}$ with $g = E_{\mathrm{obs}}/E_{\mathrm{loc}}$ being energy shift, $\mu_{\mathrm{e}} = \cos\theta_{\mathrm{e}}$ cosine of the local emission angle and $\ell = \mathrm{d}S^{\perp}_{\mathrm{obs}}/\mathrm{d}S^{\perp}_{\mathrm{loc}}$ the lensing. The power-law index Γ has different values for the energy flux ($\Gamma = 4$), the specific (per frequency) energy flux ($\Gamma = 3$), the photon flux ($\Gamma = 3$) and the specific photon flux ($\Gamma = 2$). One can see that the

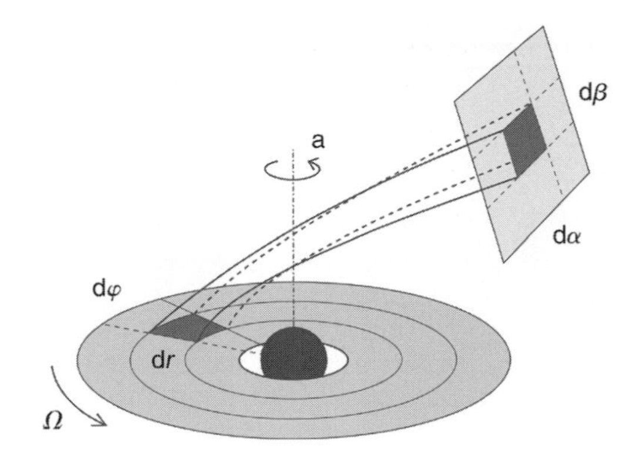

Figure 14.1 The sketch of the system with an accreting black hole.

lensing effect, light bending and aberration is in one case part of the transfer function while in the other case it is hidden in the mapping of the disc onto its image on the observer's sky, i.e. it is a part of the coordinate transformation determinant $(\mathrm{d}\alpha\mathrm{d}\beta)/(r\mathrm{d}r\mathrm{d}\varphi) = \mathrm{d}S^{\perp}_{\mathrm{obs}}/\mathrm{d}S^{\perp}_{\mathrm{loc}} \times \mathrm{d}S^{\perp}_{\mathrm{loc}}/\mathrm{d}S_{\mathrm{loc}} \times \mathrm{d}S_{\mathrm{loc}}/\mathrm{d}S = \ell \times \mu_{\mathrm{e}} \times g^{-1}$.

In the following sections we will discuss relativistic effects on the observed energy flux ($\Gamma = 4$) in the coordinates α, β.

14.3 The energy shift

The energy shift, g-factor (Figure 14.2), is the main manifestation of the relativistic effects. It reflects the change of the observed photon energy with respect to the emitted one and determines the amplification of the flux received by the detector. It is one of the main components of the transfer function, actually the only one if the transfer function is evaluated for the coordinates on the observer sky. On the approaching (receding) side of the disc the relativistic Doppler effect shifts the energy to higher (lower) values. This effect increases with the inclination of the observer. The gravitational redshift prevails close to the black hole and the value of g-factor goes to zero at the horizon. The energy shift becomes very important in spectro-polarimetry when the flux and polarization are resolved in energies. All narrow features in local spectrum and polarization become blurred (relativistically broadened) when integrated over large regions near the black hole.

14.4 The transfer function

The main meaning of the transfer function G is the overall amplification of the flux received by the detector with respect to the local emitted flux on the disc. Thus the flux and polarization from those parts of the disc where the transfer function has

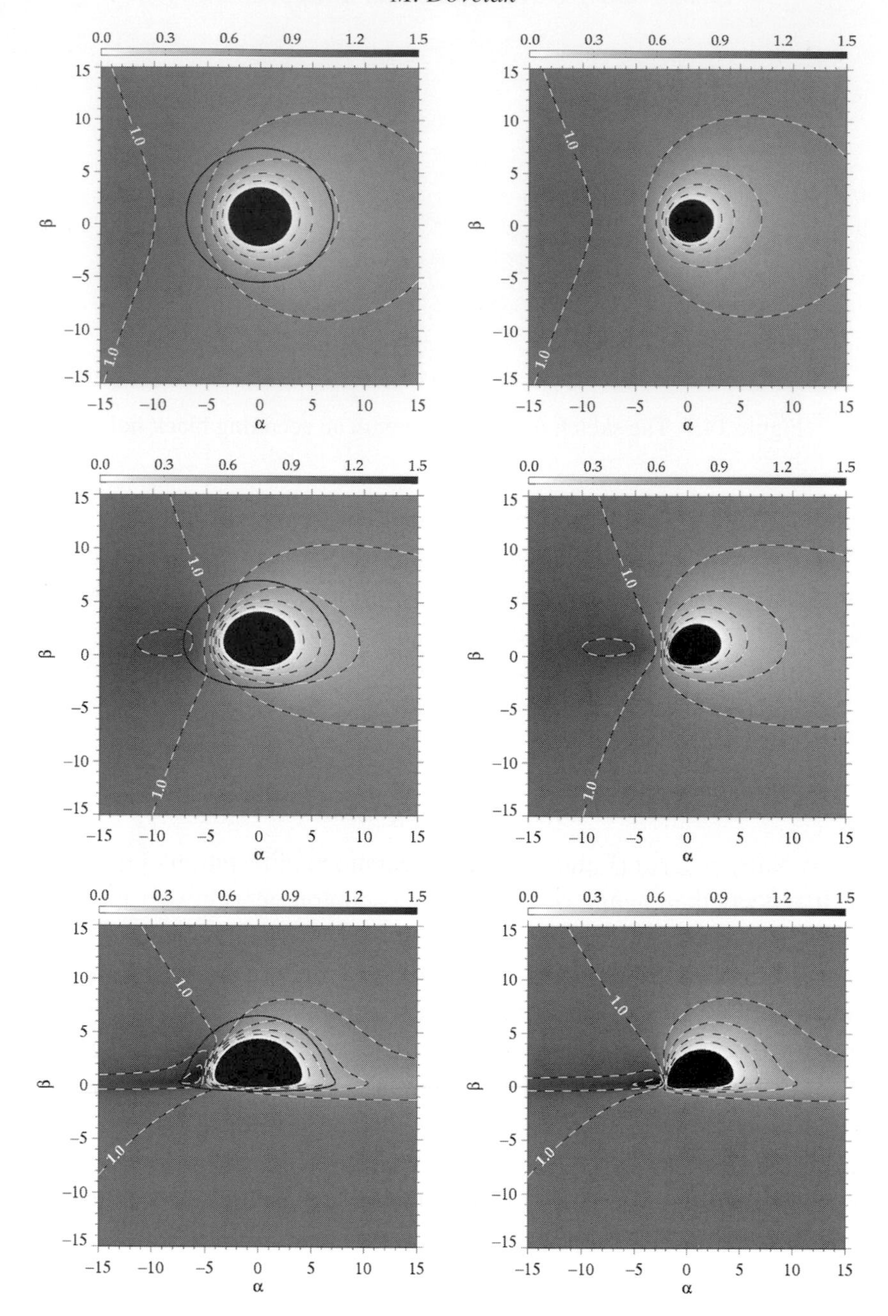

Figure 14.2 The energy shift g shown on the observer's sky with the impact parameters α, β as coordinates. Its values are represented by greyscale and by dashed contours with the step of 0.2. The inclination of the observer is $\theta_{\rm obs} = 30°$, $60°$ and $85°$ (top to bottom) and the spin of the black hole is $a = 0$ and 1 (left and right). In the left panels the last stable orbit is indicated with the solid line while on the right side it coincides with the horizon of the black hole. The accretion disc co-rotates with the black hole in the counter-clockwise direction.

the highest value will contribute to the observed quantities the most. Because of the relativistic beaming the emission from the approaching part of the disc is amplified the most, for high inclination angles more than by a factor of four.

Let us point out that not only the transfer function is important when discussing the observed flux from some part of the disc. The local emission angle may become of the same or even greater importance if the local flux and polarization depend on it, e.g. for the limb-darkening emissivity law those parts of the disc where the local emission angle is small will contribute more than others. To reach the observer the photons coming from regions far away from the black hole are to leave the disc with the local emission angle equal to the inclination. The momentum of the photons emitted from close vicinity of the horizon have to be parallel with the disc. A critical point exists for each inclination where the observed photons are emitted perpendicular to the disc. This is mainly due to the special relativistic aberration, although the exact position of the point is also influenced by the general relativistic light bending. For low inclinations this point moves farther away from the centre because lower Keplerian speeds are needed for the aberration to compensate the inclination. On the contrary, for high inclinations, the critical point moves closer to the horizon and it may be located below the marginally stable orbit.

14.5 The change of the polarization angle

From Equations (14.2) and (14.3) we can see that the change of the polarization angle ψ (Figure 14.3) is another important quantity affecting the observed polarization. The main reason why this change is nonzero is the relativistic aberration caused by fast mutual motion of the light-emitting matter and the observer. Due to this effect, the direction of the polarization vector measured in the local frame co-moving with the disc and the observer's frame at infinity is different even for a nonrotating Schwarzshild black hole. In the case of a spinning black hole, this direction changes also due to the dragging effect of the hole's gravity. The change of the polarization angle can acquire quite different values across the disc and it is heavily dependent on the inclination angle. Actually, as one can see in the Figure 14.3, it can be of any value between 0 and 180° in the region around and below the critical point. The consequence is the depolarization of the observed radiation that is more effective for lower inclinations when the depolarizing region is larger (critical point being farther away from the black hole).

Here we have to emphasize again that the dependence of the local polarization angle on the emission angle may become very important. Figure 14.3 does not include this dependence because it was assumed that the local polarization vector is perpendicular to the disc everywhere no matter what the value of the emission angle is.

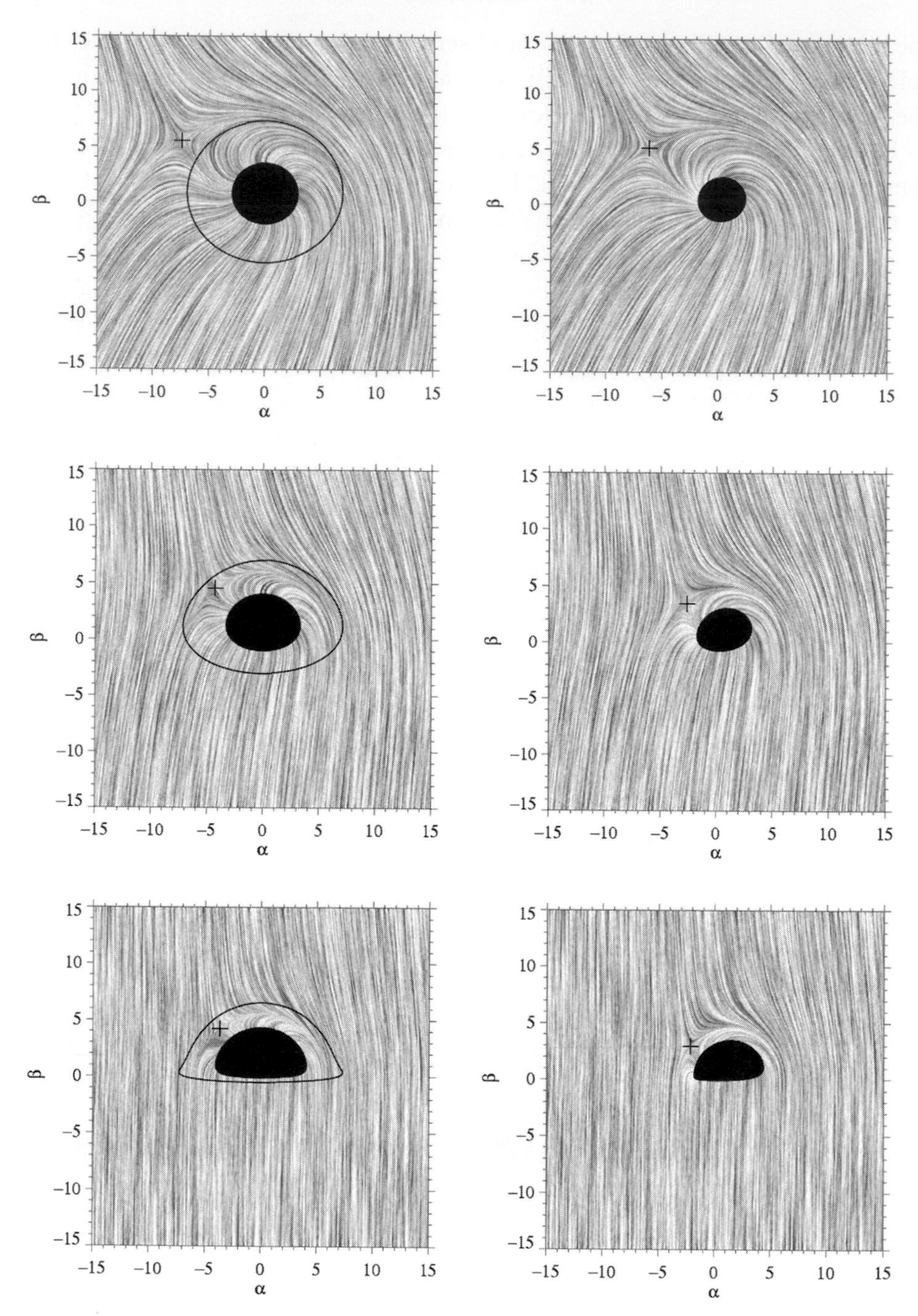

Figure 14.3 The change of the polarization angle for the Schwarzschild (left) and extremely spinning Kerr (right) black holes, the inclination of the observer is 30°, 60° and 85° (top to bottom). The flow-lines show the direction of the observed polarization vector if the local polarization vector is perpendicular to the disc. The critical point is depicted in each panel with a cross.

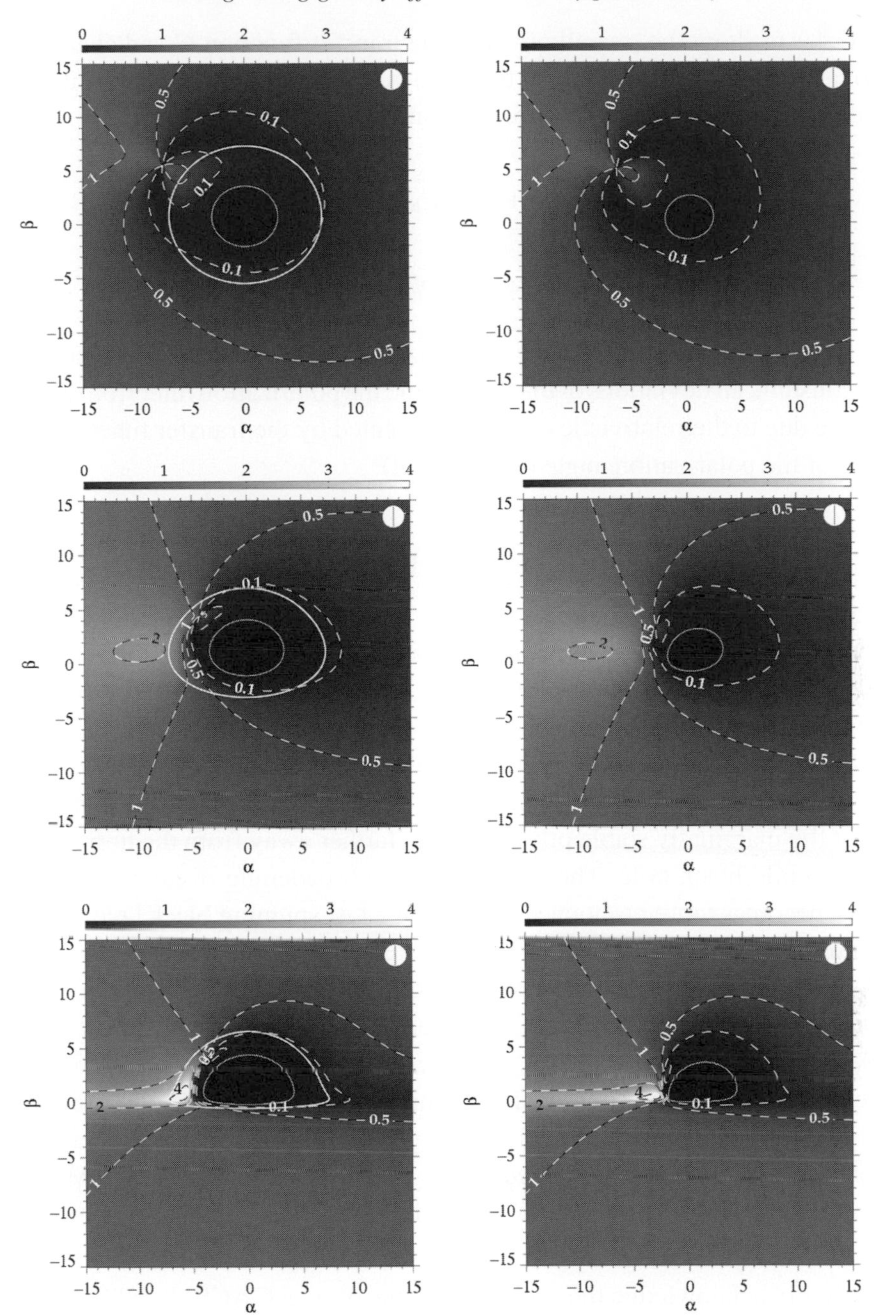

Figure 14.4 "The transfer function $G = g^4$ behind the polarization filter" for the inclination 30°, 60° and 85° (top to bottom) and the spin 0 (left) and 1 (right). The flux given by Equation (14.4) is represented by the greyscale and by dashed contours (0.1, 0.5, 1, 2 and 4). The local intensity and the local polarization degree are $I_{\rm loc} = P_{\rm loc} = 1$, the local polarization angle is $\chi_{\rm loc} = 0$ (i.e. the local polarization vector is perpendicular to the disc) and the angle of the polarization filter is $\chi_{\rm pf} = 0$ (see the upper right corner of each panel).

In order to show the overall effect of the transfer function G and change of the polarization angle ψ one can use the following equation for the intensity behind the polarization filter with the polarizing angle $\chi_{\rm pf}$

$$I_{\rm pf} = \frac{1}{2}\int_{\Sigma} {\rm d}S\ [GI_{\rm loc} + GI_{\rm loc}P_{\rm loc}\cos 2(\chi_{\rm loc}+\psi)\cos 2\chi_{\rm pf} + GI_{\rm loc}P_{\rm loc}\sin 2(\chi_{\rm loc}+\psi)\sin 2\chi_{\rm pf}]\ . \qquad (14.4)$$

If we set the local intensity, the local polarization degree and the local polarization angle constant, all deviations of the flux behind the polarization filter from a constant value are due to the relativistic effects represented by the transfer function and the change of the polarization angle (Figure 14.4).

14.6 Conclusions

We have demonstrated the effects of strong gravity on polarized radiation coming from the inner parts of the accretion disc orbiting a compact body. From the figures shown, one can notice that the relativistic effects in the case of the nonrotating and extremely rotating black holes differ only within several gravitational radii from the centre. The magnitude of these differences is not tremendous. However, the differences become much larger if we consider the accretion disc to be truncated at the marginally stable orbit, which is farther away from the horizon for the Schwarzschild black hole. Thus the relativistic broadening of all narrow spectral features becomes more pronounced in case of a fast spinning black hole where also the amplification of the flux due to the transfer function may become much larger for higher inclinations. The observed radiation is expected to be more polarized in the Schwarzschild case because a large part of the depolarizing region is hidden below the marginally stable orbit.

We would like to note that we have omitted some other relativistic effects for simplification purposes. These include different light time delays important in case of nonstationary systems, higher order images of the disc, caustics, etc. We also did not show what is the effect of self-irradiation of the disc, which may modify the local polarization.

We kindly acknowledge the support from the project GACR 207/07/0052 of the Grant Agency of the Czech Republic.

15

X-ray polarization from black holes in the thermal state

J. D. Schnittman & J. H. Krolik
Johns Hopkins University

We present new calculations of X-ray polarization from black hole accretion disks in the thermally dominated state, using a Monte-Carlo ray-tracing code in full general relativity. In contrast to many previously published studies, our approach allows us to include returning radiation that is deflected by the strong-field gravity of the BH and scatters off the disk before reaching a distant observer. Although carrying a relatively small fraction of the total observed flux, the scattered radiation tends to be highly polarized and in a direction perpendicular to the direct radiation. We show how these new features of the polarization spectra may be developed into a powerful tool for measuring black hole spin and probing the gas flow in the innermost disk.

15.1 Introduction

A recent flurry of new mission proposals has renewed interest in X-ray polarization from a variety of astrophysical sources, hopefully marking the "coming of age of X-ray polarimetry" in the very near future. The Gravity and Extreme Magnetism SMEX (*GEMS*) mission,[1] for example, should be able to detect a degree of polarization $\delta < 1\%$ for a flux of a few mCrab (e.g. [13] and these proceedings). A similar detector for the International X-ray Observatory (*IXO*) could achieve sensitivity roughly 10× greater ($\delta < 0.1\%$;[8; 5], Alessandro Brez in these proceedings). In this talk, based on our recent paper[11], we focus on the polarization signal from accreting stellar-mass black holes (BHs) in the thermal state, which are characterized by a broad-band spectrum peaking around 1 keV. The typical level of polarization from these sources should be a few percent in the 1–10 keV range, depending on BH spin and the inclination angle of the accretion disk.

X-ray Polarimetry: A New Window in Astrophysics, eds. R. Bellazzini, E. Costa, G. Matt and G. Tagliaferri. Published by Cambridge University Press.

Symmetry arguments demand that in the flat-space (Newtonian) limit, the observed polarization from the disk must be either parallel or perpendicular to the BH/disk rotation axis. However, the effects of relativistic beaming, gravitational lensing, and gravito-magnetic frame-dragging can combine to give a nontrivial net rotation to the integrated polarization vector. Early work exploring these effects[12; 3; 4] showed that they create changes in the angle and degree of polarization that are strongest for higher photon energy. Quite recently, Dovčiak *et al.*[7] investigated the effect of atmospheric optical depth on the polarization signal, and Li *et al.*[9] applied the original calculations of thermal X-ray polarization to the problem of measuring the inclination of the inner accretion disk.

Nearly all previous work has modeled the relativistic effects by calculating the transfer function along geodesics between the observer and emitter. By its very nature, this method precludes the possibility of including the effects of returning radiation – photons emitted from one part of the disk and bent by gravitational lensing so that they are absorbed or reflected from another part of the disk[6]. As described in greater detail in [11], the most important feature of our approach is that the photons are traced *from* the emitting region in all directions, either returning to the disk, scattering through a corona, getting captured by the BH, or eventually reaching a distant observer.

Using this method, we study an important new polarization feature, namely the transition between horizontal- and vertical-oriented polarization as the photon energy increases, an effect first discussed in [1]. At low energies we reproduce the "Newtonian" result of a semi-infinite scattering atmosphere emitting radiation weakly polarized in a direction parallel to the emission surface, an orientation we call *horizontal* polarization[2]. At higher energy, corresponding to the higher temperature of the inner disk, a greater fraction of the emitted photons returns to the disk and is then scattered to the observer. These scattered photons have a high degree of polarization and are aligned parallel to the disk rotation axis (*vertical*), as projected onto the image plane. At the transition point between horizontal and vertical polarization, the relative contributions of direct and reflected photons are nearly equal, and little net polarization is observed. Since the effects of returning radiation are greatest for photons coming from the innermost regions of the disk, the predicted polarization signature is strongly dependent on the behavior of gas near and inside the inner-most stable circular orbit (ISCO). Thus, polarization observations could be used to measure the spin of the black hole.

15.2 Direct radiation

In Figure 15.1(a), we show a simulated image of a Novikov-Thorne[10] accretion disk around a black hole with spin parameter $a/M = 0.9$ and luminosity $0.1 L_{\rm Edd}$,

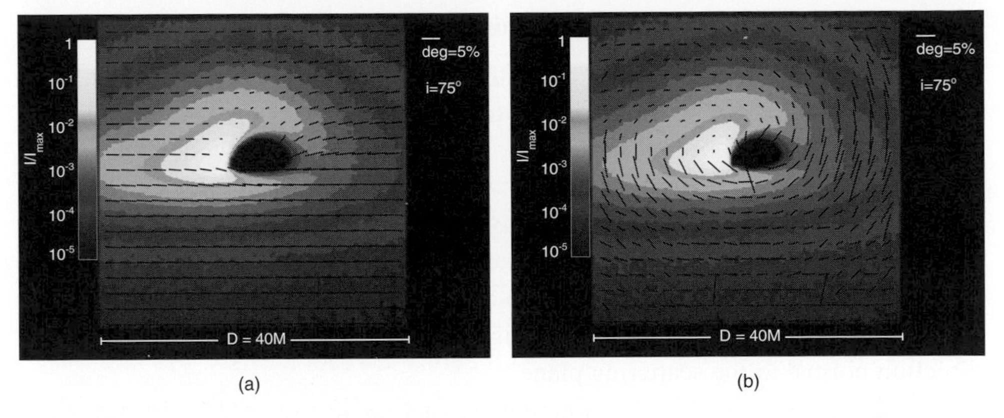

Figure 15.1 Ray-traced image of a thermal disk, including (a) only the direct, as well as (b) both direct and returning radiation. The observer is located at an inclination of 75° relative to the BH and disk rotation axis, with the gas on the left side of the disk moving towards the observer. The black hole has spin $a/M = 0.9$, mass $M = 10M_{\odot}$.

corresponding to a disk whose X-ray spectrum peaks around 1 keV for a BH mass of $10M_{\odot}$. With the observer at an inclination of $i = 75°$, significant relativistic effects are clearly apparent. The increased intensity on the left side of the disk is due to special relativistic beaming of the gas moving towards the observer, and the general relativistic light bending makes the far side of the disk appear warped and bent up towards the observer. Superposed on top of the intensity map is the polarization signature, represented by small black vectors whose lengths are proportional to the degree of polarization observed from that local patch of the disk. Far from the black hole, the polarization is essentially given by the classical result of Chandrasekhar[2] for a scattering-dominated atmosphere: horizontal orientation with $\delta \approx 4\%$ when $i = 75°$; nearer to the black hole, a variety of relativistic effects alter the polarization.

The two most prominent relativistic effects are gravitational lensing and special relativistic beaming, both lowering the net level of polarization seen by the observer. Gravitational lensing causes the far side of the disk to appear warped up towards the observer, and thus have a smaller effective inclination. Relativistic beaming causes photons emitted normal to the disk plane in the fluid frame to travel forward in the direction of the local orbital motion when seen by a distant observer; the result is a smaller effective inclination and thus degree of polarization. Naturally, these relativistic effects are most important close to the black hole, where the gas is also hottest and the photons have the highest energies. All these effects are clearly visible in Figure 15.1(a), which shows a smaller degree of polarization where the beaming is greatest (region of high intensity in the left of the image) and the lensing is strongest (just above the center of the image).

15.3 Returning radiation

When returning radiation is included, although little changes in terms of the total observed spectrum, the polarization picture (Fig. 15.1(b)) changes significantly: in much of the disk, the observed polarization rotates by 90°, even though none of the model's physical parameters has been changed at all!

This effect can be understood qualitatively in very simple fashion (see also [1]). Since the reflected flux can be roughly approximated by a single Thomson scatter, the polarization of forward-scattered photons remains unchanged, while for nearly-perpendicular scattering, the outgoing light can be close to 100% polarized, in a direction normal to the scattering plane.

For observers at high inclination angles, such as in Figure 15.1(b), returning radiation photons initially emitted from the far side of the disk (top of the image) are reflected off the near (bottom) side with a relatively small scattering angle, maintaining a moderate horizontal polarization as in Figure 15.1(a). On the other hand, photons emitted from the left side of the disk can be bent back to the right side (or *vice versa*), and then scatter at roughly 90° to reach the observer, thereby aquiring a large vertical polarization component. Although relatively small in total flux, this latter contribution can dominate the polarization because it is so strongly polarized.

Because high-energy photons from the hotter inner parts of the disk experience stronger gravitational deflection, they are more likely to return to the disk than the low-energy photons emitted at larger radii. We find that, for each value of the spin parameter, there is a characteristic "transition radius", within which the returning radiation dominates and produces net vertical polarization. Outside this point, the direct radiation dominates and produces horizontal polarization. This transition radius is in fact only weakly dependent on a/M, ranging from $R_{\rm trans} \approx 7M$ for $a/M = 0$ to $R_{\rm trans} \approx 5M$ for $a/M = 0.998$. The location and shape of the observed polarization swing can be used to infer the radial temperature profile, and thus the spin, near the transition radius.

In Figure 15.2, we compare the observed polarization as a function of energy for inclinations $i = 45, 60, 75°$, considering only direct radiation (solid curves) and also including return radiation (dashed curves). Clearly, when including the reflected photons, the polarization swings from horizontal at low energies to vertical above the thermal peak. In Figure 15.3 we show the polarization degree and angle for a range of BH spin parameters, in all cases including return radiation. As the spin increases, the ISCO moves closer to the horizon and a greater portion of the flux is emitted inside the transition radius, resulting in a higher fraction of return radiation. Thus more of the flux reaching the observer is vertically polarized, and the energy of transition moves closer to the thermal peak.

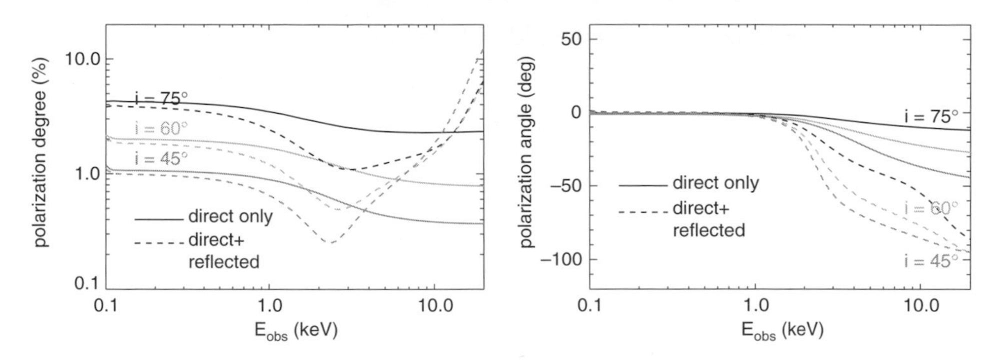

Figure 15.2 Polarization degree (*left*) and angle (*right*) for a $10M_{\odot}$ BH with spin $a/M = 0.9$ and luminosity $L = 0.1L_{\mathrm{Edd}}$, for both the direct radiation case (*solid curves*) and also when including return radiation (*dashed curves*).

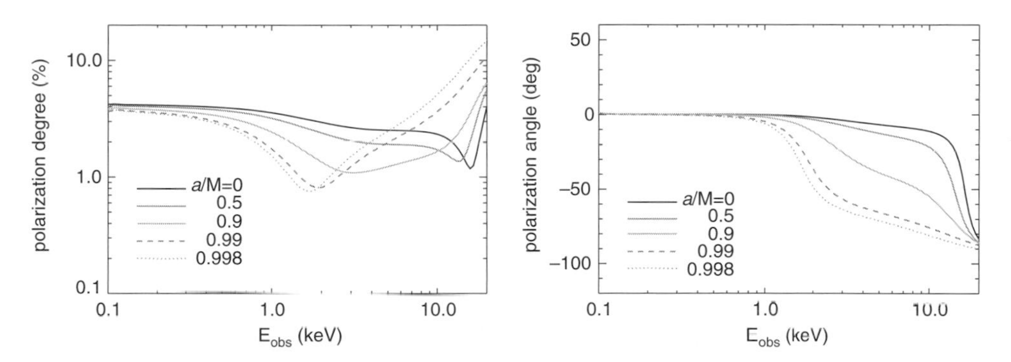

Figure 15.3 Polarization degree (*left*) and angle (*right*) for a $10M_{\odot}$ BH with luminosity $L = 0.1L_{\mathrm{Edd}}$, viewed from an inclination of 75°. The emissivity model is that of Novikov-Thorne for a range of BH spins.

15.4 Discussion

As described in [11], we have developed a new ray-tracing code to calculate the X-ray polarization from accreting black holes. In this paper we focus on polarization signatures of the thermal state in stellar-mass black holes. The emitted radiation has a diluted thermal spectrum and is weakly polarized parallel to the disk surface. The integrated polarization spectrum seen by a distant observer contains distinct energy-dependent features, as the high-energy photons from the inner disk are modified by relativistic effects such as Doppler boosting and gravitational lensing near the BH.

For radiation originating very close to the BH, the most important relativistic effect is the strong gravitational lensing that causes the photons to get bent back onto the disk and scatter towards the observer. This scattering can induce very high levels of polarization, and leads to a distinct transition in the polarization angle from horizontal at low energies to vertical above the thermal peak. Observing such a swing in the polarization angle would give the most direct evidence to date for the extreme relativistic light bending predicted around black holes. By measuring the location and shape of the polarization transition, we will be able to constrain the temperature profile of the inner disk. Assuming a Novikov-Thorne disk truncated at the ISCO, the polarization signal gives a direct measurement of the BH spin.

Notes

1. http://heasarc.gsfc.nasa.gov/docs/gems

References

[1] Agol, E., Krolik, J. H. (2000). *ApJ* **528**, 161.
[2] Chandrasekhar, S. (1960). *Radiative Transfer*. New York, Dover.
[3] Connors, P. A., Stark, R. F. (1980). *Nature* **269**, 128.
[4] Connors, P. A., Piran, T., Stark, R. F. (1980). *ApJ* **235**, 224.
[5] Costa, E., et al. (2008). *Proc. SPIE*, vol. 7011-15, [arXiv:0810.2700].
[6] Cunningham, C. T. (1976) *ApJ* **208**, 534.
[7] Dovčiak, M., Muleri, F., Goodmann, R. W., Karas, V., Matt, G. (2008). *MNRAS* **391**, 32.
[8] Jahoda, K., Black, K., Deines-Jones, P., Hill, J. E., Kallman, T., Strohmayer, T., Swank, J. (2007). [arXiv:0701090].
[9] Li, L.-X., Narayan, R., McClintock, J. E. (2009). *ApJ* **691**, 847.
[10] Novikov, I. D., Thorne, K. S. (1973) in *Black Holes*, ed. C. DeWitt and B. S. DeWitt. New York: Gordon and Breach.
[11] Schnittman, J. D., Krolik, J. H. (2009), *ApJ*, submitted. [arXiv:0902.3982]
[12] Stark, R. F., Connors, P. A. (1977). *Nature* **266**, 429.
[13] Swank, J., Kallman, T. Jahoda, K. (2008) in *Proceedings of the 37th COSPAR Scientific Assembly*, p. 3102.

16

Strong-gravity effects acting on polarization from orbiting spots

V. Karas

Astronomical Institute, Academy of Sciences, Prague

Accretion onto black holes often proceeds via an accretion disc or a temporary disc-like pattern. Variability features, observed in the light curves of such objects, and theoretical models of accretion flows suggest that accretion discs are inhomogeneous and nonaxisymmetric. Fast orbital motion of the individual clumps can modulate the observed signal. If the emission from these clumps is partially polarized, which is probably the case, then rapid polarization changes of the observed signal are expected as a result of general relativity (GR) effects.

In this contribution we will summarize the expected effects in terms of the model of bright orbiting spots. As the signal from accreting black holes peaks in X-rays, the polarimetry in this spectral band will be particularly useful to examine the strong-gravity effects that should modulate the signal originating near the horizon. We will mention similarities as well as differences between the manifestation of GR polarization change in X-rays and in other spectral bands, such as the infrared region, where the polarization measurements of the radiation flares from the immediate vicinity of the horizon are currently available and can be used to probe the Sagittarius A* supermassive black hole in the Galactic Centre.

16.1 Introduction

Polarization of light originating from different regions of a black hole accretion disc and detected by a distant observer is influenced by strong gravitational field near a central black hole. A 'spotted' accretion disc is a useful model of an interface of such an inhomogeneous medium, assuming that there is a well-defined boundary between the disc interior and the outer, relatively empty but highly curved space-time.

X-ray Polarimetry: A New Window in Astrophysics, eds. R. Bellazzini, E. Costa, G. Matt and G. Tagliaferri. Published by Cambridge University Press.

Relativistic corrections to a signal from orbiting spots can lead to large rotation in the plane of observed X-ray polarization. When integrated over an extended surface of the source, this can diminish the observed degree of polarization. Such effects are potentially observable and can be used to distinguish among different models of the source geometry and the radiation mechanisms responsible for the origin of the polarized signal. The polarization features show specific energy and time dependencies which can indicate whether a black hole is present in a compact X-ray source.

Practical implementation of the idea, originally proposed in the late 1970s by Connors et al.[7; 8] and Pineault[27], is a challenging task because the polarimetric investigations need a high signal-to-noise ratio. Also, the interpretation of the model results is often very sensitive to the assumptions about the radiation transfer in the source and the geometrical shape and orientation of the emission region. Nevertheless, the technology has achieved significant advances since the 1980s and has reached a mature state, as demonstrated also by this volume. Likewise, numerous theoretical papers have made progress in our understanding of the effects that we should look for. We assume that the gravitational field is described by a rotating black hole, and so the Kerr metric is the right model for the gravitational field. This set-up has been employed by Chen & Eardley[6] to investigate the polarization properties from an accretion disc near a Schwarzschild black hole. By using the Walker & Penrose[37] constant of motion along null geodesics in the Kerr metric, one can also determine the change of polarization angle along light rays in the case of a rotating black hole.

These dependencies suffer from some degeneracy, which could be avoided with time-resolved observations. Namely, if the source is an orbiting spot near a black hole, the time variation of the observed signal reflects the presence of strong gravity effects[3; 8]. A related problem of spots rotating on the surface of a compact star was investigated[36], as well as the case for more complicated surface patterns such as spiral waves[19; 20].

On the whole, there are some similarities as well as differences between the expected manifestation of GR polarization changes in X-rays and in other spectral bands, such as the infrared region. We will mention these interrelations and point out that the near-infrared polarization measurements of the radiation flares from the immediate vicinity of the horizon are already available for Sagittarius A* supermassive black hole in the Galactic Centre[25; 39].

16.2 Polarization from black hole accretion discs

It was realized some four decades ago that X-ray polarization studies could provide decisive clues to the physics of accreting compact objects[1; 4; 23; 29]. In the non-relativistic regime, a conceptually similar problem was discussed by Rudy[30] and

Fox[16], who studied the polarization of star light caused by an ionized circumstellar shell of free electrons.

In the context of accretion discs the effect of an electron-scattering atmosphere has been also often invoked. Further, polarization of the Comptonized radiation of accretion discs was examined as a function of various model parameters, such as the optical thickness of the disc medium, energy of scattered photons and directional angle of the emission[32; 33; 38]. Finally, relativistic effects acting on polarization from black hole accretion discs were examined[6].

Two basic schemes were proposed as being relevant for the X-ray polarization from the inner accretion disc. Firstly, the accretion disc surface lies below the scattering atmosphere and acts as a source of seed photons. Polarization of thermal radiation from a black hole accretion disc was also studied[21]. The reason for polarization is that photons from the disc are scattered by electrons within the disc atmosphere. Linear polarization should arise in the disc local co-rotating reference frame. This situation is expected to be particularly relevant for Galactic black hole candidates whose discs exhibit phases of strong multi-blackbody thermal radiation dominating over other spectral components. The early investigations were recently put forward by several groups[14; 22]. Secondly, Matt[24] and Dovčiak et al.[11] examined the polarimetric consequences of a specific model of a lamp-post-illuminated accretion disc. Within this scheme the number of reflected (polarized) photons is proportional to the incident flux arriving from the primary source.

Obviously the turbulent magnetic fields will play a role in diminishing the observed polarization degree, and that part has been neglected in the present contribution[31]. Although the geometrical effects of strong gravitational fields act on photons independently of their energy, the influence of turbulent magnetic fields, intervening via Faraday rotation, are indeed energy dependent. The impact of Faraday rotation on the observed polarization decreases with the square of photon energy, and hence it turns out to be less restrictive in the X-ray band.

In the next section we describe our model, which adopts a Keplerian, geometrically thin and optically thick disc around a Kerr black hole, assuming that the surface of the disc is covered by an orbiting spot.

16.3 Time-varying polarization from orbiting spots

The model of an orbiting bright spot[5; 9; 25; 26] has been fairly successful in explaining the observed modulation of various accreting black hole sources. Certainly not all variability patterns can be explained in this way, however, the scheme is general enough to be able to capture also the effects of spiral waves and similar kinds of transient phenomena that are expected to occur in the disc[20; 34; 35]. It can be argued that the spot lightcurves can be phenomenologically understood as a region of enhanced emission that performs a co-rotational motion closely above the innermost stable circular orbit (ISCO). For example, within the framework of the flare-spot model[10] the spots are just regions of enhanced emission on the disc surface rather than massive clumps that could suffer from fast decay due to shearing motion in the disc. The observed signal is modulated by relativistic effects. According to this idea, Doppler and gravitational lensing influence the observed radiation flux and this can be computed by ray-tracing methods. Such an approach has been extended to compute also strong gravity effects acting on polarization properties[13].

A spot on the disc surface is supposed to be intrinsically polarized (by different possible mechanisms – either by reflection of a primary flare on the disc surface or by synchrotron emission originating from an expanding blob, as detailed below). It represents a rotating surface feature which shares the bulk orbital motion of the underlying medium at sufficiently large radii above the ISCO, gradually decaying due to differential rotation of the disc.

We have applied different prescriptions for the local polarization (see References [12; 25; 39] for the detailed description of the model set-up in the individual cases that we investigated). For example, one set of models assumes the local emission to be polarized either in the direction normal to the disc plane, or perpendicular to the toroidal magnetic field. Obviously, in the case of partial local polarization the observed polarization signal will be diluted by an unpolarized fraction, and so the polarization degree of the final signal will be proportionally diminished. In another set of models we assumed a lamp-post-illuminated spot as the source of spot polarization by reflection. For the spot shape we first assumed the spot does not change its shape during its orbit, but then we also consider the spot decay with time. The relativistic effects can be clearly identified and understood with these simple (and astrophysically unrealistic) toy models, as they produce visible signatures in the observed polarization properties.

General relativistic effects present in our model can be split into two categories. Firstly, there is the symmetry breaking between the approaching and the receding part of the spot orbit. Doppler beaming as well as the light focusing contribute to the change of the observed flux, especially at high view angles when the spot orbit is seen almost edge-on. Notice that the Doppler boosting effect is off phase with respect to the light focusing effect, roughly by 0.25 of the full orbit at the corresponding radius. Here, the precise number depends on the black hole spin; it also depends on the inclination through the finite light-travel time from different parts of the spot orbit towards the observer. Also, higher-order images could be important in case of almost edge-on view of the spot.

Secondly, rotation of the polarization plane along the photon trajectory also plays a role. This effect is particularly strong for small radii of the spot orbit, in which case a critical point occurs[14]. The observed polarization angle exhibits just a small wobbling around its principal direction when the spot radius is above the critical point, whereas it starts turning around the full circle once the radius drops below the critical one. Notice that the exact location of the critical point depends on the black hole angular momentum, in principle allowing us to determine its value.

However, a caveat (and a third point on the list) is caused by sensitivity of the critical radius to the special relativistic aberration effects, especially at small view angles (i.e. when the spot is seen almost along the rotation axis). This means that the moment when the observed polarization angle starts rotating is sensitive to the underlying assumption of a perfectly planar geometry of the disc surface.

By combining the above-mentioned effects together, Dovčiak *et al.* [12] have shown that the observed polarization degree is expected to decrease (in all their models) mainly in that part of the orbit where the spot moves close to the region where the photons are emitted perpendicularly to the disc. In this situation the polarization angle changes rapidly. The decrease in the observed polarization degree for the local polarization perpendicular to the toroidal magnetic field happens also in those parts of the orbit where the magnetic field points approximately along the light ray.

For the more realistic models the resulting polarization shows a much more complex behaviour. Among persisting features is the peak in polarization degree for the extreme Kerr black hole for large inclinations, caused by the lensing effect at a particular position of the spot in the orbit where the polarization angle is changed. This is not visible in the Schwarzschild case.

The X-ray polarization lightcurves and spectra are still to be taken by future missions, but one may envision even a more challenging goal connected with imaging of the inner regions of accreting black hole sources. Obviously this is a truly distant future: imaging a black hole shadow would require in the order of ten micro-second angular resolution. However, what might be realistically foreseen is the tracking of the wobbling image centroid that a spot is supposed to produce[18; 40]. With the polarimetric resolution, the wobbling could provide excellent evidence proving the presence of the orbiting feature. See Figure 16.1 for an example of the expected form of the spot images and the corresponding centroid tracks in a simplified case of a model spot endowed with an intrinsic polarization that remains constant in the co-orbiting frame. This example assumes a spot rotating rigidly at constant radius closely above the ISCO. Orientation of the polarization filter is also fixed, as indicated in the top-right corner of the plot. Correspondingly, Figure 16.2 shows the tracks of the image centroid. Albeit the tracks are not identical in the two orientations of the polarization filter, the difference is rather subtle.

Notice that the project of detecting the centroid motion does not necessarily have to be limited to the X-ray domain. In view of recent results on Sagittarius A* flares,

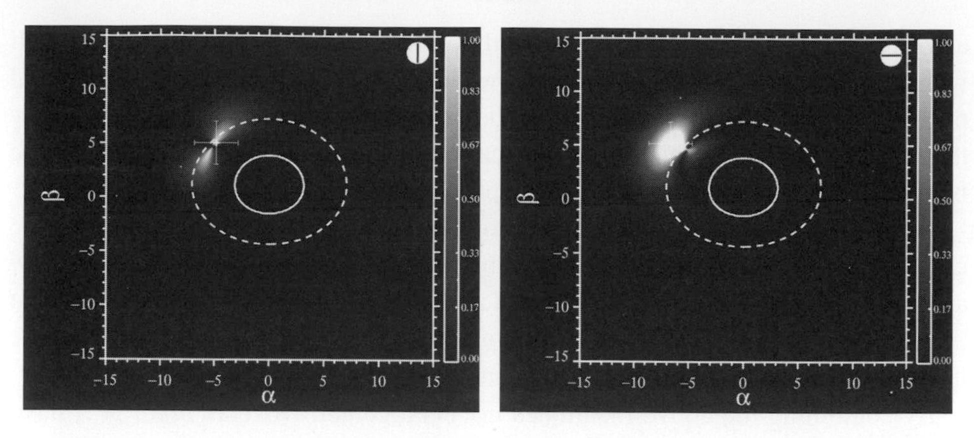

Figure 16.1 A snapshot of a spot orbiting at constant radius $r = 1.1r_{\mathrm{ISCO}}$. The image is produced by the spot emission that is assumed to be intrinsically polarized and recorded in two polarization channels, rotated by 90 degrees with respect to each other[40]. The direction of the polarization filter is indicated in the top right corner of each of the two panels. The image is shown in the observer plane (α, β), for a non-rotating black hole observed at a moderate view angle, $\theta_o = 45°$. The horizon radius (solid curve) and the ISCO (dashed curve) are also shown.

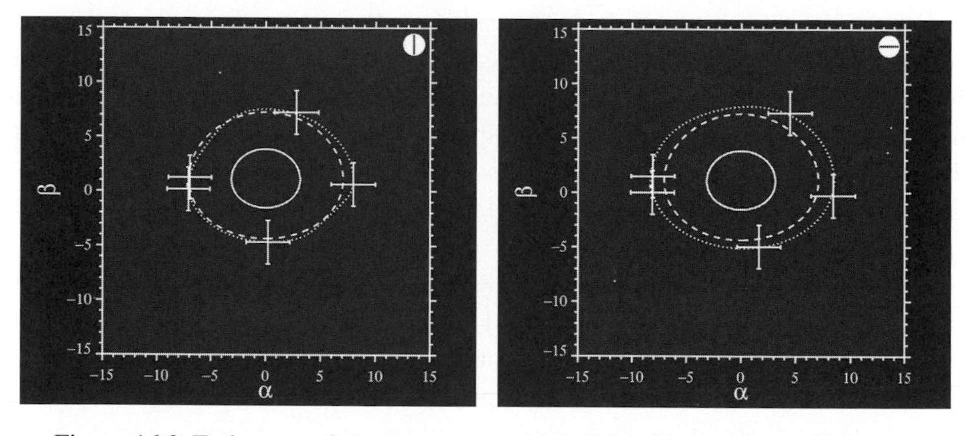

Figure 16.2 Trajectory of the image centroid during one revolution of the spot corresponding to the previous figure. The wobbling position of the image centroid is indicated by crosses at five different moments along the image track (dotted curve). Figures are courtesy of M. Zamaninasab who has applied the spot model to investigate the polarization properties of near-infrared flares from the Galactic Centre supermassive black hole[40].

which have been reported in X-rays as well as in the near infrared, submillimetre and the radio spectral bands[2; 15; 17; 28], the immediate vicinity of the black hole can be probed by various techniques. The simultaneous time-dependent measurements equipped with the polarimetric resolution seem to be a final goal of this effort.

16.4 Conclusions

Polarimetry is known to be a photon-hungry technique, and so it is not easy to identify the specific effects of general relativity that could be observed with available polarimeters (in whatever spectral band) or with those envisaged for the realistically foreseeable future. In several recent papers, and in particular in this volume, various people demonstrate that the task of detecting the relativistic effects and in this way determining the physical parameters of the black hole systems seems to be feasible. Among ways to reach the goal, time-dependent polarization profiles, such as those expected from orbiting spots, play an important role.

It may be worth reminding the reader that the KY code, employed in our computations, is publicly available, either as a part of the XSPEC package or directly from the authors[13]. The current version allows the user to include the polarimetric resolution and to compute the observational consequences of strong-gravity effects from a Kerr black hole accretion disc. Within the XSPEC notation, this polarimetric resolution is encoded by a switch defining which of the four Stokes parameters is returned in the photon count array at the moment of the output from the model evaluation. This way one can test and combine various models, and pass the resulting signal through the response matrices of different instruments.

Acknowledgements

Mohammad Zamaninasab kindly created figures for this article. The author thanks the Czech Science Foundation (ref. 202/09/0772) and the Center for Theoretical Astrophysics in Prague (ref. LC06014) for support.

References

[1] Angel J. R. P. (1969). *ApJ* **158**, 219.
[2] Baganoff F. K., Bautz M. W., Brandt W. N., et al. (2001). *Nature* **413**, 45.
[3] Bao G., Hadrava P., Wiita P. J., Xiong Y. (1997). *ApJ* **487**, 142.
[4] Bonometto S., Cazzola P., Saggion A. (1970). *A&A* **7**, 292.
[5] Broderick A. E., Loeb A. (2005). *MNRAS* **367**, 905.
[6] Chen K., Eardley D. M. (1991). *ApJ* **382**, 125.
[7] Connors P. A., Stark R. F. (1977). *Nature* **269**, 128.
[8] Connors P. A., Stark R. F., Piran T. (1980). *ApJ* **235**, 224.
[9] Cunningham C. T., Bardeen J. M. (1972). *ApJ* **173**, L137.
[10] Czerny B., Różańska A., Dovčiak M., et al. (2004). *A&A*, **420**, 1.
[11] Dovčiak M., Karas V., Matt G. (2004). *MNRAS* **355** , 1005.
[12] Dovčiak M., Karas V., Matt G. (2006). *AN* **327**, 993.
[13] Dovčiak M., Karas V., Yaqoob T. (2004). *ApJS* **153**, 205.
[14] Dovčiak M., Muleri F., Goosmann R. W., et al. (2008). *MNRAS* **391**, 32.
[15] Eckart A., Baganoff F. K., Zamaninasab M., et al. (2008). *A&A* **479**, 625.
[16] Fox G. K. (1994). *ApJ* **435**, 372.
[17] Genzel R., Schödel R., Ott T., et al. (2003). *Nature* **425**, 934.

[18] Hamaus N., Paumard T., Müller T., et al. (2009). *ApJ* **692**, 902.
[19] Karas V. (2006). *AN* **327**, 961.
[20] Karas V., Martocchia A., Šubr L. (2001). *PASJ* **53**, 189.
[21] Laor A., Netzer H., Piran T. (1990). *MNRAS* **242**, 560.
[22] Li Li-Xin, Narayan R., McClintock J. E. (2009). *ApJ* **691**, 847.
[23] Lightman A. P., Shapiro S. L. (1975). *ApJ* **198**, L73.
[24] Matt G. (1993). *MNRAS* **260**, 663.
[25] Meyer L., Eckart A., Schödel R., et al. (2006). *A&A* **460**, 15.
[26] Noble S. C., Leung Po Kin, Gammie C. F., Book L. G. (2007). *CQG* **24**, S259.
[27] Pineault S. (1977). *MNRAS* **179**, 691.
[28] Porquet D., Predehl P.. Aschenbach B., et al., (2003) *A&A* **407**, L17.
[29] Rees M. J. (1975). *MNRAS* **171**, 457.
[30] Rudy R. J. (1978). *PASP* **90**, 688.
[31] Silant'ev N. A., Gnedin Yu. N. (2008). *A&A* **481**, 217.
[32] Stark R. F. (1981). *MNRAS* **195**, 115.
[33] Sunyaev R. A., Titarchuk L. G. (1985). *A&A* **143**, 374.
[34] Tagger M., Henriksen R. N., Sygnet J. F., Pellat R. (1990). *ApJ* **353**, 654.
[35] Tagger M., Melia F. (2006). *ApJ* **636**, 33.
[36] Viironen K., Poutanen J. (2004). *A&A* **426**, 985.
[37] Walker M., Penrose R. (1970). *Comm. Math. Phys.* **18**, 265.
[38] Williams A. C. (1984). *ApJ* **279**, 401.
[39] Zamaninasab M., Eckart A., Meyer L., et al. (2008). *J. Phys.* **131**, 012008.
[40] Zamaninasab M., Eckart A., Witzel G., et al. (2009). *A&A* submitted.

17

Polarization of thermal emission from accreting black holes

M. Dovčiak & V. Karas

Astronomical Institute, Prague

R. Goosmann

Observatoire Astronomique de Strasbourg

G. Matt

Dipartimento di Fisica, Università degli Studi "Roma Tre"

F. Muleri

INAF/IASF–Roma

Multicolour black-body emission from the accretion disc around the black hole can be polarized on its way through the atmosphere above the accretion disc. We model this effect by assuming Kerr metric for the black hole, a standard thin disc for the accretion flow and Thomson scattering in the atmosphere. We compute the expected polarization degree and the angle as they can be measured for different inclinations of the observer, optical thickness of the atmosphere and different values of the black hole spin. All relativistic effects near a compact centre are taken into account. We also assess the perspectives for the next generation of X-ray polarimeters.

17.1 Introduction

We consider polarization originating from a Keplerian, geometrically thin and optically thick accretion disc near a black hole. At each radius the accretion disc emits black body radiation, the temperature of which is given by the Novikov-Thorne expression for the outer part of the standard disc. The thermal photons are scattered in the atmosphere of the disc and thus the observed radiation becomes polarized[1; 2; 3]. We assume multiple Thomson scattering with different optical depths of the disc atmosphere. The effect of hardening of the energy of photons

X-ray Polarimetry: A New Window in Astrophysics, eds. R. Bellazzini, E. Costa, G. Matt and G. Tagliaferri.
Published by Cambridge University Press.

due to scattering is taken into account via the hardening factor that increases the effective temperature.

Once the photons leave the atmosphere the polarization vector can be rotated due to strong gravity of the black hole. The energy of photons is shifted by the gravitational and Doppler effects. In our computations[1] we used the microquasar GRS 1915+105 in the thermal state as an example, assuming the mass of the central black hole $M_\bullet = 14M_\odot$, the accretion rate $\dot{M} = 1.4 \times 10^{18}$ g/s and the hardening factor $f = 1.7$.

17.2 Polarization angle and degree of polarization at detector

The local polarization induced by Thomson scattering is defined in the local frame co-moving with the disc. Due to the aberration the polarization angle for each photon is different at infinity even in the Schwarzschild space-time. In the Kerr case the rotation of the polarization vector because of the gravitational dragging is added. The dependence of the change of the polarization angle on the position of the emission from the disc was discussed in Reference [1]. Farther away from the black hole this change is not too large. Therefore the depolarizing effect of the integration over that part of the disc will be relatively small. On the other hand below the critical point, where the light is emitted perpendicularly to the disc, the change of the polarization angle can acquire any value. Thus the depolarizing effect of this region can be quite large. Therefore for the overall polarization measured at infinity it is important if this critical point is above or below the marginally stable orbit.

The decrease of the temperature with the disc radius implies that the closer to the black hole the radiation is emitted, the harder it is. As a consequence, the effects of relativity are larger at higher energies. The energy dependence of the polarization degree and angle at infinity is shown in Figure 17.1. The character of the polarization at lower energies is given mainly by the emission originating far away from the black hole. The observed polarization degree and angle are equal to the local ones. In most cases, the polarization is the highest for these energies. For high energies (above 10 keV), the polarization degree again increases. The polarization here is influenced mainly by that region of the disc where the relativistic amplification of the flux is the largest and the temperature is the highest. This area is not very large and thus the span of the change of the polarization angle is small. Therefore the depolarizing effect is not very large. Note, however, that for the same reason the flux for this interval of energy is also very small.

The dependence of the polarization degree and angle on the inclination of the observer is shown in Figure 17.2. The polarization degree in most cases increases with the inclination angle of the observer. The polarization angle does not change much with the observer's inclination, mainly for lower inclinations. However, its value depends on the spin of the black hole.

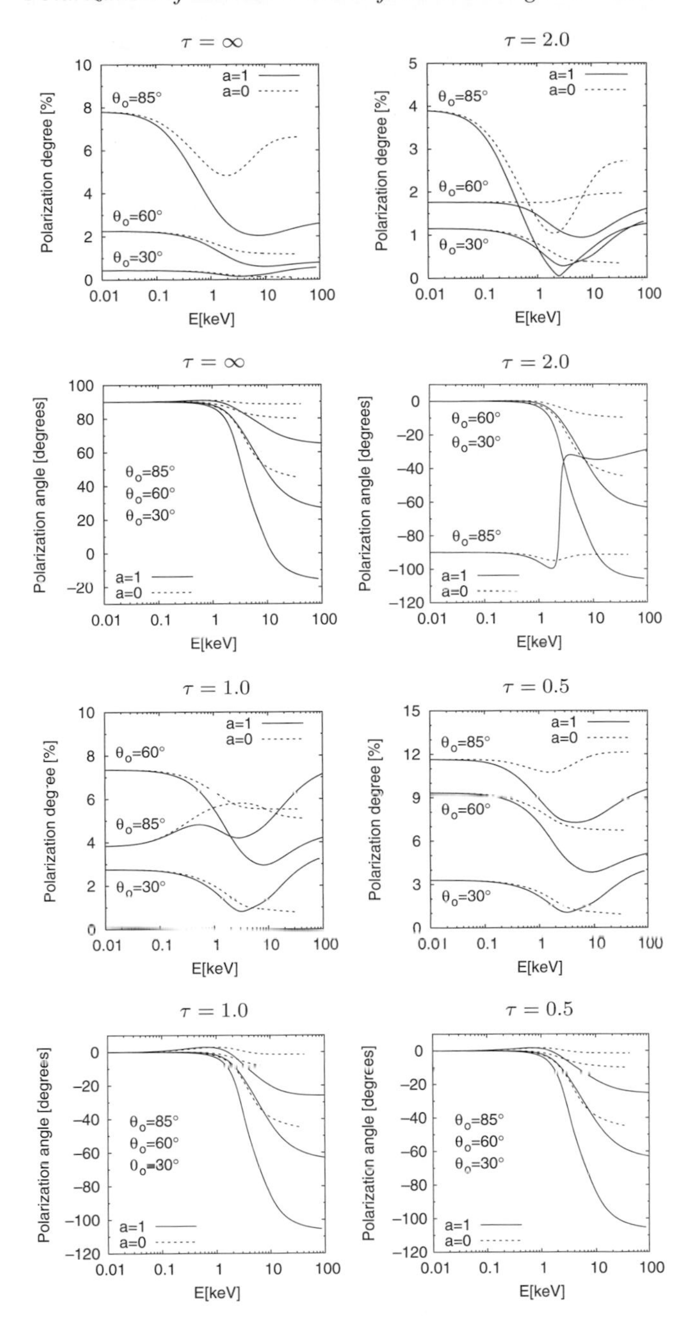

Figure 17.1 The energy dependence of the polarization degree (top) and polarization angle (bottom) as the observer at infinity would measure them for different optical depths of the atmosphere $\tau = \infty, 2.0, 1.0, 0.5$, in the case of the Schwarzschild (dashed lines) and extreme Kerr (solid lines) black holes and observer inclinations of 30°, 60° and 85°. A polarization angle of 0° represents the direction aligned with the projected symmetry axis of the disc.

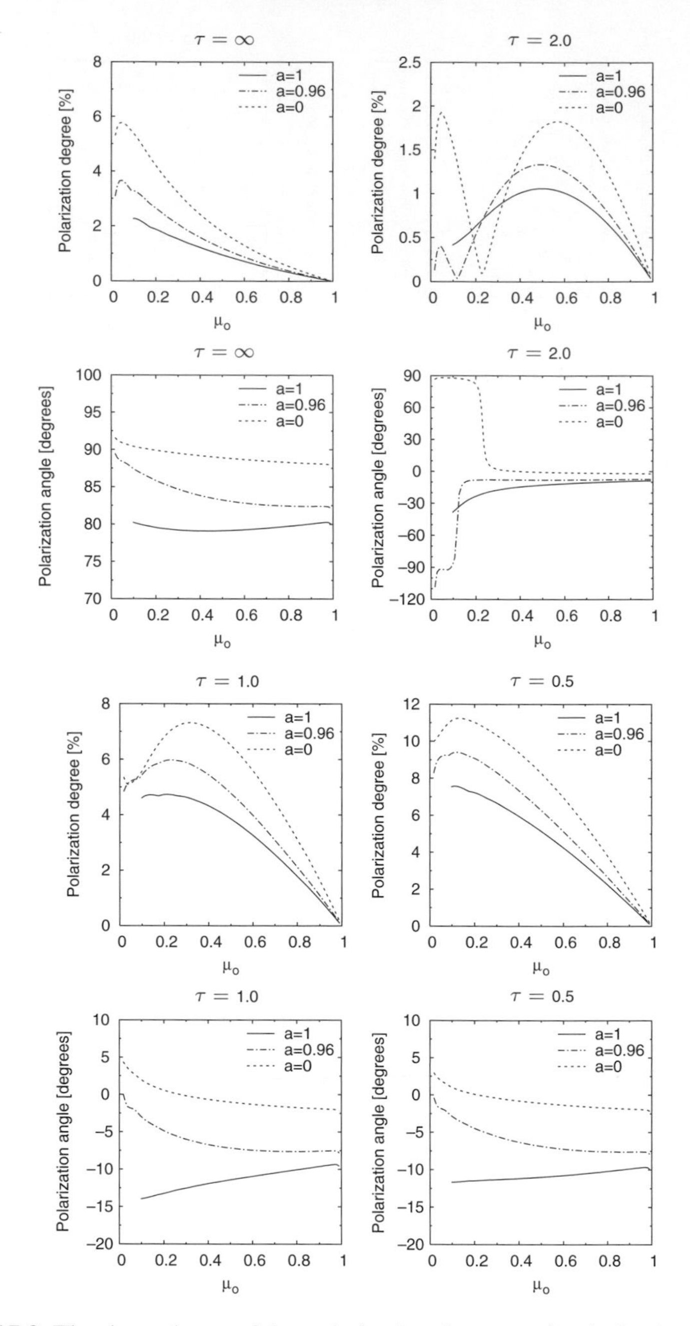

Figure 17.2 The dependence of the polarization degree and polarization angle at infinity on cosine of the observer's inclination for different optical depths of the atmosphere $\tau = \infty$, 2.0, 1.0, 0.5 in the case of the Schwarzschild (dashed lines) and extreme Kerr (solid lines) black holes. A polarization angle of $0°$ represents the direction aligned with the projected symmetry axis of the disc.

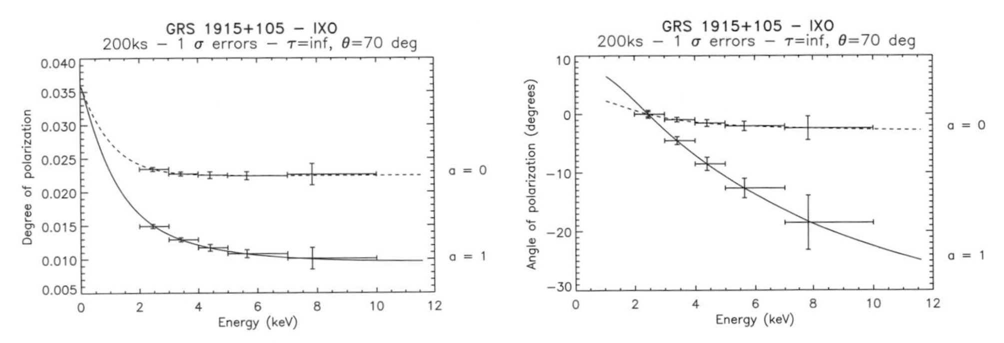

Figure 17.3 Plots of the predicted polarization for the observation carried out with the current version of IXO, assuming $\tau = \infty$, $\theta_o = 70°$, and the observation time of 200 ks. In this example the two cases of the black hole spin could be discriminated from each other.

17.3 Observational perspective

To assess the perspective of polarization observations we use the scenario of a large observatory with large collecting area – the proposed International X-ray Observatory (IXO). We consider the Gas Pixel Detector as the X-ray polarimeter. Its current design allows a large increase of performances with respect to the instruments built so far, with a sensitivity peaking at about 3 keV and a response between 2 and 10 keV. The results of simulations are shown in Figure 17.3 – the Kerr and the Schwarzschild cases can be clearly discriminated (the errors shown are at the 1-σ level).

17.4 Conclusions

The most interesting interval of energy is from 0.1 to 10 keV. Below this energy band the emission comes from far away from the disc centre, while above this range the flux rapidly decreases. Note, however, that for higher (lower) temperatures of the disc all studied energy dependences will be shifted to a higher (lower) energy.

The effects of strong gravity on the polarization state of emergent radiation will be accessible to the next missions with an X-ray polarimeter based on the Gas Pixel Detector. The energy dependence of the degree and of the angle of polarization will clearly discriminate between the Kerr and the Schwarzschild black hole cases. However, we would like to stress that the quality of the polarimetric results by IXO will increase significantly with a higher effective collecting area in the 2–10 keV band.

We acknowledge the ESA PECS project No. 98040 in the Czech Republic.

References

[1] Dovčiak, M., Muleri, F., Goosmann, R. W., et al. (2008). *MNRAS* **391**, 32.
[2] Li, Li-Xin, Narayan, R., McClintock, J. E. (2008). *ApJ* **691**, 847.
[3] Schnittman, J. D., Krolik, J. H. (2009). *ApJ*, submitted.

18

X-ray polarimetry and radio-quiet AGN

G. Matt

Dipartimento di Fisica, Università Roma Tre

Active galactic nuclei (AGN) are definitely not spherical objects, but rather have an axisymmetric morphology. This, together with the fact that the primary X-ray emission is due to Comptonization, and that there are several circumnuclear regions which may scatter the primary radiation, implies that most, if not all, of the X-ray spectral components observed in radio-quiet AGN should be significantly polarized. Moreover, the polarization properties of the radiation emitted very close to the black hole are modified by Special and General Relativity effects, which therefore can be probed with X-ray polarimetry.

18.1 Introduction

AGN have a complex morphology (see Figure 18.1). The nucleus, where a supermassive (10^6–10^9 $M_\odot$) black hole accretes matter from an accretion disc, is surrounded by optically and geometrically thick matter, possibly clumpy and probably in a toroidal (or at least axisymmetric) configuration. If the line of sight intercepts this 'torus', the nucleus is not visible and the source is classified as of type 2. For type 1 sources, instead, the line of sight does not intercept the 'torus', and the nucleus can be directly seen. Outside the 'torus', ionized matter, responsible for scattering and polarizing the nuclear radiation in type 2 sources[1] is present. A jet is also present in 'radio-loud' sources, which are not discussed in this contribution, devoted to 'radio-quiet' AGN.

The most obvious fact, when looking at Figure 18.1, is that an AGN is definitely not a spherical object, but rather an axisymmetric object. It is not surprising, therefore, that the X-ray emission (which primarily originates in the vicinity of the black

X-ray Polarimetry: A New Window in Astrophysics, eds. R. Bellazzini, E. Costa, G. Matt and G. Tagliaferri. Published by Cambridge University Press.

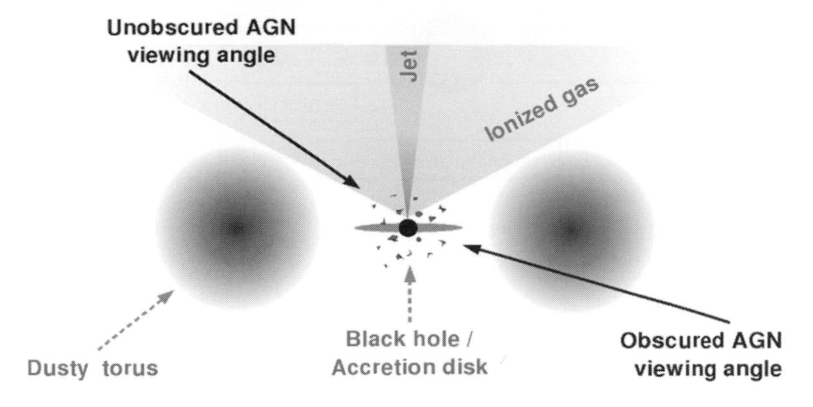

Figure 18.1 The standard picture of AGN. From [6].

hole, but which is also scattered by the various circumnuclear regions) is expected to be highly polarized, as discussed in the following.

18.2 The X-ray emission of AGN

18.2.1 The primary emission

The X-ray emission in radio-quiet AGN is largely believed to be due to Comptonization of soft photons (probably the thermal emission of the accretion disc) in a hot ($kT \approx 10^9$ K) corona. The geometry of the corona is unknown, and it may even be spherical, but the radiation field is certainly aspherical (if due to the accretion disc), so providing the asymmetry required to produce a polarized emission.

The polarization properties of course will depend on the geometry of the corona. For instance, in a plane-parallel geometry, as in the two-phase model of [10] – shown in the upper part of Figure 18.2 – the polarization degree may be very high[11; 19]. Other geometries have also been proposed for the corona (aborted jets, clouds, flaring blobs, etc.), but the polarization properties in these geometries have not been calculated yet, as well as the modifications due to strong gravity (see below).

An altogether different scenario is that of advection-dominated accretion flows (ADAF) models. In such models, the inner accreting region is very hot (electron temperature of about 10^9 K) and geometrically thick (see the lower part of Figure 18.2), and the X-ray emission is mainly via bremsstrahlung. No X-ray polarization is therefore expected in these models. The ADAF mode is often invoked for low-luminosity AGN: if this is correct, their X-ray emissions, contrary to those of 'normal' AGN, should be unpolarized.

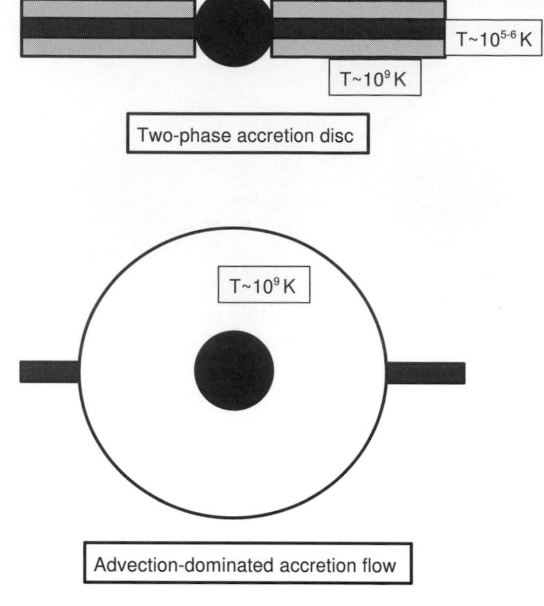

Figure 18.2 Upper: a two-phase, plane parallel geometry for the disc-corona system. Lower: the advection-dominated accretion flow scenario.

18.2.2 Probing strong gravity

Matter very close to the black hole experiences General and Special (due to the large velocities involved) Relativity (hereinafter collectively called 'strong gravity') effects. As discussed elsewhere in this volume (see contributions by M. Dovciak and J. Schnittmann), in galactic black hole systems such effects can be probed, and the black hole spin measured, by X-ray spectropolarimetry measurements of the disc thermal emission. In AGN, disc emission is much softer, peaking in the UV band. However, strong gravity effects may manifest themselves in X-rays through time-dependent, rather than energy-dependent, rotations of the polarization angle of the radiation reflected by the accretion disc. In fact, part of the primary emission illuminates the disc, where it is reprocessed and reflected, giving rise to fluorescent lines (the iron Kα line being the most prominent) and to a 'Compton reflection' (CR) continuum (e.g. [15]), which is expected to be significantly polarized[14]. The iron line and the CR are likely to be emitted very close to the black hole; strong gravity will therefore affect the transfer of radiation, giving rise to, for example, a broad and asymmetric profile of the iron line (e.g. [4] and references therein).

To explain the puzzling time variability of the iron line in the well-known Seyfert 1 galaxy MCG-6-30-15 (the best and most studied case so far for a relativistic iron line[5]), Miniutti and Fabian[17] suggested that the primary emission originates in a small region close to the black hole spinning axis (maybe an aborted jet[8]),

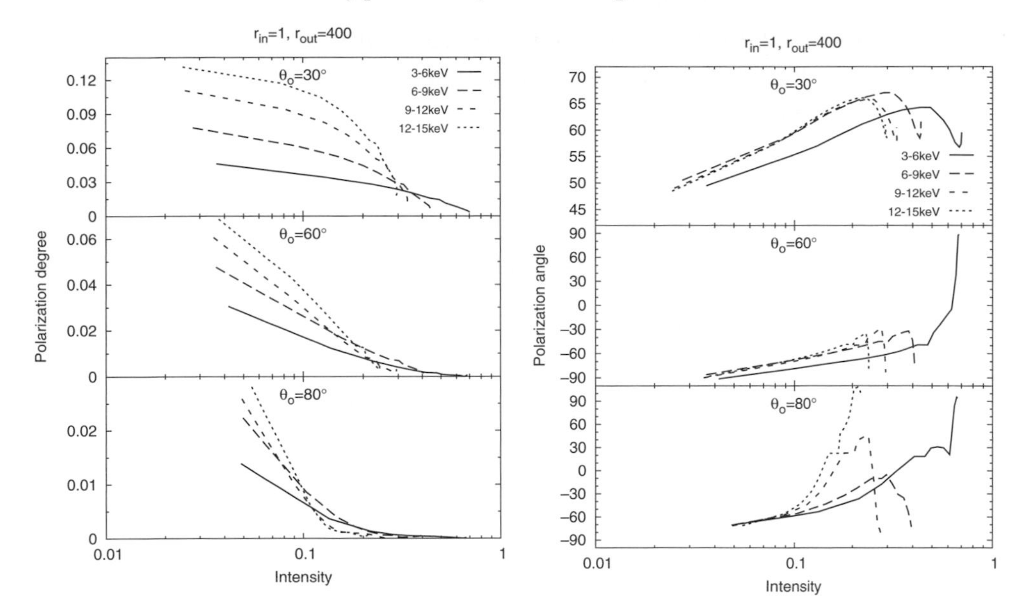

Figure 18.3 The polarization degree (left panel) and angle (right panel) of the total radiation (reflected + primary, assumed unpolarized) as a function of the flux of the primary source. A lamp-post geometry for the corona is assumed. Kindly provided by M. Dovciak.

the observed variability being due to the variation of the height of the source (the so-called 'light-bending model') and therefore of the gravitational redshift. If this is indeed the case, the polarization degree and angle of the reflected radiation must also vary in a characteristic way[3], which depends on the spin of the black hole. Polarimetry can therefore provide a further and powerful probe of radiative transfer in a strong gravity field as well as an estimate of the black hole spin.

In Figure 18.3 (kindly provided by M. Dovčiak), the polarization degree and angle of the total radiation (reflected + primary, assumed unpolarized for simplicity) as a function of the flux of the primary source (which, in the light bending model, is related to the height of the primary source) are shown for an extreme Kerr black hole. Large variations of both the polarization degree and angle with the source flux are expected, providing a test for the model.

18.3 Reflection from distant matter

As said in the Introduction, the nucleus is surrounded by both neutral and ionized matter (see Figure 18.1). Evidence for ionized matter is both direct (the ionization cones seen in the optical band) and indirect (the need for a 'mirror' which scatters and polarizes the, otherwise invisible, nuclear emission in Seyfert 2s[1]). Neutral (or low ionized) matter is almost ubiquitous: the BLR close to the nucleus, dust lanes, the 'torus', starburst regions and the galactic disc, all provide absorption if intercepting the line of sight. Among these regions, the only one which is likely to be optically thick to Compton scattering, and therefore provide significant reflection, is the pc-scale 'torus'.

But is the 'torus' really a torus?

18.3.1 The 'torus'

There are several arguments in favour of a toroidal (or at least axisymmetric) configuration of the 'torus'. The conical morphology of the 'ionization cones' seen in the optical suggests a collimation of the ionizing flux, most probably funnelled by an axisymmetric distribution of optically thick matter. Interferometric observations in nearby sources (Circinus[22]; NGC 1068[20]) seem to confirm such a scenario, even if, at least in NGC 1068, the ionization cone appears misaligned with the axis of the toroidal IR distribution[20]. It would be very important to have an independent check of these geometrical configurations, and X-ray polarimetry may provide such a check: the reflected emission from the torus should be highly polarized, with the polarization angle either parallel or perpendicular (depending on the torus half opening angle) to the torus axis [7; Goosmann & Matt, in preparation].

18.3.2 The ionized reflector

The matter which scatters and polarizes the optical/UV nuclear radiation must also scatter – and polarize – X-rays. This reflection component is indeed usually the dominant one below 1–2 keV in obscured Seyferts. Polarization degree is expected to be quite large, even if part of the radiation is due to emission lines[9], which are unpolarized (unless due to resonant scattering). For further details, and the relevant bibliography, I refer the reader to the contribution by S. Bianchi to this volume. Here I just want to remark that polarimetry may provide another way to distinguish between scattering from photoionized matter and other possible alternatives for the soft X-ray excess, as for instance emission related to starburst regions (which should be unpolarized).

18.4 The Galactic Centre: a transient AGN?

X-ray polarimetry can also test, by determining the polarization properties of the X-ray emission from the giant molecular cloud Sgr B2, the hypothesis that our own Galactic Centre was, about 300 years ago, a low-luminosity AGN.

Sgr B2 is a molecular cloud in the Galactic Centre region, located at about 100 pc (projected distance) from Sgr A*, the supermassive black hole in the center of the Galaxy. Its X-ray spectrum is well reproduced by a pure Compton reflection

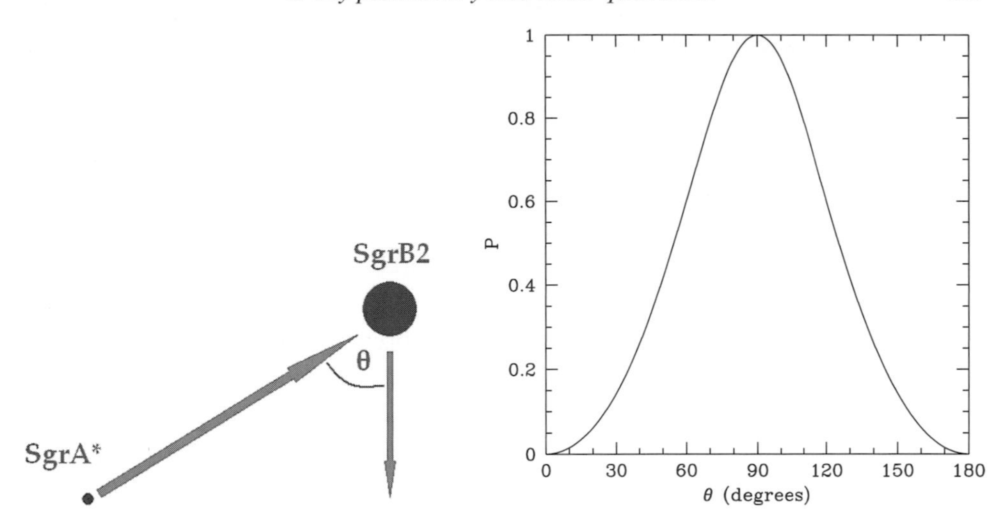

Figure 18.4 Left panel: The reflection hypothesis for Sgr B2. Right: The polarization degree from Sgr B2 as a function of the scattering angle θ.

component, indicating that Sgr B2 is reflecting the X-ray radiation emitted by a source outside the cloud[21]. Puzzling enough, there are no sufficiently bright X-ray sources in the surroundings. It has therefore been proposed that Sgr B2 is reflecting past emission of the central black hole[12], which should therefore have undergone a phase of activity about 300 years ago. If the emission from the nebula is indeed due to scattering, it should be very highly polarized[2], with the direction of the polarization vector orthogonal to the scattering plane, and therefore to the line connecting the projected position of Sgr B2 and the illuminating source (see Figure 18.4).

Polarimetry will therefore be able to confirm or disprove this hypothesis unambiguously: if the polarization vector points perpendicularly to the Sgr B2–Sgr A* projected direction, it will demonstrate that not many years ago the Galaxy was a low luminosity AGN.

It must also be noted that the flux from Sgr B2 is varying with time (unfortunately decreasing[13]). However, other X-ray reflection nebulae are present around the central black hole, which are also varying with time[18]. The brightest of them, when an X-ray polarimeter will be operating, will of course be chosen for this important experiment.

18.5 Observational perspectives

Can the effects described above be realistically observed in the not-too-distant future? As described elsewhere in this volume, several proposed missions, usually making use of photoelectric polarimeters, are at present (May 2009) performing, or have just finished, phase A studies (*Polarix, GEMS, IXO*). It is with reference to these missions that I now briefly discuss, in a qualitative way, the observational perspectives.

The **primary emission** is expected to be strongly polarized, and it is of course the brightest component in unobscured AGN. It looks a quite easy measurement to do. The problem here is interpretation: only polarization from a plane-parallel geometry has been calculated so far, and much work is needed before having a comprehensive view of the polarization properties expected in the variouus geometries, indispensable to fully exploit the diagnostic capability of X-ray polarimetry.

Observations of **strong gravity effects** will instead be quite challenging even for *IXO*. Reflected radiation is important above 7 keV or so, where the efficiency of photoelectric polarimeters goes down. If the disc is significantly ionized, the increased albedo at low energies would help significantly.

Polarization of the reflected radiation from the **'torus'** should be easily observable, at least in the brightest Compton-thick sources (the Circinus galaxy and NGC 1068) thanks to the high degree of polarization expected.

Large polarizations are also expected for the **soft X-ray emission** in obscured AGN. Here the problem is the sensitivity of soft X-ray polarimeters, as photoelectric polarimeters work at too high energies for this emission to be studied.

Finally, reflected emission from **Sgr B2** and other X-ray reflection nebulae is rather weak, but the polarization degree is expected to be very high. Moreover, the interpretation here would be straightforward and model-independent.

To summarize, X-ray polarimetry will probably have an impact on our understanding of almost every aspect of radio-quiet AGN, from very close to the black hole event horizon up to distant matter, as well as unveiling the history of the accretion onto the black hole in our own Galaxy.

References

[1] Antonucci, R. R. F., Miller, J. S. (1985). *ApJ* **297**, 621–632.
[2] Churazov, E., Sunyaev, R., Sazonov, S. (2002). *MNRAS* **330**, 817–820.
[3] Dovčiak, M., Karas, V., Matt, G. (2004). *MNRAS* **355**, 1005–1009.
[4] Fabian, A. C., Iwasawa, K., Reynolds, C. S., Young, A. (2000). *PASP* **112**, 1145–1161.
[5] Fabian, A. C., et al. (2002). *MNRAS* **335**, L1–5.
[6] Gandhi, P. (2006). In *Asian Journal of Physics (AJP) special issue '21st century astrophysics'*, ed. S. K. Saha and V. K. Rastogi.
[7] Ghisellini, G., Haardt, F., Matt, G. (1994). *MNRAS* **267**, 743–754.
[8] Ghisellini, G., Haardt, F., Matt, G. (2004). *A&A* **413**, 535–545.
[9] Guainazzi, M., Bianchi, S. (2007). *MNRAS* **374**, 1290–1302.
[10] Haardt, F., Maraschi, L. (1991). *ApJ* **380**, L51–54.
[11] Haardt, F., Matt, G. (1993). *MNRAS* **361**, 346–352.
[12] Koyama, K., et al. (1996). *PASJ* **48**, 249–255.

[13] Koyama, K., Inui, T., Matsumoto, H., Tsuru, T. (2008). *PASJ* **60**, 201–206.
[14] Matt, G., Costa, E., Perola, G. C., Piro, L. (1989). In *The 23rd ESLAB Symposium on Two Topics in X Ray Astronomy. Volume 2: AGN and the X Ray Background*, pp. 991–993.
[15] Matt, G., Perola, G. C., Piro, L. (1991). *A&A* **247**, 25–34.
[16] Matt, G. (1993). *MNRAS* **260**, 663–674.
[17] Miniutti, G., Fabian, A. (2004). *MNRAS* **349**, 1435–1448.
[18] Muno, M. P., Baganoff, F. K., Brandt, W. N., Park, S., Morris, M. R. (2007). *ApJ* **656**, L69–72.
[19] Poutanen, J., Vilhu, O. (1993). *A&A* **275**, 337–344.
[20] Raban, D., Jaffe, W., Röttgering, H., Misenheimer, K., Tristram, K. R. W. (2009). *MNRAS* **394**, 1325–1337.
[21] Sunyaev, R., Markevitch, M., Pavlinsky, M. (1993). *ApJ* **407**, 606–610.
[22] Tristram, K. R. W. et al. (2007). *A&A* **474**, 837–850.

19

The soft X-ray polarization in obscured AGN

S. Bianchi, G. Matt & F. Tamborra
Università degli Studi Roma Tre, Italy

M. Chiaberge
Space Telescope Science Institute, USA

M. Guainazzi
XMM-Newton SOC, ESAC, ESA, Spain

A. Marinucci
Università degli Studi Roma Tre, Italy

The soft X-ray emission in obscured active galactic nuclei (AGN) is dominated by emission lines, produced in a gas photoionized by the nuclear continuum and probably spatially coincident with the optical narrow-line region (NLR). However, a fraction of the observed soft X-ray flux appears like a featureless power law continuum. If the continuum underlying the soft X-ray emission lines is due to Thomson scattering of the nuclear radiation, it should be very highly polarized. We calculated the expected amount of polarization assuming a simple conical geometry for the NLR, combining these results with the observed fraction of the reflected continuum in bright obscured AGN.

19.1 Introduction

The presence of a 'soft excess', i.e. soft X-ray emission above the extrapolation of the absorbed nuclear emission, is very common in low-resolution spectra of nearby X-ray obscured active galactic nuclei (AGN)[8; 20]. It is generally very difficult to discriminate between thermal emission, as expected by gas heated by shocks induced by AGN outflows or episodes of intense star formation, and emission from a gas photoionized by the AGN primary emission. However, the high energy and spatial resolution of XMM-*Newton* and *Chandra* have allowed us to make important progress in the last few years.

X-ray Polarimetry: A New Window in Astrophysics, eds. R. Bellazzini, E. Costa, G. Matt and G. Tagliaferri. Published by Cambridge University Press.

19.2 The photoionization signatures

The high-resolution spectra of the brightest obscured AGN, made available by the gratings aboard *Chandra* and XMM-*Newton*, revealed that the 'soft excess' observed in CCD spectra was due to the blending of strong emission lines, mainly from He- and H-like transitions of light metals and L transitions of Fe (see Figure 19.1, e.g. [17; 18; 9; 4; 19; 3]). The presence of narrow radiative recombination continua (RRC) features from carbon and oxygen, whose width indicates typical plasma temperatures of the order of a few eV, are unequivocal signatures of photoionized spectra[13]. Moreover, the intensity of higher-order series emission lines, once normalized to the Kα, are larger than predicted by pure photoionization, and are consistent with an important contribution by photoexcitation (resonant scattering[1; 14; 11; 15]). All these pieces of evidence agree that the observed lines should be produced in a gas photoionized by the AGN, with little contribution from any collisionally ionized plasma.

These results have been confirmed to be common in a large catalog of Seyfert 2 galaxies (CIELO-AGN: Catalog of Ionized Emission Lines in Obscured AGN[7]), which presented results of a high-resolution soft X-ray (0.2–2 keV) spectroscopic study on a sample of 69 nearby obscured AGN observed by the Reflection Grating Spectrometer (RGS) on board XMM-*Newton*. Their analysis confirmed the dominance of emission lines over the continuum in the soft X-ray band of these sources, the presence of narrow RRC and the important contribution from higher-order series lines. Therefore, this study allows us to confirm that the results extracted from the

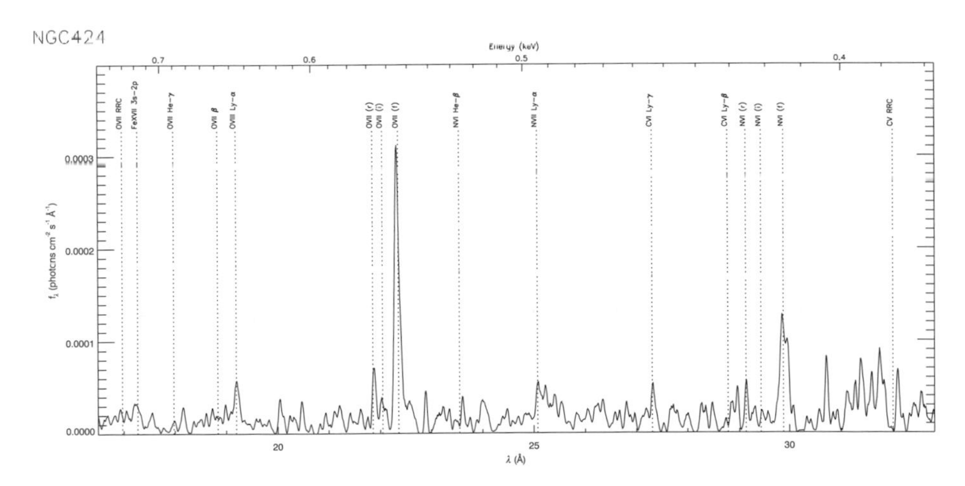

Figure 19.1 RGS spectrum of NGC 424 (a.k.a. Tololo 0109-383). Spectra of the two RGS cameras have been merged and smoothed with a five-channel wide triangular kernel for illustration purposes only. The positions of the line transitions measured in CIELO-AGN are labelled. From [Marinucci *et al.*, in preparation].

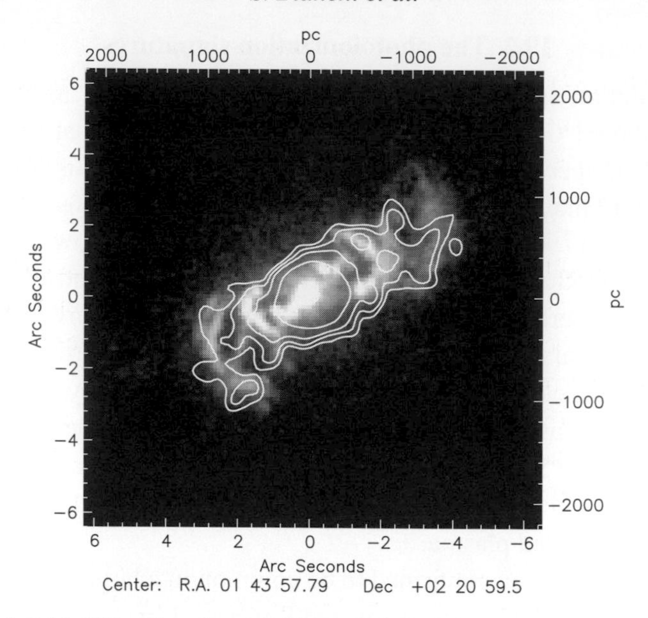

Figure 19.2 Mrk 573: *Chandra* soft X-ray (0.2-2 keV) contours superimposed on the *HST* [O III] image. The contours refer to 0.001, 0.1, 0.2, 0.4 and 0.95 levels with respect to the brightest pixel. North is up, east to the left. From [Bianchi et al., in preparation].

detailed study of high-quality spectra of the brightest objects can be extended to the whole population of nearby obscured AGN.

19.3 The coincidence with the NLR: the same medium?

Thanks to the unrivaled spatial resolution of *Chandra*, it has been possible to resolve the soft X-ray emission of Seyfert 2 galaxies, which appears to be strongly correlated with that of the Narrow Line Region (NLR), as mapped by the [O III] λ5007 *HST* images (see Figure 19.2, e.g. [21; 2; 12]).

The possibility that the NLR, which is also believed to be gas photoionized by the AGN, is the same material responsible for the soft X-ray emission was investigated in detail by [2]. They found that such a simple scenario is tenable. Moreover, the observed [O III] to soft X-ray flux ratio remains fairly constant up to large radii, thus requiring that the density decreases roughly as r^{-2}, similarly to what is often found for the NLR (e.g. [10; 6]).

19.4 Polarization

High-resolution spectroscopy of soft X-ray emission in obscured AGN reveals that it is dominated by strong emission lines. The common spatial coincidence between the soft X-ray emission and the NLR suggests they can also be one and the same medium, photoionized by the central AGN. In this scenario, a fraction of the soft X-ray flux is still expected to be constituted by a featureless continuum, due to Thomson scattering of the primary radiation. This is indeed what is found in low-resolution, CCD spectra, when all the flux detected in emission lines is taken into account.

We are systematically analysing all the brightest obscured AGN observed by the XMM-*Newton* gratings (RGS) and CCD (EPIC), in order to disentangle the fraction of soft X-ray flux in the continuum from that of the emission lines [Tamborra et al., in preparation]. First results include the Phoenix Galaxy (60%, based on *Suzaku* data[16]), NGC 424 (30%, XMM-*Newton* RGS and EPIC pn data [Marinucci et al., in preparation]), and Mrk 573 (35%, XMM-*Newton* RGS and *Chandra* ACIS-S [Bianchi et al., in preparation]).

The fraction of the soft X-ray emission constituted by the continuum should be very highly polarized. Assuming a conical shape and homogeneous density for the soft X-ray emitting region, we estimated the expected polarization degree for this radiation (see Figure 19.3). The results shown in the figure have been calculated following[5], who derived the formulae for the flux and polarization of the Thomson scattered radiation in axisymmetric configurations of matter illuminated by a

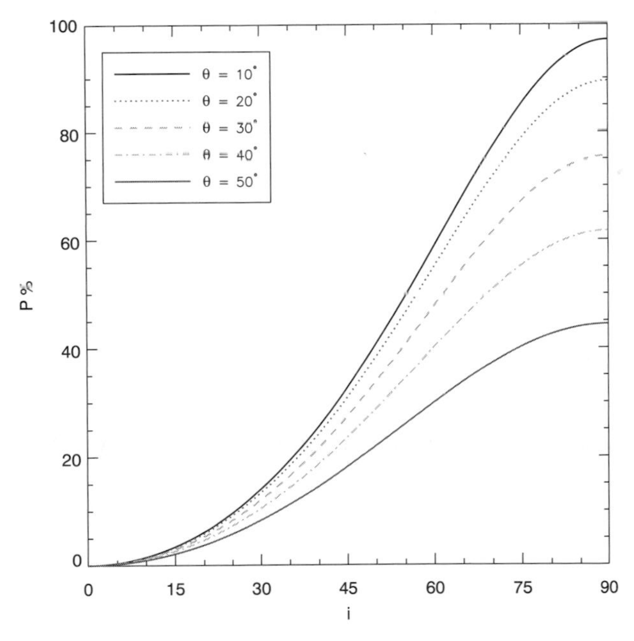

Figure 19.3 Expected polarization degrees for the continuum component of the soft X-ray emission in obscured AGN as a function of the inclination angle i between the line of sight and the axis of symmetry, and for different values of the half-opening angle θ of the emission cone.

point-like source. The formulae are valid in the optically thin case, i.e. in the single scattering approximation. Unpolarized primary radiation is assumed for simplicity. For symmetry reasons, the polarization vector is orthogonal to the axis of the cone. The polarization degree increases with the inclination angle (i.e. the angle between the line of sight and the axis of the cone), because of the well-known dependence of the polarization degree on the scattering angle α, i.e. $P = (1-\cos^2\alpha)/(1+\cos^2\alpha)$. The polarization degree decreases with the half-opening angle of the cone because of the decreasing level of asymmetry. For typical values of the half–opening angles (30°–50°) the polarization degree is very high, up to 45–70% for an edge-on view.

With the continuum fraction measured for the Phoenix galaxy, NGC 424, and Mrk 573, these results would correspond to net polarization degrees of around 20% for the first source (total $F_{0.5-2\,\mathrm{keV}} = 2.7 \times 10^{-13}$ erg cm^{-2} s^{-1}), and 10% for the others (total $F_{0.5-2\,\mathrm{keV}} = 1.9 \times 10^{-13}$ and 2.9×10^{-13} erg cm^{-2} s^{-1}, respectively), assuming $i = 50°$ and $\theta = 30°$. These estimates assume that line emission is completely unpolarized, i.e. polarization of resonant scattering lines is not considered.

19.5 Conclusions

The soft X-ray emission in obscured AGN is likely due to continuum and line emission from photoionized matter in an axisymmetric configuration. Even after dilution by (mostly unpolarized) line emission, high degrees of polarization are expected. Obscured AGN are therefore among the best candidates for any polarimeter working in the soft X-ray band.

References

[1] Band, D. L., Klein, R. I., Castor, J. I., Nash, J. K. (1990). *ApJ* **362**, 90–99.
[2] Bianchi, S., Guainazzi, M., Chiaberge, M. (2006). *A&A* **448**, 499–511.
[3] Bianchi, S., Miniutti, G., Fabian, A. C., Iwasawa, K. (2005). *MNRAS* **360**, 380–389.
[4] Brinkman, A. C. et al. (2002). *A&A* **396**, 761–772.
[5] Brown, J. C, McLean, I. S. (1977). *A&A* **57**, 141–149.
[6] Collins, N. R. (2005). *ApJ* **619**, 116–133.
[7] Guainazzi, M., Bianchi, S. (2007). *A&A* **374**, 1290–1302.
[8] Guainazzi, M., Matt, G., Perola, G. C. (2005). *A&A* **444**, 119–132.
[9] Kinkhabwala, A. et al. (2002). *ApJ* **575**, 732–746.
[10] Kraemer, S. B. (2000). *ApJ* **531**, 278–295.
[11] Krolik, J. H., Kriss, G. A. (1995). *ApJ* **447**, 512.
[12] Levenson, N. A. et al. (2006). *ApJ* **648**, 111–127.
[13] Liedahl, D. A., Paerels, F. (1996). *ApJ* **468**, L33.
[14] Matt, G. (1994). *MNRAS* **267**, L17–L20.
[15] Matt, G., Brandt, W. N, Fabian, A. C. (1996). *MNRAS* **280**, 823–834.
[16] Matt, G., et al. (2009). *A&A* **496**, 653.

[17] Sako, M., Kahn, S. M., Paerels, F., Liedahl, D. A. (2000). *ApJ* **543**, L115–L118.
[18] Sambruna, R. M., et al. (2001). *ApJ* **546**, L13.
[19] Schurch, N. J., Warwick, R. S., Griffiths, R. E., Kahn, S. M. (2004). *MNRAS* **350**, 1–9.
[20] Turner, T.J., George, I. M., Nandra, K., Mushotzky, R. F. (1997). *ApJS* **113**, 23.
[21] Young, A. J., Wilson, A. S., Shopbell, P. L. (2001). *ApJ* **556**, 6–23.

20

The polarization of complex X-ray sources

R. W. Goosmann

Observatoire Astronomique de Strasbourg, France

The radiative transfer code STOKES was extended to allow for X-ray polarimetry modelling. The physical mechanisms of Compton scattering, photo-absorption, and the production of iron K lines were added and are illustrated by modelling the X-ray polarization spectrum of irradiated, cold matter disks. These models confirm that the orientation of the polarization position angle is related to the size of the disk. Although strongly diminishing the spectral flux, an obscuring torus around a small irradiated disk significantly increases the polarization at intermediate viewing angles. Our modelling shows that the polarization can be very sensitive to the radiative coupling between different reprocessing regions.

20.1 Introduction

Polarimetry and spectropolarimetry are an important extension of photometry and spectroscopy techniques. In addition to the spectral intensity, i.e. the first Stokes parameter, the linear polarization percentage, P, and position angle, ψ, can provide further information about the geometry and the dynamics of a given object. To decode this information, accurate modelling is necessary. For this purpose I started the development of the radiative transfer code STOKES[3] that is based on the Monte-Carlo method and that computes the polarization spectrum due to multiple photon–matter interactions in geometrically complex environments. The code is publicly available on the web.[1] New versions of STOKES will be subsequently provided for download after they have been carefully tested and documented. The code has recently been extended to include polarization effects that are relevant in the X-ray range. I implemented the prescription of Compton scattering, the effects of photo-absorption, and the production of iron Kα and Kβ lines.

X-ray Polarimetry: A New Window in Astrophysics, eds. R. Bellazzini, E. Costa, G. Matt and G. Tagliaferri. Published by Cambridge University Press.

In this proceedings note the X-ray capabilities of STOKES are illustrated by modelling the polarization of irradiated, cold, constant density disks of different sizes. Adopting this setup as a very simplified approximation to an AGN accretion disk in a lamp-post geometry, I then add an equatorial, cold matter torus. For each model setup I briefly discuss the resulting flux and polarization spectra as a function of different viewing angles.

20.2 X-ray polarization spectra of cold reprocessing disks

A point-like radiation source is located on the symmetry axis of a circular, geometrically thin and optically thick disk with radius r. The source is put at the height h from the disk surface and isotropically emits a flux spectrum with a power law shape according to $F(E) \propto E^{-1}$, E being the spectral energy in the range of 1–100 keV. The disk has a constant density of neutral matter with cosmic abundances. The energy-dependent photo-absorption cross-sections of hydrogen and helium are implemented and Compton scattering is taken into account. The production of iron Kα and Kβ fluorescent lines is also included. Two different horizontal sizes of the disk are examined: an 'infinitely wide' disk with $h/r = 10^{-4}$, and a smaller disk with the ratio $h/r = 1$.

The results for the large disk are plotted in Figure 20.1 considering only the reprocessed component and ignoring the direct radiation from the point source. The total flux spectrum confirms previous results (see, e.g., [1]) by showing soft X-ray absorption, strong neutral iron Kα and Kβ lines together with the K-shell absorption edge around 7.1 keV, and a broad Compton hump centered at roughly 30 keV. The lines reveal a Compton shoulder, i.e. a small excess in the red wing that is due to subsequent Compton down-scattering of line photons.

The polarization spectrum of the large disk for a given viewing direction does not vary much below 6.4 keV. The absolute value of P decreases strongly across the iron K lines and less strongly across the Compton hump. The fluorescent lines are intrinsically unpolarized and the drop in $|P|$ is caused by simple dilution. The Compton hump is due to multiple electron scattering, again causing depolarization. The absolute normalization of P rises with the viewing angle i and can reach more than 30% for an edge-on view at low spectral energies. These high values are due to the strong absorption in the soft X-ray band. Most often, the escaping photons only undergo a single electron scattering event so that depolarization by multiple scattering is not efficient. Note that for all viewing angles the polarization position angle is aligned with the projected axis of the disk.

It is useful to also plot the polarized flux spectrum $P \times F$ to estimate the amount of radiation that is affected by scattering. Since for most of the energy range considered here P does not vary much with E, the shape of the polarized flux spectra is rather

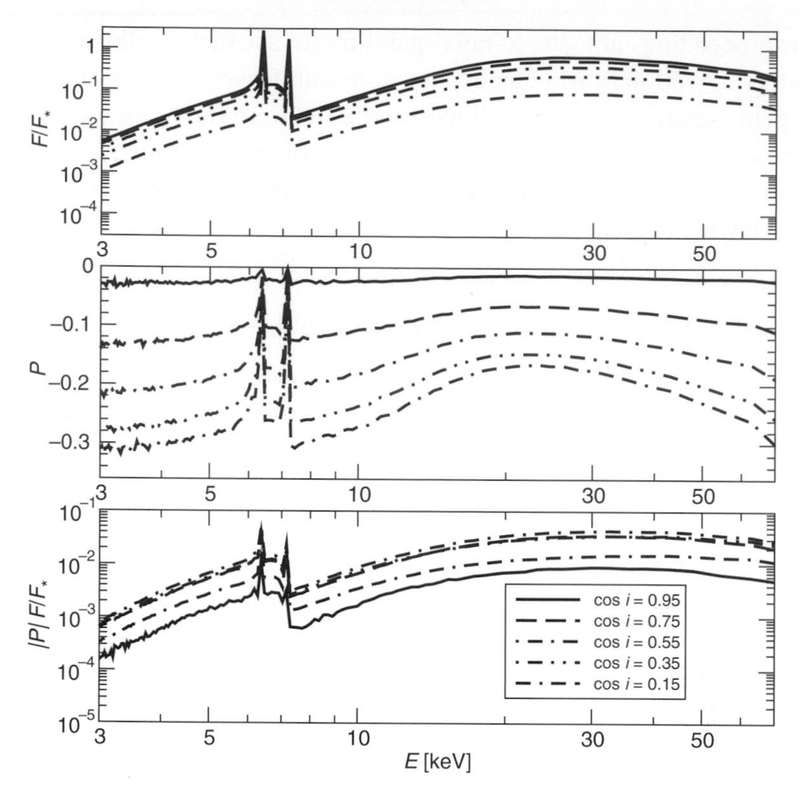

Figure 20.1 Model results for the cold X-ray reprocessing by an irradiated, large constant density disk with $h/r = 10^{-4}$. Top: reprocessed flux, F, normalized to the emitted source flux, F_*. Middle: polarization percentage, P. Bottom: normalized polarized flux, $P \times F/F_*$. All values are given as a function of photon energy, E, and for different viewing angles, i. The viewing angle i is measured with respect to the disk's symmetry axis. The negative values of P indicate that the polarization position angle is aligned with the projected symmetry axis of the disk.

similar to the one of the total flux. It is worth noting, though, that the normalization of $P \times F$ does not vary monotonically with i. Furthermore, it is interesting that at most viewing angles the iron K lines still appear as an emission feature in $P \times F$. The weak polarization of the lines is due to electron scattering of the intrinsically unpolarized fluorescent line photons.

The results for the smaller disk size are shown in Figure 20.2. The most important difference to the large disk is in the orientation of ψ. For a small disk, the position angle of the polarization is oriented perpendicularly to the symmetry axis. The orientation of ψ as a function of the disk size was investigated before in a slightly different geometry[4]. It can be understood by the fact that the net polarization results from integrating over the disk surface: the geometrical relation between the source, the disk surface and the observer, as well as the fact that a maximum polarization is induced for perpendicular scattering angles, explains a strong

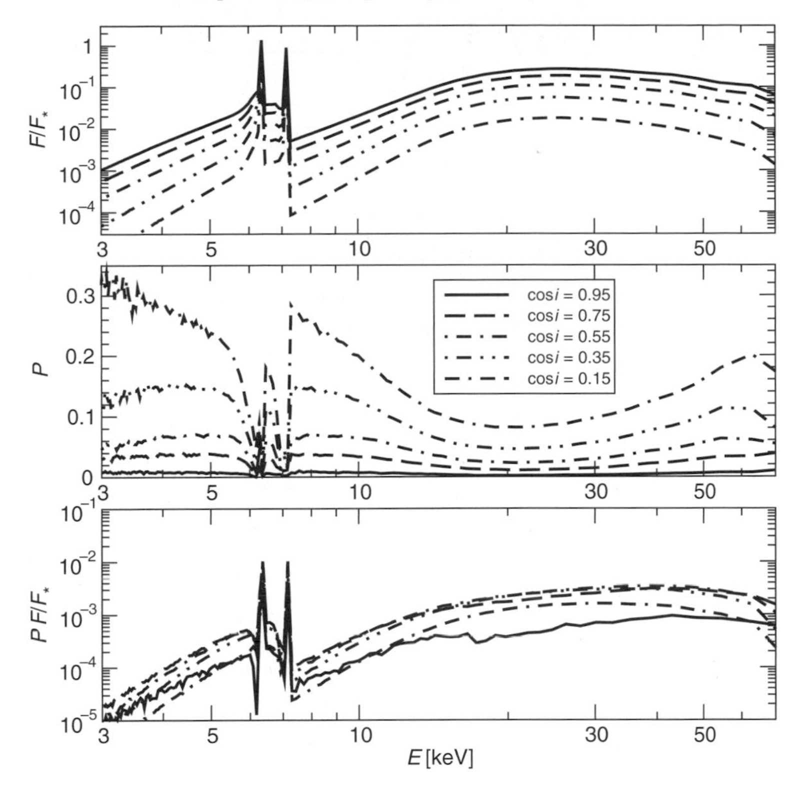

Figure 20.2 Modelling results for the case of an irradiated small disk with $h/r = 1$. The denotations are as in Figure 20.1. The positive values of P denote that the polarization position angle is oriented perpendicularly to this axis.

contribution of parallel polarization by the outer parts of the disk. In the case of a large disk these dominate over the inner disk parts that preferentially produce perpendicular polarization.

20.3 A small disk and an obscuring torus

The STOKES code allows several emission and scattering regions to be combined inside the same model space and then computes the radiative transfer and the polarization coherently. Having the uniform scheme of AGN in mind, I add an optically thick, equatorial torus to the irradiated small, constant density disk that is interpreted as a very simple approximation of an accretion disk. The disk is located at the center of the torus funnel. The torus half-opening angle is set to 30°. The isotropically emitting X-ray point-source thus irradiates the underlying disk and the inner surface of the torus, for which again cold reprocessing is assumed.

This model setup, which extends previous work[2], is evaluated in Figure 20.3. The total flux spectra and the polarization percentage reveal a significant difference

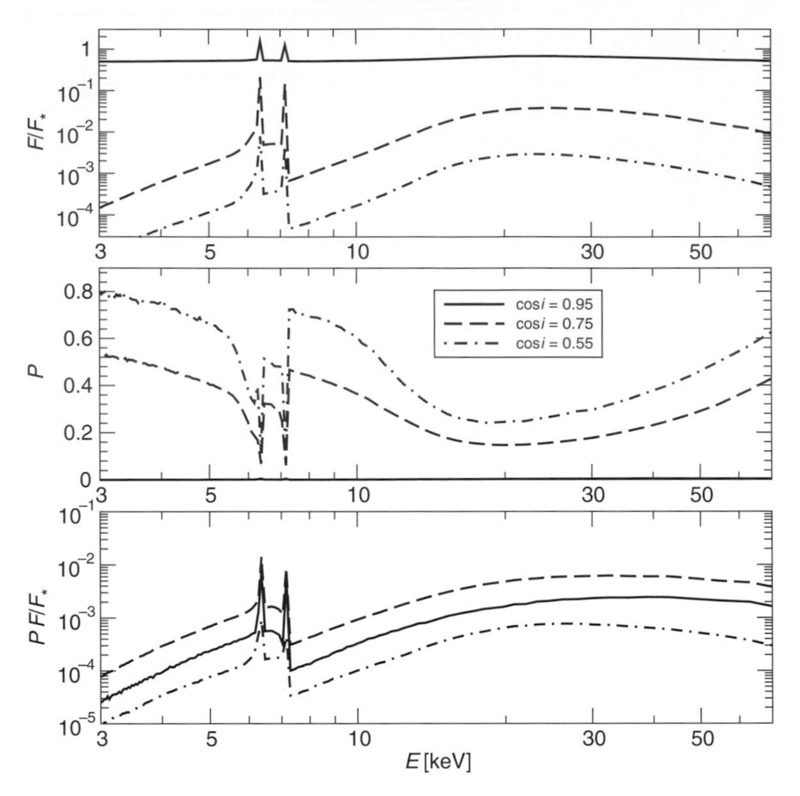

Figure 20.3 Modelling results for the case of an irradiated small disk with $h/r = 1$ at the center of an obscuring torus with a half-opening angle of 30°. The denotations are as in Figure 20.2.

between the cases of $i < 30°$ and $i > 30°$. When looking directly at the source and the accretion disk, as it is the case for type-1 AGN, the spectral flux is high and the polarization very low. At higher viewing angles, as for type-2 AGN, only scattered radiation is seen. This scattered flux decreases quickly with increasing i, especially in the soft X-ray band where absorption either in the disk or in the torus is very likely. On the other hand, the polarization degree for type-2 viewing angles rises with i and for a given viewing angle the values exceed the results for a scattering small disk alone. In this particular model setup, the torus therefore seems to have a collimation effect on the polarization.

20.4 Summary and perspectives

The latest version of the STOKES code was applied to model the X-ray polarization spectrum for some simple reprocessing scenarios in AGN. It turns out that the radiative coupling between different components of a given object can have an important impact on the resulting polarization. For the analysis of future X-ray polarimetry data it will be necessary to further extend the multi-component modelling and also to include lower energy wavebands. Multi-wavelength polarimetry observations should then provide important constraints on the complex inner geometry of AGN.

Notes

1. `http://www.stokes-program.info`

References

[1] George, I. M. & Fabian, A. C. (1991). *MNRAS* **249**, 352–367.
[2] Ghisellini, G., Haardt, F., & Matt, G. (1994). *MNRAS* **267**, 743–754.
[3] Goosmann, R. W. & Gaskell, C. M. (2007). *A&A* **465**, 129–145.
[4] Matt, G. (1993). *MNRAS* **260**, 663–674.

21
Polarization of Compton X-rays from jets in AGN

A. McNamara & Z. Kuncic
University of Sydney
K. Wu
University College London

We investigate the polarization of Compton scattered X-rays from relativistic jets in active galactic nuclei (AGN) using Monte-Carlo simulations. We consider three scenarios: scattering of photons from an accretion disk, scattering of cosmic microwave background (CMB) photons, and synchrotron self-Comptonization (SSC) within the jet. For Comptonization of thermal disk photons or CMB photons the maximum linear polarization attained is slightly over 20% at viewing angles close to 90°. The value decreases with the viewing inclination. For SSC, the maximum value may exceed 80%. The angle dependence is complicated, and it varies with the photon injection sites. Our study demonstrates that X-ray polarization, in addition to multi-wavelength spectra, can distinguish certain models for emission and particle acceleration in relativistic jets.

21.1 Introduction

Observations of extended jets in AGN by *Chandra* have revealed that the origins of their X-ray emission is less trivial than previous thought (see [9; 14] for X-ray jet surveys). The X-rays may arise from various processes. The polarization in the radio and optical bands suggests that the emission is generated by the synchrotron process[7]. Thus, synchrotron and synchrotron self-Compton (SSC) emission are candidates for the X-ray continuum emission[8]. However, the X-rays can also be generated from external Comptonization (EC) of disk black-body radiation[15] or of the CMB[4]. It has been suggested that X-ray polarization measurements are able to discriminate these competing emission mechanisms. To date only a few approximate analytical predictions have been made for SSC[2; 1; 3] or EC polarizations

X-ray Polarimetry: A New Window in Astrophysics, eds. R. Bellazzini, E. Costa, G. Matt and G. Tagliaferri. Published by Cambridge University Press.

[13]. Here we show results of our X-ray polarization calculations for photons scattered by energetic electrons in jets at relativistic bulk speeds. We consider Compton scattering of thermal photons emitted from an underlying accretion disk and the scattering of polarized synchrotron photons emitted within the jet. We also consider Compton scattering of CMB photons by the bulk flow of jet electrons.

21.2 Models

We consider a conical jet with a bulk Lorentz factor $1 < \Gamma_{\rm j} < 10$, launched at a height z_0 above the disk midplane (Figure 21.1). The jet base has a radius $r_{\rm b}$, and the jet power is $P_{\rm j}$. The electrons have a number density distribution, $N_{\rm e}(\gamma) = K\gamma^{-p}$, where γ is the Lorentz factor, p is the energy spectral index, and K is obtained from $N_{\rm e} = \int_{\gamma_{\rm min}}^{\gamma_{\rm max}} N_{\rm e}(\gamma)\, d\gamma$. Along the jet $N_{\rm e}$ falls off according to $N_{\rm e}(z) = N_0(z/z_0)^{-2}$, where

$$N_0 = N_{\rm e}(z_0) \approx \frac{P_{\rm j}}{\beta_{\rm j}\Gamma_{\rm j}(\Gamma_{\rm j}-1)m_{\rm p}c^3\pi r_{\rm b}^2}\,. \tag{21.1}$$

Polarized synchrotron photons are emitted isotropically by the electrons with Lorentz factors $\gamma_{min} \le \gamma \le \gamma_{\rm max}$ in the plasma rest frame. The emission is Lorentz boosted by the relativistic bulk flow of the jet. The accretion disk emission is thermal, with the photons generated according to the flux

$$F(R) = \frac{3GM\dot{M}_{\rm a}}{8\pi R^3}\left[1 - \left(\frac{R}{R_{\rm i}}\right)^{-1/2}\right], \tag{21.2}$$

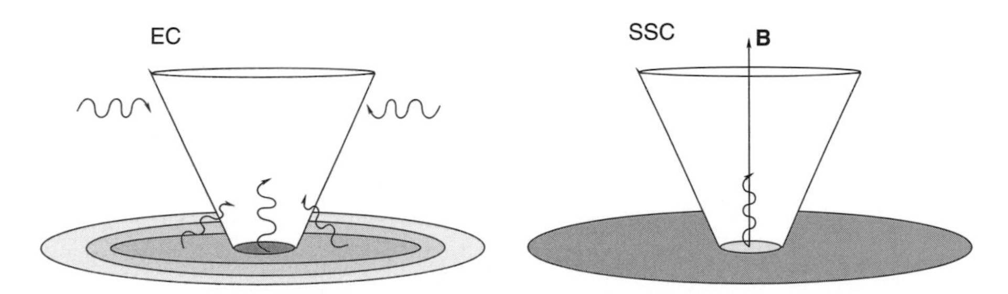

Figure 21.1 Schematic illustrations of the geometrical set-up of the calculations for the cases of external Comptonization (EC) (left) and SSC (right). For the EC case thermal photons are either injected from an accretion disk or injected isotropically around the conical jet. For the SSC case, linearly polarized synchrotron photons are injected at specific locations inside the jet. A magnetic field **B** is permeated in the jet along the bulk flow direction.

where $R_{\rm i}$ is the inner disk radius and $\dot{M}_{\rm a}$ is mass accretion rate of the disk. The CMB photons are generated from a single-temperature black-body with $T_{\rm bb} = 2.8\,(1+z)$ K. They are injected isotropically and uniformly around the jet. In this work, we set the redshift $z = 2$. The numerical values of the model parameters are the same as those in [12].

The non-linear Monte-Carlo algorithm given in [5; 6] is used to determine the scattering event and the photon transport in the jet. The polarization is calculated in each scattering event, and the emergent photons are summed, following the procedures as described in [10; 11]. In the simulations, three different co-ordinate frames – the electron rest frame, the plasma rest frame and the observer's frame – are used. The transformation of the polarization of the photons between the frames is given in [12].

21.3 Results and discussions

The results of our simulations are shown in Figure 21.2. For the EC case with photons injected from an accretion disk, the degree of linear polarization $P_{\rm L}$ increases with the viewing inclination angle θ. At $\theta = 10°$, $P_{\rm L} \approx 3\%$, and at $\theta = 80°$, $P_{\rm L} \approx 21\%$. The angle dependence of the polarization for the case with CMB photon injection is very similar, with $P_{\rm L} \approx 3\%$ at $\theta = 10°$, increasing to 24% at $\theta = 80°$. In these two cases, the average polarization of the seed photons is zero. Large linear polarization results when large-angle scattering events dominate. As the jet

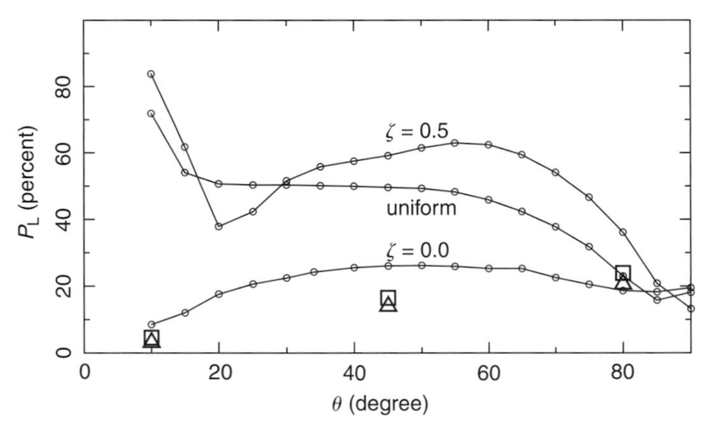

Figure 21.2 Linear polarization $P_{\rm L}$ of Compton-scattered photons with energies between 1 and 10 keV as a function of the viewing inclination angle θ. The solid lines corresponding to the SSC cases with seed photons injected at the jet base ($\zeta = 0$), in the middle of the jet ($\zeta = 0.5$) and uniformly along the jet. The triangles correspond to the EC case with seed photons generated from an accretion disk. The squares correspond to the EC case with CMB photon injection.

plasma has bulk relativistic motion, the incoming photons are essentially seen as head-on in the electron rest frame. Regardless of the initial angle distribution of the seed photons, most photons emerging at large θ have undergone a large scattering event immediately before escaping from the jet. Thus, $P_{\rm L}$ increases with θ, and the degrees of polarization are similar for the two EC cases with thermal photon injection.

The polarization is substantially higher in the SSC cases than in the EC cases – above the 20% level, except when the photons are injected at the jet base and when the viewing inclination is low (small θ). It reaches 70% or higher if photon injection occurs at substantial heights or uniformly along the jet. The high polarization is partially due to the fact that the seed photons are polarized. Scattering randomizes the photon polarization angles. Multiple scattering does not always increase the net average polarization, thus the angle dependence is more complicated in the SSC cases than in the EC cases, where net polarization is generated by multiple scatterings of unpolarized thermal photons. The polarization values are similar for all cases when θ is close to 90°. This is due to the fact that the corresponding incident photons have similar properties, in the rest frame of the electrons, at the last scattering before they escape from the jet.

Our study has shown that X-ray polarization can distinguish whether the X-rays are generated from EC or from SSC. It can also discriminate between the particle acceleration sites in the case of SSC emission.

References

[1] Begelman, M. C., Sikora, M. (1987). *ApJ* **322**, 650–661.
[2] Bjornsson, C. -I., Blumenthal, G. R. (1982). *ApJ* **259**, 805–819.
[3] Celotti, A., Matt, G. (1994). *MNRAS* **268**, 451–458.
[4] Celotti, A., Ghisellini, G., Chiaberge, M. (2001). *MNRAS* **321**, L1–L5.
[5] Cullen, J. G. (2001a). PhD Thesis, University of Sydney.
[6] Cullen, J. G. (2001b). *JCoPh* **173**, 175–186.
[7] Jorstad, S. G., et al. (2007). *ApJ* **134**, 799–824.
[8] Maraschi, L., Ghisellini, G., Celotti, A. (1992). *ApJ* **397**, L5–L9.
[9] Marshall, H. L. et al. (2005). *ApJS* **156**, 13–33.
[10] McNamara, A. L., Kuncic, Z., Wu, K. (2008a). *MNRAS* **386**, 2167–2172.
[11] McNamara, A. L., Kuncic, Z., Wu, K., Galloway, D. K., Cullen, J. G. (2008b). *MNRAS* **383**, 962–970.
[12] McNamara, A. L., Kuncic, Z., Wu, K. (2009). *MNRAS* **395**, 1507–1514.
[13] Poutanen, J. (1994). *ApJS* **92**, 607–609.
[14] Sambruna, R., et al. (2004). *ApJ* **608**, 698–720.
[15] Wagner, S. J. (1995). *A&A* **298**, 688–698.

22

Polarization of X-ray lines from galaxy clusters and elliptical galaxies

I. V. Zhuravleva,

MPA, Garching, Germany

E. M. Churazov, S. Yu. Sazonov & R. A. Sunyaev

MPA, Garching, Germany; IKI, Moscow, Russia

W. Forman

Harvard-Smithsonian Center for Astrophysics, Cambridge, USA

K. Dolag

MPA, Garching, Germany

We study the impact of gas motions on the polarization of bright X-ray emission lines from the hot intercluster medium (ICM). The polarization naturally arises from the resonant scattering of the emission lines owing to a quadrupole component in the radiation field produced by a centrally peaked gas-density distribution. If differential gas motions are present in the cluster then the line can leave the resonance, affecting both the degree and the direction of polarization. The changes in the polarization signal are, in particular, sensitive to the gas motions *perpendicular* to the line of sight. We calculate the expected degree of polarization for several patterns of gas motions, including a slow inflow expected in the simplest version of a cooling flow model and a fast outflow in an expanding spherical shock wave. In both cases the effect of nonzero gas velocities is found to be minor. We also calculate the polarization signal for a set of clusters, taken from large-scale structure simulations, and evaluate the impact of the gas bulk motions on the polarization signal. We argue that the expected degree of polarization is within reach of the next generation of space X-ray polarimeters.

22.1 Introduction

Owing to quadrupole anisotropy in the radiation field, the scattered emission in certain resonance X-ray lines should be polarized. Anisotropy can be due to *A*) the

X-ray Polarimetry: A New Window in Astrophysics, eds. R. Bellazzini, E. Costa, G. Matt and G. Tagliaferri. Published by Cambridge University Press.

centrally concentrated gas distribution and B) the gas bulk motions. The sole effect of small-scale random (turbulent) motions is to make lines broader, decreasing optical depth and reducing the effect of resonant scattering and polarization[4], i.e. $P \sim c/\sqrt{V_{\rm therm}^2 + V_{\rm turb}^2}$. The effect of the bulk motions is different, since the line can leave the resonance along the direction of motion leading to the change of scattering ability of a given region in the cluster. Changes in the polarization degree in this case are defined by the shift of the photon energy relative to the Gaussian line profile $P \sim \exp[-V_{\rm bulk}^2/(V_{\rm therm}^2 + V_{\rm turb}^2)]$. This opens a unique possibility to reconstruct both radial and tangential velocity components by observing the line energy, which is connected with the radial velocity component, and polarization, which primarily depends on the tangential component.

22.2 Spherically symmetric problems

We have performed calculations for Perseus and Virgo/M87 clusters, as examples of cooling flow clusters. The expected polarization degree in the Perseus cluster in the He-like iron line at 6.7 keV increases with the distance from the cluster center and the maximum polarization is $\sim$10% at a distance $\sim$1 Mpc. For Virgo cluster the highest polarization degree is achieved in the Fe XXIII 1.129 keV line, reaching a maximum of 3.5% at a distance $\sim$15 kpc from the center and falling off from this maximum value when moving away from the center.

We then consider a velocity field expected in the simplest 'cooling flow' model[5]. For Perseus and M87/Virgo gas velocities do not exceed $\sim$100 km/s outside the central $\sim$1 kpc region and they do not change the degree of polarization since the flow velocities have an appreciable effect only in the center of clusters where the degree of polarization is very small anyway.

A case of an expanding shock wave was also considered in M87[3]. In this case, polarization diminishes noticeably in the central region owing mostly to the changes in number density and temperature rather than due to gas motions.

22.3 Three-dimensional problem

For three-dimensional simulations the density, temperature distributions and the velocity field of galaxy clusters were taken from a set of high-resolution Gadget-2 simulations[2]. All data were adaptively smoothed and placed in a cube with the half size 1000 kpc and cell size 3 kpc. We made a number of radiative transfer simulations varying the Mach number M and the characteristic amplitude of gas

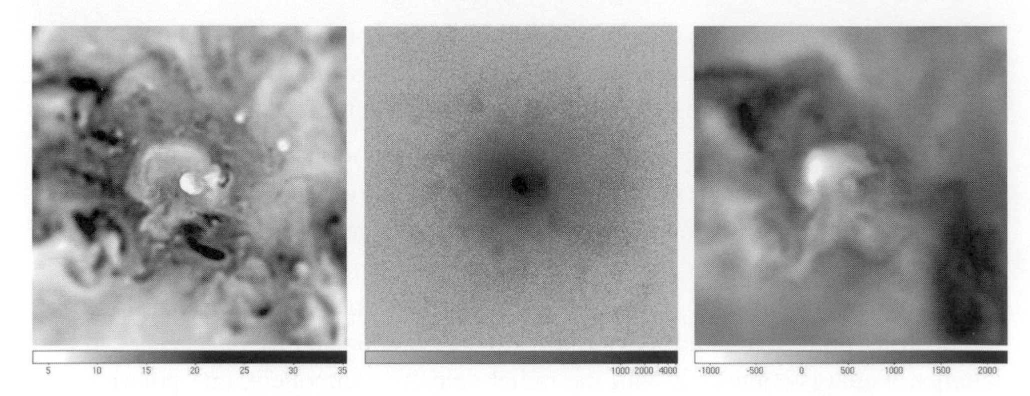

Figure 22.1 Simulated cluster g8. Slices of temperature in keV are shown in the left panel, surface brightness (ph/s/cm^2/arcmin^2) in the middle panel and the x component of velocity (km/s) in the right panel. In the left panel temperature ranges from 3 to 35 keV.

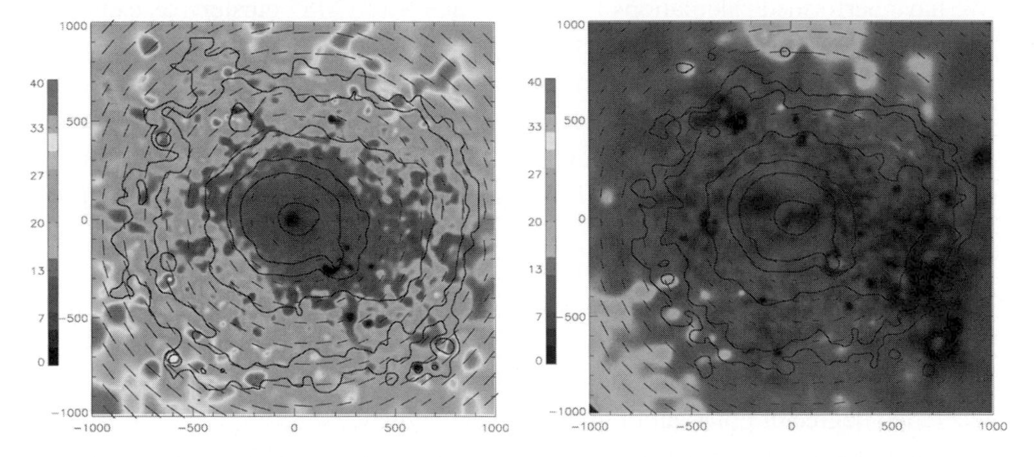

Figure 22.2 Polarization degree in g8 cluster calculated in Fe XXV line at 6.7 keV. The left panel shows the case (f=0, M=0) and the right panel the case (f=1, M=0). The shading shows the polarization degree, short lines show the orientation of the electric vector and contours the X-ray surface brightness.

velocities, introducing a multiplicative factor f which is used to scale all velocities obtained from the hydrodynamical simulations, e.g. $(\mathrm{f}=0, \mathrm{M}=0)$ is the case of no motions, $(\mathrm{f}=1, \mathrm{M}=0.25)$ shows the case when bulk motions and turbulent motions with Mach number 0.25 are included. Our results for g72 and g8 (see Figure 22.1) clusters are presented in Table 22.1 and in Figure 22.2.

Table 22.1 *Mean temperature, RMS velocities and polarization degree at a distance of 500 kpc in g72 and g8 clusters for different cases of gas motions.*

Cluster	T_{mean}	V_{rms}	f = 0, M = 0	f = 1, M = 0	f = 0, M = 0.25	f = 1, M = 0.25
g8	15	1000	25%	10%	10%	7%
g72	7	600	25%	10%	12%	5%

22.4 Requirements for the future X-ray polarimeters

1. It is desirable to have good energy resolution in order to avoid contamination of the polarized line flux by unpolarized continuum or nearby lines. For the 6.7 keV iron line, polarization will decrease by 50% if the energy resolution is degraded to 50 eV. For lines with energy around 1 keV a much better resolution ∼10 eV is needed since there is a 'forest of lines' in the spectrum at low energies.
2. The angular resolution has to be better than ∼arcminute. The polarization largely vanishes when integrated over the whole cluster, we need to be able to resolve regions where the polarization signal does not change much.
3. Large effective area of the telescope and field of view are necessary to detect polarization signals from clusters, which are relatively faint X-ray sources.

Currently, there are discussions of two X-ray polarimeters: EXP on-board HXMT[1] and X-ray polarimeter on-board IXO[6]. It should be possible to measure the polarization in galaxy clusters with the IXO polarimeter since it has large effective area (at 6 keV it is 6500 cm^2), angular resolution is ∼5 arcsec and spectral resolution is 2.5 eV. With the IXO polarimeter and an exposure time of 10^5 s, the measured MDP for the g8 and g72 clusters varies from 0.1% and 0.5% at their centers, respectively, to 2% at a distance of 500 kpc from the center.

References

[1] Costa, E., et al. (2007). *UV, X-Ray and Gamma-Ray Space Instrumentation for Astronomy XV*, ed. H. O. Siegmund (Proc. of the SPIE).
[2] Dolag, K., Borgani, S., Murante, G., Springel, V. (2008). [arXiv:0808.3401].
[3] Forman, W., et al. (2005). *ApJ* **635**, 894.
[4] Gilfanov, M. R., Sunyaev, R. A., Churazov, E. M. (1987). *SvAL* **13**, 3.
[5] Sarazin, C. L. (1996). *UV and X-ray Spectroscopy of Astrophysical and Laboratory Plasmas: Proc. of the Eleventh Colloquium on UV and X-ray*, ed. K. Yamashita and T. Watanabe.
[6] White, N. E., Parmar, A., Kunieda, H., International X-ray Observatory Team (2009). *AAS* **41**, 357.

23

Polarization characteristics of rotation-powered pulsars

A. K. Harding

NASA Goddard Space Flight Center

Polarization measurements of rotation-powered pulsars have been a very powerful diagnostic in the radio band and promise to be at least as useful a diagnostic in the X-ray band. Since the relativistic particles that radiate pulsar high-energy emission tightly beam the radiation along the pulsar magnetic field, phase-resolved polarimetry has the potential to map the emission patterns. Fermi observations of young pulsars at gamma-ray energies have disfavored polar-cap models where emission takes place near the neutron star surface, and strongly favor outer-magnetosphere models where emission takes place close to the light cylinder or beyond. Since the different outer-magnetosphere models predict similar gamma-ray light curves, it is difficult to discriminate between them at gamma-ray energies. But X-ray polarization has the potential to provide this discrimination, since the models predict distinct polarization signatures and optical detections will only be possible for a small number of pulsars.

23.1 Introduction

Rotating systems are particularly interesting to study with polarization because the rotation provides changing views of the emitting regions. Rotation-powered pulsars have the additional advantage that the emission is radiated by highly relativistic particles moving along magnetic field lines and therefore the emission angle measures the field direction to within an angle of $1/\gamma$, where γ is the particle Lorentz factor. The magnetic field structure is also able to be derived independently, using either retarded-vacuum[12], force-free (e.g. [28]) or pair-starved[21] solutions. The measured polarization position angle can therefore truly map the emission pattern in the magnetosphere, one of the "holy grails" of pulsar astrophysics. We have one

X-ray Polarimetry: A New Window in Astrophysics, eds. R. Bellazzini, E. Costa, G. Matt and G. Tagliaferri. Published by Cambridge University Press.

additional recent advantage with the launch of the Fermi Gamma-Ray Space Telescope, which is discovering new γ-ray pulsars at an unprecedented rate, including discovery of new pulsars in γ-rays using blind-search techniques[6]. Fermi has increased the number of known γ-ray pulsars by nearly a factor of ten in the first six to nine months of operation, providing a variety of light curves and spectral measurements. These measurements have been able to definitively rule out the near-surface polar-cap models as contributing the bulk of the γ-ray emission. The choices have been narrowed down to the outer-magnetosphere models: the outer-gap, slot-gap or striped-wind models. In any case, we are dealing with emission from field lines that are not dipole but are severely distorted by relativistic effects and by particle currents.

23.2 The rotating vector model

Radio astronomers developed this model to understand the patterns of position angle variation with pulse phase, assuming that radio-emitting particles produce radiation polarized either parallel or perpendicular to a dipole magnetic field[24]. According to this model, when our line-of-sight crosses the radio beam we see a sweep in the polarization position angle (PA) on the sky if the radio beam is centered on the magnetic pole as most studies of radio profiles suggest. In the outer edges of the beam, the PA changes slowly but near the center of the pulse as our line-of-sight approaches the magnetic pole, the PA changes quickly and reverses direction as the pole is crossed. This forms the classic S-shaped pattern that is observed in many radio pulsars. As our line-of-sight sweeps nearer the magnetic pole (i.e. small impact angle β), the sweep of the PA becomes more pronounced. Additionally, if our line-of-sight crosses the radio beam on the other side of the magnetic pole, the PA sweep pattern is flipped in sign. It is therefore possible to deduce the sign of β, which is negative for lines-of-sight between the spin axis and magnetic axis and positive otherwise. Therefore the Rotating Vector Model (RVM) is able to derive the value and sign of β fairly precisely, but cannot derive the value of the magnetic inclination to the rotation axis, α, very accurately. Often, only upper or lower limits on α can be given for a particular pulsar.

The RVM only works well if the radio emission takes place at altitudes that are a small fraction of the pulsar light cylinder distance. Otherwise, relativistic effects of aberration and retardation will distort the emission pattern. This is precisely why we cannot apply this model to the high-energy outer-magnetosphere emission models.

23.3 Intrinsic polarization of radiation

The models for high-energy emission usually invoke curvature, synchrotron or inverse-Compton radiation, all of which produce highly polarized radiation for relativistic particles. In the case of curvature and synchrotron emission, the percent polarization (where $P_{||}$ and $P_{\perp}$ denote the percent of linear polarization parallel and perpendicular to the magnetic field) is

$$\Pi \approx \frac{P_{\perp} - P_{||}}{P_{\perp} + P_{||}}$$

which reaches 75% for single electron emission. For CR, the electric vector is parallel to the magnetic field direction while for SR it is perpendicular to the magnetic field. For a power law electron distribution, the polarization percent is somewhat lower,

$$\Pi(p) = \frac{p+1}{p+\frac{7}{3}}$$

which gives 60–75% for power law indices $p = 1-3$.

In the case of inverse Compton emission, off-beam viewing where the lines-of-sight are at angles around $1/\gamma$, where γ is the emitting particle Lorentz factor, gives a high degree of polarization (~90%). However, particle beams that are broad or have a distribution of Lorentz factors will lower the degree of polarization[7].

23.4 High-energy emission models

The models for high-energy pulsar emission divide into three main regimes: polar-cap models, where the emission occurs within a few stellar radii of the neutron star surface, slot-gap and outer-gap models, where emission occurs in the outer magnetosphere near the light cylinder, and striped-wind models, where emission occurs in the pulsar wind outside the light cylinder. The inner- and outer-magnetosphere emission models give different polarization predictions. Polar-cap models[5; 10; 11] assume that particles begin accelerating near the neutron star (NS) surface and that γ-rays result from a curvature radiation or inverse-Compton-induced pair cascade in a strong magnetic field. The slot-gap accelerator, a recently explored part of the polar-cap model[19; 20], extends to high altitude along the last open magnetic field line, producing curvature, synchrotron and inverse-Compton components in the outer magnetosphere[16]. Outer-gap models[8; 25; 15] assume that acceleration occurs in vacuum gaps that form between the null charge surfaces and the light cylinder in the outer magnetosphere and that γ-rays result from curvature radiation and γ–γ pair-production-induced cascades. Both the slot-gap, which is a class of two-pole-caustic (TPC), models[14] and outer-gap (OG) models produce caustics in their emission. These caustics[18] form as a result of relativistic aberration and travel-time delays at large altitudes that cancel the phase delays of radiation along trailing dipole field lines. Striped-wind models[22] explore the possibility of pulsed emission in the pulsar wind that forms beyond the light cylinder. This wind has alternating toroidal magnetic field of opposite sign, which forms the current sheet in pulsars with high inclination. If the magnetic field undergoes reconnection to convert magnetic energy into particle energy, and if this dissipation region is small enough ($\Delta r \leq \Gamma^2 r_{LC}$, where Γ is the particle Lorentz factor and r_{LC} is the light cylinder radius), then double-peaked light curves can result.

23.5 Fermi observations and gamma-ray pulsars

Since its launch in June 2008, Fermi has been revolutionizing our understanding of pulsar high-energy emission. Among the early results was the discovery of the pulsar in the young supernova remnant CTA 1[1] using only γ-ray photons. Not only did this mark the first time that a pulsar has been discovered through γ-ray timing alone, but it was the first of many pulsars to be discovered by Fermi. There are now at least 16 new pulsars discovered by Fermi by blind search[2] and the majority of these are not detected at radio wavelengths. These findings strongly imply that the γ-ray beam is much larger than the radio beam, so that the γ-ray beam is seen at most viewing angles while the narrower radio beam is seen at a more limited range of viewing angles. In the first few months of Fermi operation, the phase-averaged spectrum of the Vela pulsar was measured precisely enough to rule out a super-exponential cutoff, as predicted by polar cap models due to magnetic pair production attenuation[10; 11], to a significance of 16σ[3]. In the first six months, Fermi has discovered γ-ray pulsations from many pulsars, including 18 radio-selected normal pulsars, 16 discovered in γ rays by blind search[2], and 8 radio-loud millisecond pulsars[4]. The γ-ray light curves of most of these show two sharp peaks with wide (0.4–0.5) phase separations. Among the radio-loud pulsars, the first γ-ray peak lags the radio pulse by typically 0.1–0.2 in phase. Furthermore, their spectra are all best described by power laws with simple exponential cutoffs. All of these results are characteristic of the outer-magnetosphere models and are not compatible with polar-cap models. The early Fermi results have therefore ruled out near-surface polar-cap emission for the young pulsars and even the millisecond pulsars detected so far.

23.6 Polarization in high-energy emission models

Now that Fermi has firmly established that pulsar high-energy emission comes from the outer magnetosphere, we expect the polarization characteristics of the high-energy emission to be quite different from that of the lower altitude radio emission. Indeed, the measured optical polarization of the Crab pulsar[27] does not show the pattern expected of the RVM, such as the S-shaped sweep of PA.

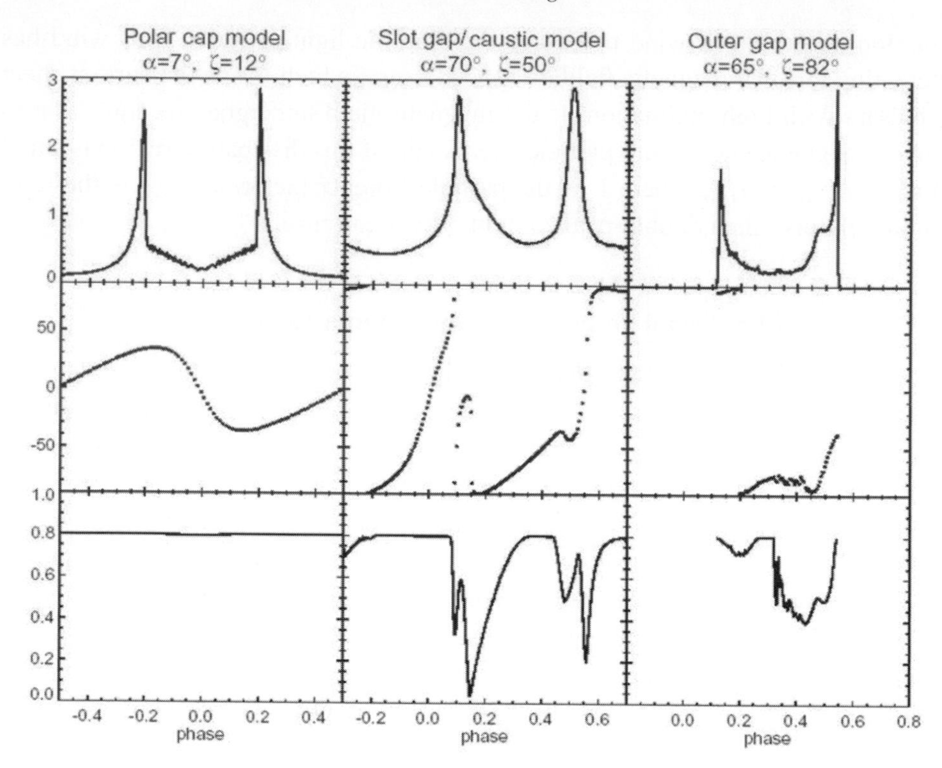

Figure 23.1 Light curve (top), position angle (middle) and percent polarization (bottom) in polar-cap (left), slot-gap (middle) and outer-gap (right) models[13].

In fact the PA shows very rapid sweeps through the main peak and interpulse, and there is strong depolarization associated with each peak. These features are actually expected of the caustic emission that is common in both two-pole-caustic models like the slot-gap and in outer-gap models. Predictions of polarization of the Crab for these high-energy emissions were first derived by Romani & Yadigaroglu[26] for the outer-gap, and later by [13] for the TPC and SG models. A comparison of TPC/SG, OG and PC model predictions is shown in Figure 23.1. The PC model shows the S-shaped sweep of PA that is definitely not in agreement with the optical polarization of the Crab. The TPC model shows large and rapid PA sweeps through each peak as well as significant drop in percent polarization just following each peak, properties that are in reasonable agreement with the observations. The OG model shows a large PA sweep at the interpulse but only a small sweep at the main peak. This is because the main peak in the classic OG model is formed by overlapping field lines from opposite poles, while in the TPC model this peak and the interpulse are the caustics associated with the two poles. In order to achieve a full PA sweep, it is necessary that some emission from the main peak come from below the null charge surface. Takata & Chang[29] have made note of this in their revised OG model, inspired by the 2D OG solutions of Hirotani[17] who found that the OG can be extended below the null surface if currents are allowed to flow into the OG from its lower boundary. In fact, such external currents are required to produce sufficient flux from the OG to account for the Crab observed flux. Takata & Chang demonstrate that a large PA sweep in the main pulse can be produced with this revised OG model, and it is due to extending the emission below the null charge surface. Such a geometry also produces the trailing emission at the two peaks that is absent in classic OG model geometry. However, their OG model produces depolarization through the entire interpeak region, not just at the peaks. This is due to the geometry of the overlapping field lines that extend between the two peaks. It seems that the measured optical polarization of the Crab is defining the high-energy emission geometry of SG and OG models where emission takes place within the corotating magnetosphere. However, Petri & Kirk[22] have shown that the Crab polarization characteristics can be almost equally well reproduced in emission in the striped pulsar wind outside the light cylinder, if a small dissipation region $\Delta r \leq \Gamma^2 r_{LC}$ is assumed. They also showed that the background polarization of the striped wind is at a PA of 120°, the same as the PA of the rotation axis. This PA direction would form the background for all models since we cannot yet resolve the pulsar from its wind. This agrees quite well with both the Optima[27] and INTEGRAL[9] results.

23.7 Summary

It is clear that phase-resolved polarimetry is a powerful tool for understanding the nature of high-energy pulsar emission. The beautiful optical polarization measurements of the Crab have demonstrated how well these measurements are constraining the high-energy emission models. X-ray polarimetry can play an equally powerful role. Although the spatial and phase resolution may not be as high as can be achieved in the optical, the Crab is an exceptionially bright optical pulsar. Most other pulsars are extremely faint at optical wavelengths and only a handful have optical pulsed detections. Some 60–70 rotation-powered pulsars have detected X-ray pulsations, and for the vast majority of these, X-ray polarimetry is the only option.

References

[1] Abdo, A. A. et al. (2008). *Science* **322**, 1218.
[2] Abdo, A. A. et al. (2009). *Science*, in press.
[3] Abdo, A. A. et al. (2009). *ApJ* **696**, 1084.
[4] Abdo, A. A. et al. (2009). *Science*, in press.
[5] Arons, J. (1983). *ApJ* **266**, 215.

[6] Atwood, W. B. et al. (2006). *ApJL* **652**, L49.
[7] Behelman, M. C. & Sikora, M. (1987). *ApJ* **322**, 650.
[8] Cheng, K. S., Ho, C. & Ruderman, M. A. (1986). *ApJ* **300**, 500.
[9] Dean, A. J. et al. (2008). *Science* **321**, 1183.
[10] Daugherty, J. K. & Harding, A. K. (1982). *ApJ* **252**, 337.
[11] Daugherty, J. K. & Harding A. K. (1996). *ApJ* **458**, 278.
[12] Deutsch, A. (1955). *Annales d'Astrophysique* **18**, 1.
[13] Dyks, J., Harding, A. K. & Rudak, B. (2004). *ApJ* **606**, 1125.
[14] Dyks, J. & Rudak, B. (2003). *ApJ* **598**, 1201.
[15] Hirotani, K. & Shibata, S. (2001). *MNRAS* **325**, 1228.
[16] Harding, A. K., Stern, J. V., Dyks, J. & Frackowiak, M. (2008). *ApJ* **680**, 1378.
[17] Hirotani, K. (2006). *ApJ* **652**, 1475.
[18] Morini, M. (1983). *MNRAS* **303**, 495.
[19] Muslimov, A. G. & Harding, A. K. (2003). *ApJ* **588**, 430.
[20] Muslimov, A. G. & Harding, A. K. (2004). *ApJ* **606**, 1143.
[21] Muslimov, A. G. & Harding, A. K. (2004). *ApJ* **617**, 471.
[22] Petri, J. & Kirk, J. G. (2005). *ApJ* 627L, 37.
[23] Petrova, S. A. (2003). *A&A* **408**, 1057.
[24] Radhakrishnan & Cooke (1969). *ApL* **3**, 225.
[25] Romani, R. W. (1996). *ApJ* **470**, 469.
[26] Romani, R. W. & Yadigaroglu, I. A. (1995). *ApJ* **438**, 314.
[27] Slowikowska, A., Kanbach, G., Kramer, M. & Stefanescu, A. (2009). *MNRAS* **397**, 103.
[28] Spitkovsky, A. (2006). *ApJ* **648**, L51.
[29] Takata, J. & Chang, H. K. (2007). *ApJ* **670**, 677.

24

Polarized X-rays from magnetized neutron stars

D. Lai
Cornell University
W. C. G. Ho
University of Southampton
M. van Adelsberg
Kavli Institute for Theoretical Physics, UCSB
C. Wang
NAOC and Cornell University
J. S. Heyl
University of British Columbia

We review the polarization properties of X-ray emission from highly magnetized neutron stars, focusing on emission from the stellar surfaces. We discuss how X-ray polarization can be used to constrain neutron star magnetic field and emission geometry, and to probe strong-field quantum electrodynamics and possibly constrain the properties of axions.

24.1 Introduction

One of the most important advances in neutron star (NS) astrophysics in the last decade has been the detection and detailed studies of surface (or near-surface) X-ray emission from a variety of isolated NSs[15; 7]. This has been made possible by X-ray telescopes such as *Chandra* and *XMM-Newton*. Such studies can potentially provide invaluable information on the physical properties and evolution of NSs (e.g. equation of state at super-nuclear densities, cooling history, surface magnetic fields and compositions, different NS populations). The inventory of isolated NSs with detected surface emission includes: (i) radio pulsars: e.g. the phase-resolved spectroscopic observations of the 'three musketeers' revealed the geometry of the

X-ray Polarimetry: A New Window in Astrophysics, eds. R. Bellazzini, E. Costa, G. Matt and G. Tagliaferri. Published by Cambridge University Press.

NS polar caps; (ii) magnetars (AXPs and SGRs): e.g. the quiescent emission of magnetars consists of a black body at $T \sim 0.5$ keV with a power-law component (index 2.7–3.5), plus significant emission up to ∼100 keV; (iii) central compact objects (CCOs) in SNRs: these now include six to eight sources, several have $P, \dot{P}$ measurements and two have absorption lines[3]; (iv) thermally-emitting isolated NSs: these are a group of seven nearby (≲1 kpc) NSs with low ($\sim 10^{32}$ erg s^{-1}) X-ray luminosities and long (3–10 s) spin periods, and recent observations have revealed absorption features in many of the sources[27; 14].

In the coming decade, the most important goals in NS astrophysics include: (i) understanding how these different types of NSs evolve and relate to each other; (ii) elucidating the different observational manifestations (e.g. radiative processes in NS atmospheres and magnetospheres); (iii) using NSs to probe physics under extreme conditions. An obvious message of this paper is that in addition to imaging, timing and spectroscopy, X-ray polarimetry provides a window to study NSs: e.g. even when the spectrum or light curve is 'boring', polarization can still be interesting and very informative. Recent advances in detector technology suggest that polarimetry study of X-ray sources holds great promise in the future (see contributions by E. Costa, J. Swank, and M. Weisskopf in these proceedings).

24.2 Polarized X-rays from NSs: basics

The surface emission from magnetized NSs (with $B \gtrsim 10^{12}$ G) is highly polarized[5; 23] for the following reason. In the magnetized plasma that characterizes NS atmospheres, X-ray photons propagate in two normal modes: the ordinary mode (O-mode, or ∥-mode) is mostly polarized parallel to the **k**-**B** plane, while the extraordinary mode (X-mode, or ⊥-mode) is mostly polarized perpendicular to the **k**-**B** plane, where **k** is the photon wave vector and **B** is the external magnetic field. This description of normal modes applies under typical conditions, when the photon energy E is much less than the electron cyclotron energy $E_{Be} = \hbar eB/(m_e c) = 11.6 B_{12}$ keV (where $B_{12} = B/(10^{12}\,\mathrm{G})$), E is not too close to the ion cyclotron energy $E_{Bi} = 6.3 B_{12}(Z/A)$ eV (where Z and A are the charge number and mass number of the ion), the plasma density is not too close to the vacuum resonance (see below) and θ_B (the angle between **k** and **B**) is not close to zero. Under these conditions, the X-mode opacity (due to scattering and absorption) is greatly suppressed compared to the O-mode opacity, $\kappa_X \sim (E/E_{Be})^2 \kappa_O$. As a result, the X-mode photons escape from deeper, hotter layers of the NS atmosphere than the O-mode photons, and the emergent radiation is linearly polarized to a high degree[23; 10; 11; 26]. Thus, if we were to put a polarimeter on the NS surface we would measure high-polarization X-rays.

To translate this 'surface measurement' into the observed signals at infinity, we need to understand how a photon (including its polarization state) evolves as it travels from the emission point to the observer. This will involve considering the geometry of the emission region (including magnetic field), light bending, and polarization evolution. Before discussing these issues, we summarize some of the general expected X-ray polarization characteristics:

(i) The X-ray polarization vector is either $\perp$ or $\parallel$ to the $\mathbf{k}$-$\boldsymbol{\mu}$ plane, depending on the photon energy and surface field strength, even when the surface field is non-dipole! (Here $\boldsymbol{\mu}$is the magnetic dipole axis.)

(ii) As the NS rotates, we will obtain a 'linear polarization sweep' (as in the rotating vector model of radio pulsars). This will provide a constraint on the dipole magnetic field geometry.

Thus, measurements of X-ray polarization, particularly when phase-resolved and measured in different energy bands, could provide unique constraints on the NS magnetic field (both the dipole component and the 'total' strength) and geometry. There is only a modest dependence on M/R of the NS, but as we discuss in the next section, quantum electrodynamics (QED) plays an important role.

Other Issues: (i) For sufficiently low T and high B, there is a possibility that the NS surface is in a condensed (metallic) form, with negligible 'vapour' above it[20]. This has been suggested in the case of the TINS RX J1856.5-3754[27; 13]. Radiation from a condensed surface is certainly quite different from an atmosphere, with distinct X-ray polarization characteristics[25]. (ii) In the case of magnetars, Compton scatterings by mildly relativistic $e^{\pm}$ in the NS magnetosphere/corona are important in determining the X-ray spectra at $E \gtrsim 2$ keV[4; 22]. How such scatterings affect the polarization signals of surface emission has not been studied.

24.3 QED effects on X-ray polarization signals

It has long been predicted from quantum electrodynamics (QED) that in a strong magnetic field the vacuum becomes birefringent[1; 8]. While this vacuum polarization effect makes the photon index of refraction deviate from unity only when $B \gtrsim 300 B_{\rm Q}$, where $B_{\rm Q} = m_e^2 c^3/(e\hbar) = 4.414 \times 10^{13}$ G is the critical QED field strength, it can significantly affect the spectra of polarization signals from magnetic NSs in more subtle way, at much lower field strengths. In particular, the combined effects of vacuum polarization and magnetized plasma give rise to a 'vacuum resonance': a photon may convert from the high-opacity mode to the low-opacity one and vice versa when it crosses the vacuum resonance region in the inhomogeneous NS atmosphere[17]. For $B \gtrsim 7 \times 10^{13}$ G (see below), this vacuum resonance phenomenon tends to soften the hard spectral tail due to the non-greyness of the atmospheric opacities and to suppress the width of absorption lines, while for $B \lesssim 7 \times 10^{13}$ G the spectrum is unaffected[11; 12; 18; 19; 26].

The QED-induced vacuum birefringence influences the X-ray polarization signals from magnetic NSs in two ways. (i) *Photon mode conversion in the NS atmosphere:* Since the mode conversion depends on photon energy and magnetic field strength, this vacuum resonance effect gives rise to a unique energy-dependent polarization signal in X-rays: For 'normal' field strengths ($B \lesssim 7 \times 10^{13}$ G), the plane of linear polarization at the photon energy $E \lesssim 1$ keV is perpendicular to that at $E \gtrsim 4$ keV, while for 'superstrong' field strengths ($B \gtrsim 7 \times 10^{13}$ G), the polarization planes at different energies coincide[19; 26]. (ii) *Polarization mode decoupling in the magnetosphere:* The birefringence of the magnetized QED vacuum decouples the photon polarization modes, so that as a polarized photon leaves the NS surface and propagates through the magnetosphere, its polarization direction follows the direction of the magnetic field up to a large radius (the so-called polarization limiting radius). The result is that although the magnetic field orientations over the NS surface may vary widely, the polarization directions of the photon originating from different surface regions tend to align, giving rise to large observed polarization signals[9; 26; 29].

24.3.1 QED effect in NS atmospheres

In a magnetized NS atmosphere, both the plasma and vacuum polarization contribute to the dielectric tensor of the medium[6; 21]. The vacuum polarization contribution is of the order $10^{-4}(B/B_Q)^2 f(B)$ (where $f \sim 1$ is a slowly varying function of B), and is quite small unless $B \gg B_Q$. The plasma contribution depends on $(\omega_p/\omega)^2 \propto \rho/E^2$. The 'vacuum resonance' arises when the effects of vacuum polarization and plasma on the polarization of the photon modes 'compensate' each other. For a photon of energy E (in keV), the vacuum resonance occurs at the density

$$\rho_V \simeq 0.964\, Y_e^{-1} B_{14}^2 E^2 f^{-2} \text{ g cm}^{-3}, \tag{24.1}$$

where Y_e is the electron fraction[17]. Note that ρ_V lies in the range of the typical densities of a NS atmosphere. For $\rho \gtrsim \rho_V$ (where the plasma effect dominates the dielectric tensor) and $\rho \lesssim \rho_V$ (where vacuum polarization dominates), the photon modes are almost linearly polarized – they are the usual O-mode and X-mode described above; at $\rho = \rho_V$, however, both modes become circularly polarized as a result of the 'cancellation' of the plasma and vacuum polarization effects (Figure 24.1). When a photon propagates outward in the NS atmosphere, its polarization state will evolve adiabatically if the plasma density variation is sufficiently gentle. Thus the photon can convert from one mode into another as it traverses the

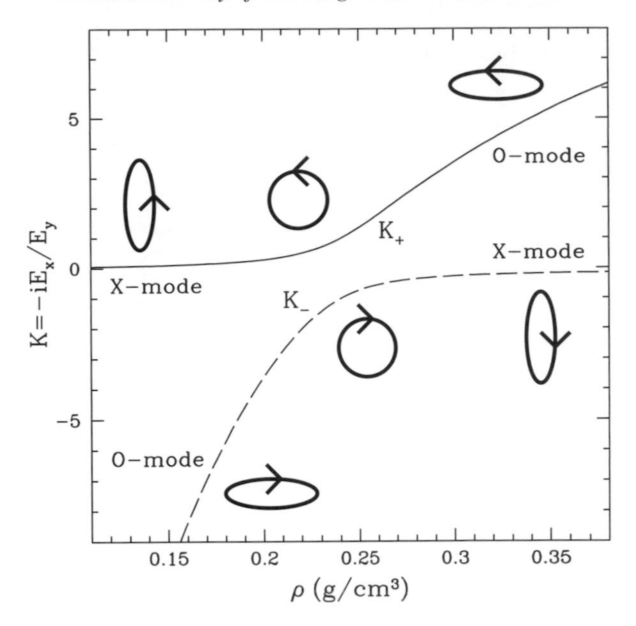

Figure 24.1 Vacuum resonance in a NS atmosphere: The polarization ellipticities of the two photon modes are shown as a function of density, for $B = 10^{13}$ G, $E = 5$ keV, $Y_e = 1$ and $\theta_B = 45°$.

vacuum resonance. For this conversion to be effective, the adiabatic condition must be satisfied:

$$E \gtrsim E_{\rm ad} = 1.49\left(f\ \tan\theta_B|1-u_i|\right)^{2/3}\left(5\,{\rm cm}/H_\rho\right)^{1/3}\ {\rm keV}, \qquad (24.2)$$

where θ_B is the angle between $\mathbf{k}$ and $\mathbf{B}$, $u_i = (E_{Bi}/E)^2$ (E_{Bi} is the ion cyclotron energy), and $H_\rho = |ds/d\ln\rho|$ is the density scale height (evaluated at $\rho = \rho_V$) along the ray. For a typical atmosphere density scale height ($\sim$1 cm), adiabatic mode conversion requires $E \gtrsim 1$–2 keV[17; 18].

The location of vacuum resonance relative to the photospheres of X-mode and O-mode photons is important. For magnetic field strengths satisfying[18; 12]

$$B \gtrsim B_l \simeq 6.6\times10^{13}\,T_6^{-1/8}E_1^{-1/4}S^{-1/4}{\rm G}, \qquad (24.3)$$

where $T_6 = T/(10^6\ {\rm K})$ and $S = 1 - e^{-E/kT}$, the vacuum resonance density lies between the X-mode and O-mode photospheres for typical photon energies, leading to suppression of spectral features and softening of the hard X-ray tail characteristic of the atmospheres. For 'normal' magnetic fields, $B \lesssim B_l$, the vacuum resonance lies outside both photospheres, and the emission spectrum is unaltered by the vacuum resonance, although the observed polarization signals are still affected[19]. See Figure 24.2.

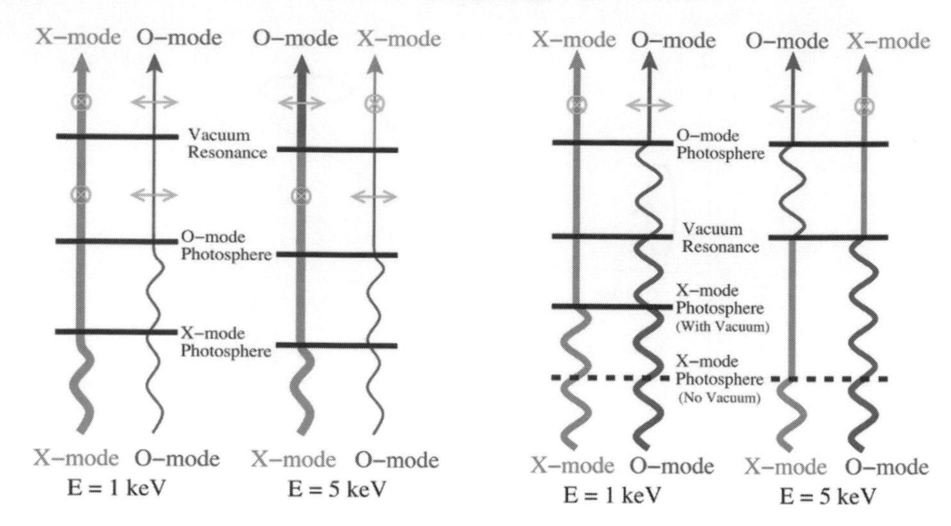

Figure 24.2 A schematic diagram illustrating how vacuum polarization affects the polarization state of the emergent radiation from a magnetized NS atmosphere. The left panel is for $B \lesssim B_l \simeq 7 \times 10^{13}$ G, and the right panel for $B \gtrsim B_l$. The photosphere is defined where the optical depth (measured from outside) is 2/3.

Figures 24.3 and 24.4 give some examples of the polarization signal of a NS hot spot. We see that for $B \lesssim B_l$, the sign of the F_Q Stokes parameter is opposite for low- and high-energy photons; this implies that the planes of polarization for low- and high-energy photons are perpendicular. For $B \gtrsim B_l$, the planes of linear polarization at different E values coincide. This is a unique signature of vacuum polarization.

24.3.2 QED effect in magnetospheres: polarization evolution

Consider radiation from a large patch of the NS, with **B** varying significantly across the emission region. Recall that locally at the NS surface, the emergent radiation is dominated by one of the two modes. If the photon polarization were parallel-transported to infinity, then the net polarization (summed over the observable patch of the NS) may be significantly reduced. However, this is incorrect[9; 19; 26]. The reason is that, even in vacuum, the photon polarization modes are decoupled near the NS surface due to QED-induced birefringence, so that parallel transport does not apply.[1]

It is straightforward to obtain the observed polarized X-ray fluxes (Stokes parameters) from the fluxes at the emission region, at least approximately, without integrating the polarization evolution equations in the magnetosphere[19; 26]. For a given (small) emission region of projected area (this is the area perpendicular to the ray at the emission point – General Relativistic light bending effect can be easily included in this), $\Delta A_\perp$, one needs to know the intensities of the two photon

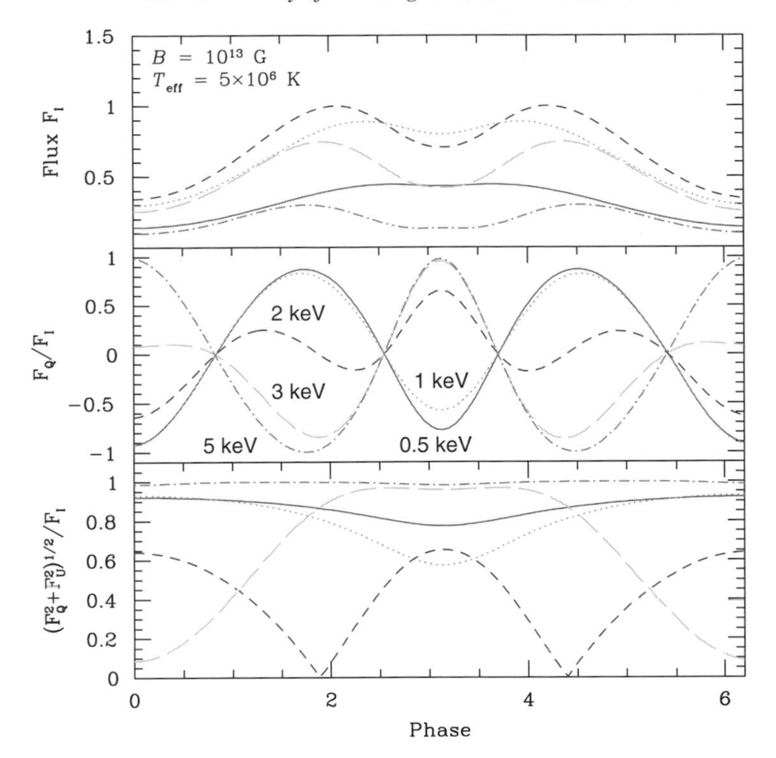

Figure 24.3 Lightcurve and polarization as a function of rotation phase for a NS hot spot with $B = 10^{13}$ G, $T_{\rm eff} = 5 \times 10^6$ K. The angle of the spin axis relative to the line of sight is $\gamma = 30°$, and the inclination of the magnetic axis relative to the spin axis is $\eta = 70°$.

modes at emission, $I_\perp$ and $I_\parallel$. In the case of thermal emission, these can be obtained directly from atmosphere/surface models. As the radiation propagates through the magnetosphere, the photon mode evolves adiabatically, following the variation of the magnetic field, until the polarization limiting radius $r_{\rm pl}$. This is where the two photon modes start recoupling to each other, and is determined by the condition $(\omega/c)\Delta n = 2|d\phi_B/ds|$ (where Δn is the difference of the indices of refraction of the two modes, ϕ_B specifies the direction of $\mathbf{B}$ along the ray), giving

$$\frac{r_{\rm pl}}{R_*} \simeq 70 \left(E_1 B_{*13}^2 P_1\right)^{1/6}, \tag{24.4}$$

where R_* is the NS radius and B_{*13} is the polar magnetic field at the stellar surface in units of 10^{13} G (see Reference [26] for a more detailed expression). Beyond $r_{\rm pl}$, the photon polarization state is frozen. Thus the polarized radiation flux at $r \gtrsim r_{\rm pl}$ is $F_Q = (I_\parallel - I_\perp)\Delta A_\perp/D$, and $F_U \simeq F_V \simeq 0$,[2] where D is the distance of the source, F_Q and F_U are defined in the coordinate system such that the stellar magnetic field

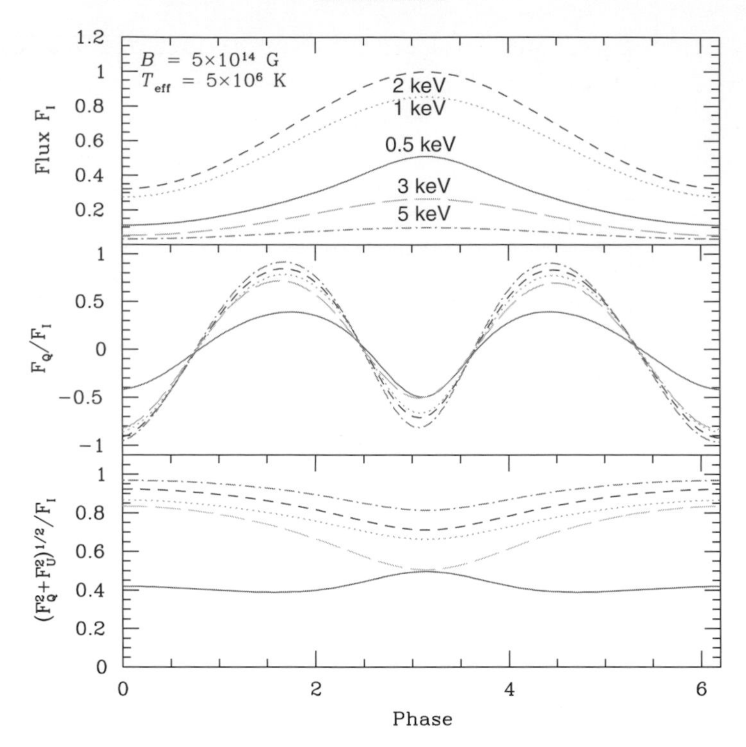

Figure 24.4 Same as Figure 24.3, but for $B = 5 \times 10^{14}$ G.

at $r_{\rm pl}$ lies in the XZ plane (with the Z-axis pointing towards the observer). Since $r_{\rm pl}$ is much larger than the stellar radius, the magnetic fields as 'seen' by different photon rays are aligned and are determined by the dipole component of the stellar field, one can simply add up contributions from different surface emission areas to F_Q to obtain the observed polarization fluxes.

Note that the description of the polarization evolution in the last paragraph is valid regardless of the possible complexity of the magnetic field near the stellar surface. This opens up the possibility of constraining the *surface* magnetic field of the neutron star using X-ray polarimetry. For example, the polarization light curve (particularly the dependence on the rotation phase) depends only on the dipole component of the magnetic field, while the intensity lightcurve of the same source depends on the surface magnetic field. On the other hand, the linear polarization spectrum (i.e., its dependence on the photon energy) depends on the magnetic field at the emission region[19; 26]; thus it is possible that a NS with a weak dipole field ($\lesssim 7 \times 10^{13}$ G) may exhibit X-ray polarization spectrum characteristic of a $B \gtrsim 7 \times 10^{13}$ G NS.

One complication arises from the *quasi-tangential* (QT) effect: as the photon travels through the magnetosphere, it may cross the region where its wave vector is aligned or nearly aligned with the magnetic field (i.e., θ_B is zero or small). In such a QT region, the two photon modes ($\|$ and $\perp$ modes) become (nearly) identical, and can temporarily recouple, thereby affecting the polarization alignment[29]. This QT effect gives rise to partial mode conversion: after passing through the QT point, the mode intensities change to $\bar{I}_\| \neq I_\|$ and $\bar{I}_\perp \neq I_\perp$. The observed polarization flux is then $\bar{F}_Q = (\bar{I}_\| - \bar{I}_\perp)\Delta A/D \neq F_Q$.

In the most general situations, to account for the QT propagation effect, it is necessary to integrate the polarization evolution equations in order to obtain the observed radiation Stokes parameters. However, Wang & Lai [29] showed that for generic near-surface magnetic fields, the effective region where the QT effect leads to significant polarization changes covers only a small area of the NS surface. For a given emission model (and the size of the emission region) and magnetic field structure, Wang & Lai derived the criterion to evaluate the importance of the QT effect. In the case of surface emission from around the polar cap region of a dipole magnetic field, they quantified the effect of QT propagation in detail and provided a simple, easy-to-use prescription to account for the QT effect in determining the observed polarization fluxes. The net effect of QT propagation is to reduce the degree of linear polarization, so that $\bar{F}_Q/F_Q < 1$, with the reduction factor depending on the photon energy, magnetic field strength, geometric angles, rotation phase and the emission area. The largest reduction is about a factor of two, and occurs for a particular emission size. Obviously, for emission from a large area of the stellar surface, the QT effect is negligible.

24.4 Probing axions with polarized X-rays

The axion is a hypothesized pseudoscalar particle, introduced in the 1980s to explain the absence of strong CP violation. The axion is also an ideal candidate for cold dark matter, with the allowed axion mass m_a in the range of $10^{-6} \lesssim m_a \lesssim 10^{-3}$ eV.

A general property of the axion is that it can couple to two photons (real or virtual) via the interaction

$$\mathcal{L}_{a\gamma\gamma} = -\frac{1}{4} g\, a F_{\mu\nu}\tilde{F}^{\mu\nu} = g\, a\, \mathbf{E} \cdot \mathbf{B}, \tag{24.5}$$

where a is the axion field, $F_{\mu\nu}$ ($\tilde{F}^{\mu\nu}$) is the (dual) electromagnetic field strength tensor, and g is the photon–axion coupling constant. Accordingly, in the presence of a magnetic field, a photon (the $\|$ component) may oscillate into an axion and vice versa. Exploiting such photon–axion oscillation, various experiments and astrophysical considerations have been used to put constraint on the allowed values of g and m_a[24; 2].

Magnetic NSs can serve as a useful laboratory to probe axion–photon coupling. Lai & Heyl [16] presented the general methods for calculating the axion–photon conversion probability during propagation through a varying magnetized vacuum as well as across an inhomogeneous atmosphere. Partial axion–photon conversion may take place in the vacuum region outside the NS. Strong axion–photon mixing occurs due to a resonance in the atmosphere, and depending on the axion coupling strength and other parameters, significant axion–photon conversion can take place at the resonance. Such conversions may produce observable effects on the radiation spectra and polarization signals from the star. More study is needed in order to determine whether it is possible to separate out the photon–axion coupling effect from the intrinsic astrophysical uncertainties of the sources. See Reference [16] for more details and references.

Notes

1. For photons with frequencies much higher than radio, vacuum birefringence dominates over the plasma effect for all reasonable magnetosphere plasma parameters[28].
2. Note that F_V is not exactly zero because of the NS rotation and because mode recoupling does not occur instantly at $r_{\rm pl}$; see[26].

References

[1] Adler, S. L. (1971). *Ann. Phys.* **67**, 599.
[2] CAST collaboration (2009). [arXiv:0905.4273].
[3] De Luca, A. (2008). in *40 years of Pulsars: Millisecond Pulsars, Magnetars and More*, eds. C. Bassa et al. (NY: AIP), 311.
[4] Fernandez, R. & Thompson, C. (2007). *ApJ* **660**, 615.
[5] Gnedin, Yu. N., Sunyaev R. A. (1974). *A&A* **36**, 379.
[6] Gnedin, Yu. N., Pavlov, G. G., & Shibanov, Yu. A. (1978). *Sov. Astro. Lett.* **4**, 117.
[7] Harding, A. K., & Lai, D. (2006). *Rept. Prog. Phys.* **69**, 2631.
[8] Heyl, J. S., & Hernquist, L. (1997). *Phys. Rev.* **D55**, 2449.
[9] Heyl, J. S., & Shaviv, N. J. (2002). *Phys. Rev.* **D66**, 023002.
[10] Ho, W. C. G., & Lai, D. (2001). *MNRAS* **327**, 1081–1096.
[11] Ho, W. C. G., & Lai, D. (2003). *MNRAS* **338**, 233.
[12] Ho, W. C. G., & Lai, D. (2004). *ApJ* **607**, 420.
[13] Ho, W. C. G., et al. (2007). *MNRAS* **375**, 821.
[14] Kaplan, D. L. (2008). in *40 years of Pulsars: Millisecond Pulsars, Magnetars and More*, eds. C. Bassa et al. (NY: AIP), 331.
[15] Kaspi, V. M., Roberts, M., & Harding, A. K. (2006). in *Compact Stellar X-ray Sources*, eds. W. Lewin & M. van der Klis (Cambridge Univ. Press).
[16] Lai, D., & Heyl, J. (2006). *Phys. Rev.* **D74**, 123003.
[17] Lai, D., & Ho, W. C. G. (2002). *ApJ* **566**, 373.
[18] Lai, D., & Ho, W. C. G. (2003). *ApJ* **588**, 962.
[19] Lai, D., & Ho, W. C. G. (2003). *Phys. Rev. Lett.* **91**, 071101.
[20] Medin, Z., & Lai, D. (2007). *MNRAS* **382**, 1833.
[21] Mészáros, P. & Ventura, J. (1979). *Phys. Rev.* **D19**, 3565.

[22] Nobili, L., Turolla, R., & Zane, S. (2008). *MNRAS* **386**, 1527.
[23] Pavlov, G. G. & Zavlin, V. E. (2000). *ApJ* **529**, 1011.
[24] Raffelt, G. G. (2008). *Lect. Notes. Phys.* **741**, 51 [arXiv:hep-ph/0611350].
[25] van Adelsberg, M., et al. (2005). *ApJ* **628**, 902.
[26] van Adelsberg, M., & Lai, D. (2006). *MNRAS* **373**, 1495.
[27] van Kerkwijk, M. H., & Kaplan, D. L. (2007). *Ap&SS* **308**, 191.
[28] Wang, C. & Lai, D. (2007). *MNRAS* **377**, 1095.
[29] Wang, C. & Lai, D. (2009). *MNRAS* in press [arXiv:0903.2094].

25

Polarization properties of X-ray millisecond pulsars

J. Poutanen

Astronomy Division, Department of Physics, University of Oulu, Finland

Radiation of X-ray bursts and of accretion shocks in weakly magnetized neutron stars in low-mass X-ray binaries is produced in plane-parallel atmospheres dominated by electron scattering. We first discuss polarization produced by single (non-magnetic) Compton scattering, in particular the depolarizing effect of high electron temperature, and then the polarization due to multiple electron scattering in a slab. We further predict the X-ray pulse profiles and polarization properties of nuclear- and accretion-powered millisecond pulsars. We introduce a relativistic rotation vector model, which includes the effect of rotation of polarization plane due to the rapid motion of the hot spot as well as the light bending. Future observations of the X-ray polarization will provide a valuable tool to test the geometry of the emission region in pulsars and its physical characteristics.

25.1 Introduction

Polarization has proved to be a valuable tool in determining the geometry of the emission region in radio pulsars[1]. For X-ray pulsars, the data are not yet available, but their interpretation in any case is not going to be easy, because of the strong magnetic field effects on the radiation transport. Discovery of millisecond coherent pulsations during X-ray bursts in nearly 20 low-mass X-ray binaries (so-called nuclear-powered millisecond pulsars (NMSP) see [23]) and in the persistent emission of at least eight sources (accretion-powered millisecond pulsars (AMSP) see [25; 14; 15]) opens a completely new range of possibilities. The emission in these cases is produced at the surface of a rapidly spinning, weakly magnetized neutron star. Thus the magnetic field does not affect the radiation transport and much more reliable predictions for the radiation pattern from the surface can be

X-ray Polarimetry: A New Window in Astrophysics, eds. R. Bellazzini, E. Costa, G. Matt and G. Tagliaferri. Published by Cambridge University Press.

obtained. The observed pulse profiles and polarization are affected not only by general, but also special relativistic effects[17; 24; 16]. The pulse profile alone, however, does not allow us to determine uniquely the pulsar geometry, while the phase dependence of the polarization angle (PA) is a powerful tool to distinguish between the models and to put strong constraints on the geometry. In this review, we first discuss the physics of polarization produced by Compton scattering. Then we apply these results to predict the polarization properties of NMSP and AMSP.

25.2 Polarization properties of Compton scattering

25.2.1 Polarization in single Compton scattering

Radiation Thomson-scattered once by cold electrons becomes linearly polarized with the PD depending strongly on the scattering angle Θ: $P(\Theta) = \sin^2\Theta/(1+\cos^2\Theta)$. With increasing electron temperature the PD drops because of random aberrations and corresponding random rotations of the polarization plane due to the thermal motions of the electrons[12; 13] (see Figure 25.1). For isotropic relativistic electrons of arbitrary energies, the exact analytical expressions for the scattering redistribution matrix for Stokes parameters was derived in [11]. For $T_e = 50$ keV, polarization is smaller by 50% and for 100 keV it drops by a factor of two compared to the Thomson scattering case. As electrons in accreting black holes and neutron stars often reach such temperatures, the depolarization effect has to be accounted

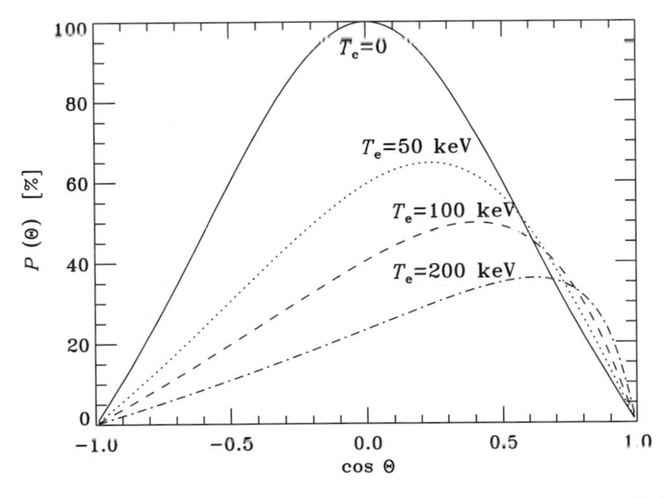

Figure 25.1 PD at 1 keV of 0.1 keV black-body photons scattered by the hot isotropic electrons of various temperatures T_e as a function of the scattering angle (adapted from[13]). For the dependence of the PD on photon energy and for cases of initially polarized radiation, see [12].

for. For isotropic power-law distribution of relativistic electrons, the net scattered polarization is zero if the incident polarization is unpolarized[2; 11].

25.2.2 *Scattering in optically thick atmosphere*

In a classical problem of radiative transfer in a plane-parallel semi-infinite atmosphere with opacity dominated by Thomson scattering[3; 21], the PD reaches the maximum of $\approx 11.7\,\%$ at $\mu = 0$ (where μ is the cosine of the angle from the normal) and decreases to zero at $\mu = 1$ due to the symmetry:

$$P \approx -\frac{1-\mu}{1+3.582\mu} 11.71\%. \tag{25.1}$$

The dominant direction of the electric vector oscillations is parallel to the slab plane.

25.2.3 *Comptonization in optically thin atmosphere*

The detailed study of the transport of polarized radiation in plane-parallel atmospheres with absorption and Thomson scattering included was considered in [8; 9]. The absorption causes the rotation of the polarization plane by 90° (Nagirner effect,[10]) even in optically thick atmospheres. The same happens when the scattering optical depth becomes small $\tau_T \lesssim 1$[22; 6; 24]. We are often interested in

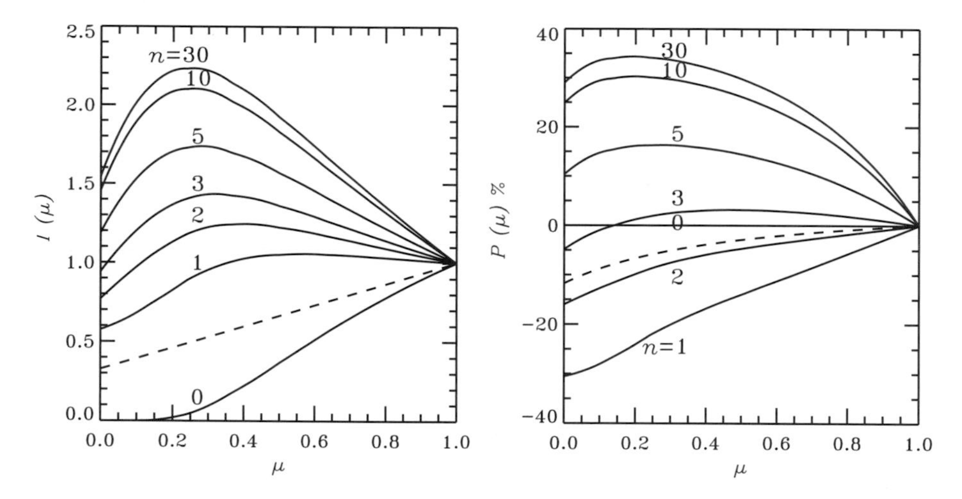

Figure 25.2 Angular dependence of the intensity and polarization of the radiation escaping from a slab of $\tau_T = 1$ for different scattering orders n. Positive P corresponds to the plane of polarization parallel to the slab normal. The unpolarized seed photons with $I(\mu)$ =const were injected from the slab bottom. The dashed curves correspond to the classical results of Chandrasekhar-Sobolev for $\tau_T = \infty$. From [24].

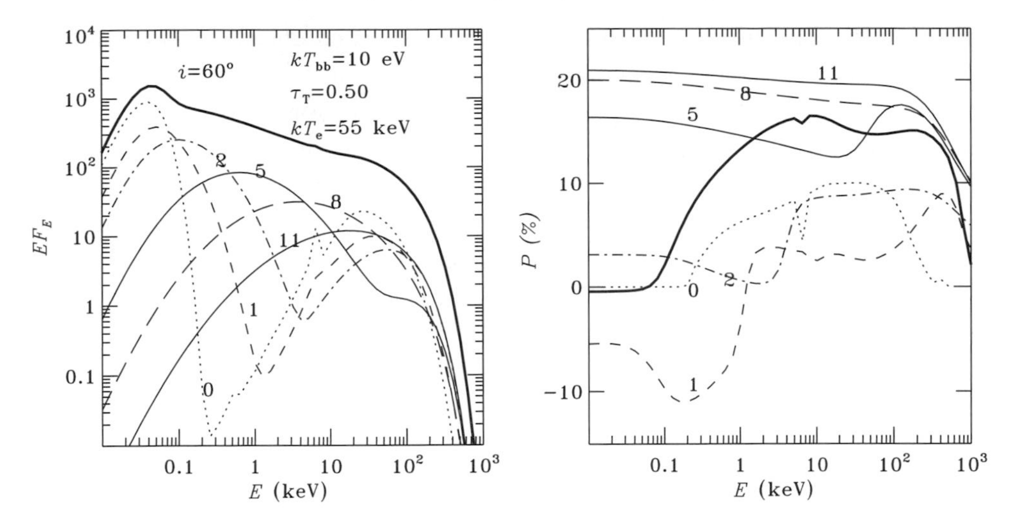

Figure 25.3 The flux and the PD emergent from a slab of $\tau_T = 0.5$ and electron temperature $kT_e = 55$ keV at inclination of 60°. The incident photons are black body of temperature 10 eV. Different scattering orders are marked on the corresponding curves. The thick solid curves correspond to the total radiation. The zeroth order radiation consists of unscattered black body and a polarized Compton reflection component at high energies (computed using Green's matrix from [19]). Adapted from [18].

polarization of photons undergoing a certain number of scattering events (because photons gain energy every time they are scattered by hot electrons). The dependence of intensity and PD on angle relative to the normal for various scattering orders is shown in Figure 25.2 for $\tau_T = 1$.

Usually polarization was considered in Thomson scattering approximations, while, as we have seen in Section 25.2.1, the polarization does depend on the electron temperature. Calculations of the PD in a slab using a fully relativistic kernel from [11] is shown in Figure 25.3. One sees that radiation scattered once is polarized perpendicular to the slab normal (as for optically thick atmosphere), while at higher scattering orders polarization changes the sign.

25.3 Polarization of accreting neutron stars

25.3.1 Relativistic rotating vector model

When considering polarization of radiation from coherently oscillating sources, we can assume that the emission originates in one or two (in the case of AMSP) spots at the neutron star surface. In AMSP the hard X-rays are produced by thermal Comptonization in a plane-parallel slab (accretion shock) of optical depth of order unity[17; 15]. The emission during the X-ray bursts is produced in a semi-infinite electron-scattering-dominated atmosphere. It is natural to assume that the radiation pattern has azimuthal symmetry in the co-moving frame of a spot. Once we have the Stokes parameters in spot frame, we can transform them to the observer frame. First, we make the Lorentz transformation to the non-rotating frame accounting for Doppler boosting and relativistic aberration and then follow photon trajectories to the observer at infinity in Schwarzschild space-time. Deviations from the Schwarzschild metric and from sphericity of the star due to the stellar rotation have a small effect and are usually neglected. For pulsar rotational frequencies of $\nu \gtrsim 400$ Hz, we also need to account for time delays, which in the extreme cases can reach about 5–10 per cent of the pulsar period.

Lorentz transformation and gravitational light bending do not change the PD and the observed value corresponds to the polar angle α' at which a photon is emitted in the spot co-moving frame. Lorentz transformation gives $\cos\alpha' = \delta\cos\alpha$, where δ is the Doppler factor and α is the angle in the nonrotating frame, which is related to the position angle ψ by the usual gravitational bending formula (see Figure 25.4 and[17; 24; 16] for details).

For a slowly rotating star, the polarization vector lies in the plane formed by the spot normal and the line of sight. The PA (measured from the projection of the spin axis on the plane of the sky in the counter-clockwise direction) is given by

$$\tan\chi_0 = -\frac{\sin\theta\ \sin\phi}{\sin i\ \cos\theta - \cos i\ \sin\theta\ \cos\phi}, \tag{25.2}$$

as in the rotating vector model of Radhakrishnan & Cooke[20]. Here ϕ is the pulsar phase, θ is the spot colatitude, and i is the observer's inclination (see Figure 25.4). The effect of the stellar spin on rotation of the polarization vector is discussed in [4; 5]. Viironen & Poutanen[24] presented the correction to the expression for

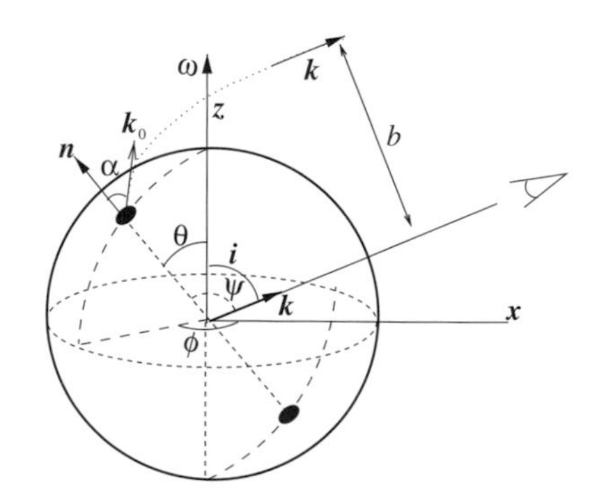

Figure 25.4 Geometry of the emitting regions for AMSP.

the PA which includes light bending and aberration:

$$\tan \chi_c = \beta_{eq} \cos\alpha \sin\theta \frac{\cos i \, \sin\theta - \sin i \, \cos\theta \, \cos\phi}{\sin\alpha \, \sin\psi + \beta_{eq} \sin\theta \sin i \, \sin\phi}, \tag{25.3}$$

where β_{eq} is the equatorial stellar velocity in units of speed of light. The total polarization angle for each spot is then $\chi = \chi_0 + \chi_c$.

25.3.2 Polarization properties of NMSP

The X-ray bursts often show coherent oscillations[23]. The energy dissipation takes place deep in the atmosphere, and thus we can assume that optical depth is infinite. At sub-Eddington luminosities the atmosphere is pinned down to the neutron star surface, and thus a plane-parallel approximation is valid. At effective temperatures of about 2 keV, most of the material is ionized and scattering dominates the opacity. The radiation escaping from the surface then can be described by Chandrasekhar-Sobolev formulae from Section 25.2.2. The predicted pulse profiles and behavior of PD and PA are shown in Figure 25.5 for a small bright spot. The polarization is increasing with the spot colatitude θ, reaching the maximum of $\sim$12% close

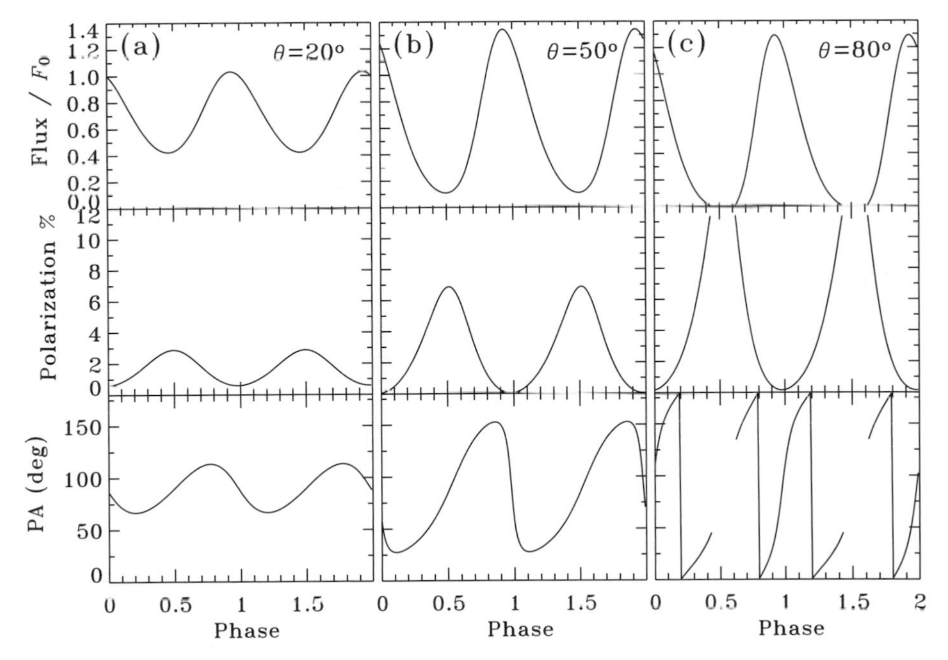

Figure 25.5 Light curves, PD and PA expected from an X-ray burst happening at various colatitudes θ. Semi-infinite electron scattering atmosphere is considered. The neutron star mass is $M = 1.4\mathrm{M}_{\odot}$, radius $R = 10.3$ km, rotational frequency $\nu = 400$ Hz, and the observer inclination $i = 60^{\circ}$. From [24].

to the eclipses. The PA varies around 90° as the electric vector is predominantly perpendicular to the meridional plane and its variability amplitude grows with θ. The larger is the spot, the smaller is the polarization.

25.3.3 Polarization properties of accreting millisecond pulsars

The spectra of accreting millisecond pulsars can be represented as a sum of a black-body-like emission and a Comptonized tail (see review [15] and references therein). The black body photons play also a role of seed photons for Comptonization. This interpretation is supported by the shape of the pulse profile above 7 keV which is consistent with being produced by the hotspot with the angular pattern characteristic for an optically thin slab[17]. Because below a few keV the contribution of the black body is large, the expected polarization degree is small. Scattering in the hot electron slab modifies significantly the angular distribution of radiation and produces polarization signal as described in Section 25.2.3. The polarization degree is a strong function of the scattering order for small n (see Figure 25.3) and therefore of the photon energy also. After a few scatterings polarization saturates. In Figure 25.6 we show the prediction for the polarization signal for $n = 7$ (in Thomson approximation), which would correspond to about 10 keV for a typical

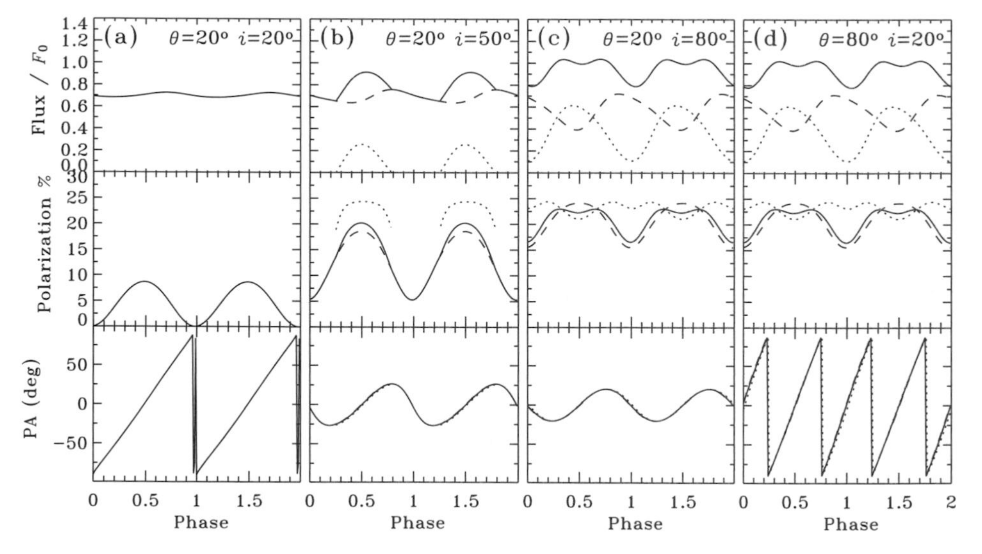

Figure 25.6 Normalized pulse profiles, PD and PA from two antipodal point-like spots for various inclinations and magnetic inclinations. The intrinsic radiation is scattered seven times in a slab of Thomson optical depth $\tau_T = 1$. Dashed curves correspond to the primary spot, dotted curves to the antipodal spot, and solid curves to the total signal. The stellar parameters are $M = 1.4\mathrm{M}_\odot$, $R = 10.3$ km, $\nu = 300$ Hz. From [24].

electron temperature of 50 keV and the seed photon temperature of 0.6 keV. We choose the slab optical depth $\tau_T = 1$ which is consistent with the spectra of SAX J1808.4−3658[17; 7]. We see that exchanging i and θ does not affect the pulse shape and PD, but the PA changes dramatically.

25.4 Conclusions

Polarization properties of non-magnetic Compton scattering, which is the main emission mechanism in weakly magnetized accreting neutron stars, have been studied in detail. Light bending in the gravitational field of the neutron star and the rotation of the polarization plane because of relativistic effects are also well understood. Thus the theory of the polarization from millisecond X-ray pulsars is rather well developed by now and is ready to be applied to the data. The launch of an X-ray polarimeter will hopefully start a new era in studies of these sources and will provide a powerful tool for understanding the geometry and the physics of pulsars.

References

[1] Blaskiewicz, M., Cordes, J. M. & Wasserman, I. (1991). *ApJ* **370**, 643–669.
[2] Bonometto, S., Cazzola, P. & Saggion, A. (1970). *A&A* **7**, 292–304.
[3] Chandrasekhar, S. (1960). *Radiative transfer* New York, Dover.
[4] Ferguson, D. C. (1973). *ApJ* **183**, 977–986.
[5] Ferguson, D. C. (1976). *ApJ* **205**, 247–260.
[6] Haardt, F. & Matt, G. (1993). *MNRAS* **261**, 346–352.
[7] Ibragimov, A. & Poutanen, J. (2009). *MNRAS*, submitted [arxiv:0901.0073].
[8] Loskutov, V. M. & Sobolev, V. V. (1981). *Astrofizika* **17**, 535–546.
[9] Loskutov, V. M. & Sobolev, V. V. (1982). *Astrofizika* **18**, 81–91.
[10] Nagirner, D. I. (1962). *Tr. Astron. Obs. Leningrad Univ.* **19**, 79.
[11] Nagirner, D. I. & Poutanen, J. (1993). *A&A* **275**, 325–336.
[12] Nagirner, D. I. & Poutanen, J. (1994). *Astrophys. Space Phys. Rev.* **9**, 1–83.
[13] Poutanen, J. (1994). *ApJS* **38**, 2697–2703.
[14] Poutanen, J. (2006). *Adv. Sp. Res* **38**, 2697–2703.
[15] Poutanen, J. (2008). In *AIP Conf. Proc. 1068, A Decade of Accreting Millisecond X-ray Pulsars*, ed. R. Wijnands et al., Melville, NY, AIP.
[16] Poutanen, J. & Beloborodov, A. M. (2006). *MNRAS* **373**, 836–844.
[17] Poutanen, J. & Gierliński, M. (2003). *MNRAS* **343**, 1301–1311.
[18] Poutanen, J. & Svensson, R. (1996). *ApJ* **470**, 249–268.
[19] Poutanen, J., Nagendra K. N. & Svensson, R. (1996). *MNRAS* **283**, 892–904.
[20] Radhakrishnan, V. & Cooke, D. J. (1969). *Ap. Letters* **3**, 225–904.
[21] Sobolev, V. V. (1963). *A treatise on radiative transfer* Princeton, NJ, Van Nostrand.
[22] Sunyaev, R. A. & Titarchuk, L. G. (1985). *A&A* **143**, 374–388.
[23] Strohmayer, T. & Bildsten, L. (2006). In *Compact stellar X-ray sources*, ed. W. Lewin & M. van der Klis, Cambridge, Cambridge University Press.
[24] Viironen, K. & Poutanen, J. (2004). *A&A* **426**, 985–997.
[25] Wijnands, R. (2006). In *Trends in pulsar research*, ed. J. A. Lowry, New York, Nova Science Publishers.

26

X-ray polarization signatures of neutron stars

P. Ghosh

Tata Institute of Fundamental Research

We explore expected polarization signatures in thermal X-ray emission from magnetized neutron stars. We study the interplay between the photospheres of ordinary and extraordinary modes, and the vacuum resonance. We consider propagation in the neutron star magnetosphere. We identify distinct regimes of magnetic field strengths, and summarize their polarization signatures.

26.1 Introduction

We are discussing the state of the art of X-ray polarization detection techniques in this conference, so that it is important to remind ourselves of the expected X-ray polarization properties of various astrophysical objects. In this paper, we give a brief overview of the expected X-ray polarization signatures of magnetic neutron stars found in diverse situations, e.g. in accretion-powered pulsars, low-mass X-ray binaries (LMXBs), recycled pulsars, isolated neutron stars and finally the fascinating magnetars. We concentrate here only on some aspects of the basic physics of radiation propagation around magnetized neutron stars which lead to some basic, expected polarization features in the X-rays which we consider relatively robust. Accordingly, our discussion here is qualitative. Quantitative aspects of a few of these features have been described by other participants of the conference, and detailed calculations on some other aspects will be reported elsewhere.

The X-ray emission we are concerned with here is basically thermal emission from the surface of the neutron star, powered by accretion or otherwise. This radiation propagates through the neutron-star atmosphere, then through the accretion columns over the magnetic poles of the neutron star if it is an accreting one, and

X-ray Polarimetry: A New Window in Astrophysics, eds. R. Bellazzini, E. Costa, G. Matt and G. Tagliaferri. Published by Cambridge University Press.

finally through the neutron-star magnetosphere. In all of these locations, the plasma is magnetized by the enormous magnetic field of the neutron star, and such plasma is birefringent, leading to two distinct modes of propagation for the radiation, namely, the ordinary or O mode and the extraordinary or X mode. These modes have their directions of polarization perpendicular to each other, so that their interplay decides the overall polarization of the escaping radiation. Generally (though not always, as we shall see below) the X mode comes from deeper and hotter layers of emission because of the very different opacities of the two modes in magnetized media, and so dominates the polarization properties. The way in which this is normally quantified is in terms of the photospheres of the two modes, where the optical depth counted from a distant observer is approximately unity. So, the photosphere of the X mode generally lies deeper than that of the O mode.

A key concept in this problem is that of vacuum resonance: in the presence of strong magnetic fields, even the vacuum is birefringent because of vacuum polarization due to virtual e^+e^- pairs. At a critical point, these vacuum effects cancel the plasma effects described above, leading to resonant features in the opacities: this is the vacuum resonance point V. It is the interplay between the locations of the O and X mode photospheres and the V point that determine the key polarization signatures near the surface of the neutron star. Subsequent propagation through the magnetosphere of the neutron star causes further modification in the polarization. We consider each in turn.

26.2 Polarization near neutron-star surface

We show in Figure 26.1 the density ρ_O at the photosphere of the O mode, the density ρ_X at the photosphere of the X mode, and the density ρ_V at the vacuum resonance as functions of the magnetic field strength B. The essential scalings of these densities with B and the photon energy E are given by the approximate expressions[1]:

$$\rho_O \approx 0.4E^{3/2} \text{ g cm}^{-3}, \tag{26.1}$$

$$\rho_X \approx 5E^{1/2}B_{12} \text{ g cm}^{-3}, \tag{26.2}$$

$$\rho_V \approx 10^{-4}E^2B_{12}^2 \text{ g cm}^{-3}. \tag{26.3}$$

Here B_{12} is B in units of 10^{12} G, and E is in units of 1 keV.

The crossover field strengths $B_l \approx 7 \times 10^{13}$ G (where O and V curves cross), $B_h \approx 5 \times 10^{16}$ G (where X and V curves cross) and $B_{XO} \approx 9 \times 10^{10}$ G (where X and O curves cross) lead to several interesting signatures in various regimes due to the phenomenon of *mode conversion* at higher photon energies between O and X modes at the vacuum resonance point V[1]. In the regime $B_{XO} < B < B_l$, which we can call the *normal* regime, and where the accretion-powered pulsars with relatively high

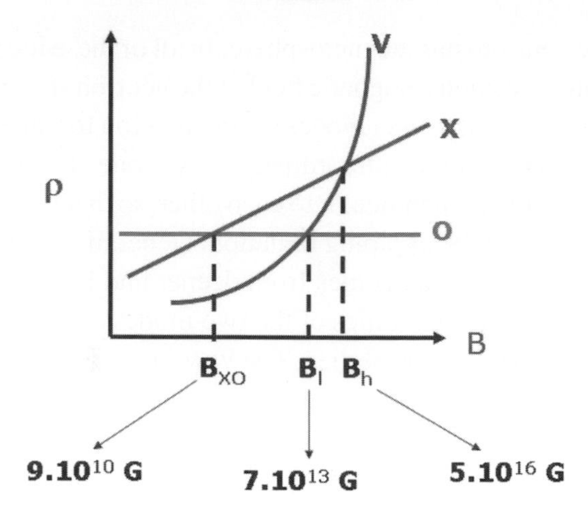

Figure 26.1 Scaling with magnetic field of the photospheric densities of O and X modes, and the density of vacuum resonance.

magnetic fields are, the vacuum resonance lies at a lower density (and therefore at a larger radius in any density profile that decreases outward) than the photospheres of both O and X modes. Accordingly, the X mode dominates at lower photon energies E, but the O mode dominates at higher E, so that the linear polarization rotates by 90° in going from low to high E, and the Stokes parameter changes sign. However, in the regime $B_l < B < B_h$, where magnetars are believed to be, and which we can call the *magnetar* regime, the vacuum resonance lies between the photospheres of the O and X modes, and the same mode-conversion phenomenon now leads to a different signature[1]. The X mode dominates at both low and high E, so that the Stokes parameter does not change sign, and the linear polarization does not change direction.

What happens when $B < B_{XO}$, which we can call the *low* regime, and where the O mode photosphere has a higher density, and so lies deeper, than the X mode photosphere? The same mode-conversion arguments now show that the O mode dominates at low E and the X mode dominates at high E. So the linear polarization does rotate by 90° in going from low to high X-ray photon energies, but now in the *reverse* direction from what happens in the normal regime, as shown in Figure 26.2. At exactly $B = B_{XO}$, what do we expect? Pending further investigation, the simple answer is of course that linear polarization would be replaced by circular polarization.

Given a radial density profile $\rho(r)$, the radii r_O, r_X and r_V of the O and X mode photospheres and the vacuum resonance can be readily found. However, this profile depends on the nature of the neutron-star source. For isolated neutron stars (and possibly for magnetars) the relevant profile is that within the neutron-star

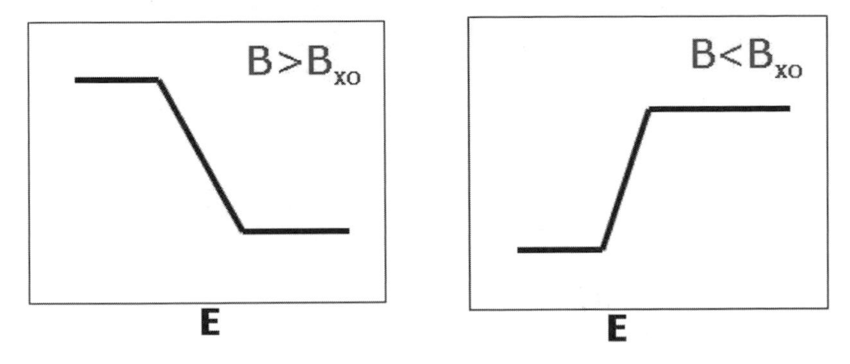

Figure 26.2 Rotation of linear polarization by 90° above and below B_{XO}.

atmosphere, the scale height being extremely small, ~1–10 cm. For accretion-powered neutron stars, the relevant profile is that within the accretion columns, with scale heights typically ~100 m. Detailed results for the latter case will be reported elsewhere. The essential point to be noted here is that in all cases the interplay between r_O, r_X and r_V occurs *near* the neutron-star surface.

26.3 Magnetospheric effects

We now consider the propagation of the above polarized radiation through the magnetosphere, in the case of accreting neutron stars. Much larger scale lengths now enter the problem, since the scale size r_m of the magnetosphere, given approximately by

$$r_m \approx 3 \times 10^8 B_{12}^{4/7} \dot{M}_{-9}^{-2/7} \text{cm}, \tag{26.4}$$

is two orders of magnitude larger than the size of the neutron star R_{ns} for typical accretion rates $\dot{M}$ for bright X-ray sources. Here, $\dot{M}_{-9}$ is $\dot{M}$ in units of $10^{-9} M_\odot$ yr^{-1}.

This is shown in Figure 26.3, where the collection of lines close to the neutron-star radius R_{ns} are the above radii r_O, r_X and r_V, shown schematically to stress the very different scale of r_m for canonical, accretion-powered, ~10^{12} G neutron stars. As an X-ray photon propagates outward through decreasing magnetic field in the magnetosphere, its polarization state evolves adiabatically, until the polarization is frozen beyond a limiting radius r_{pl}[2; 3]. Using the prescription given in the latter reference, we obtain

$$r_{pl} \approx 3 \times 10^7 E^{1/6} B_{12}^{1/3} \text{cm}, \tag{26.5}$$

which is about a tenth of the magnetospheric radius and about 30 times the neutron-star radius at canonical field strengths 10^{12} G.

Table 26.1 *Linear polarization strength.*

B	q
10^{10} G	10%
10^{9} G	1%

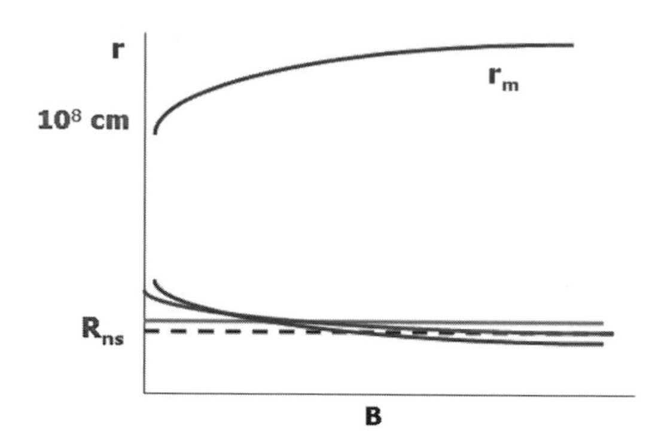

Figure 26.3 Schematic scaling with B of the magnetospheric radius r_m, and the radii r_O, r_X and r_V close to the neutron-star radius R_{ns} (see text).

Note, however, that r_{pl} scales more slowly than r_m with B and so overtakes the latter at very low field strengths, the crossover point being at $B_{vl} \sim 10^8$ G. If we wish, we can use this last crossover field to define the very-low end of the "low" regime introduced above, i.e. $B_{vl} < B < B_{XO}$. This end is populated by LMXBs and strongly recycled pulsars, including accretion-powered millisecond pulsars, with $B \sim 10^8$–10^9 G. By contrast, the upper end of the low regime with $B \sim 10^{10}$–10^{11} G is populated by accretion-powered pulsars with lower fields and mildly recycled pulsars (e.g. those recycled in massive X-ray binaries). It can be readily seen that in the above 'very-low' regime, not only are r_{pl} and r_m close to each other, but both are close to R_{ns} and therefore also to r_O, r_X and r_V. The whole picture of separation between the two regions fails at this point.

But more important from the observational point of view is the fact that the level of polarization becomes very low at such field strengths. This is easily seen with the aid of the expression $q \sim 12B_{12}/E(\text{keV})$ for a rough measure[4] of the linear polarization strength, and displayed in Table 26.1. Thus LMXBs and accretion-powered millisecond pulsars would be expected to have very low levels of linear polarization from the neutron star magnetic field.

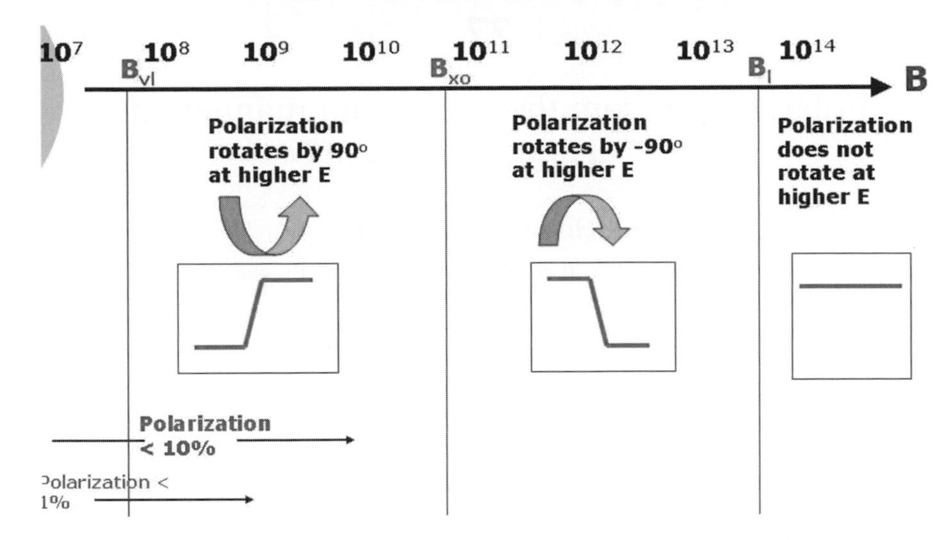

Figure 26.4 Polarization signatures in different regimes of B.

26.4 Conclusions

X-ray emission from magnetized neutron stars is expected to show strong linear polarization (∼10–50%) for canonical field strengths of accretion-powered pulsars or magnetars. We have identified, above, distinct regimes of B which should have different signatures, as summarized in Figure 26.4.

References

[1] Lai, D. and Ho, W. (2003). *ApJ* **588**, 962–974.
[2] Heyl, J. and Shaviv, N. (2002). *Phys. Rev. D* **66**, 023002.
[3] van Adelsberg, M. and Lai, D. (2006). *MNRAS* **373**, 1495–1522.
[4] Pavlov, G. and Zavlin, V. (2000). *ApJ* **529**, 1011–1018.

27

Polarization from the oscillating magnetized accretion torus

J. Horák & M. Bursa

Astronomical Institute, Academy of Sciences, Prague

We study oscillations of the accretion torus with azimuthal magnetic field. For several lowest-order modes we calculate eigenfrequencies and eigenfunctions and calculate corresponding intensity and polarization light curves using advanced ray-tracing methods.

27.1 Introduction and model description

In addition to spectroscopy, polarimetry provides us with further information about a source of radiation, namely about its geometry. In particular, time-resolved polarimetry may be a useful tool in identifying various types of oscillations in non-stationary sources. In this note we explore this idea for a system consisting of a rotating black hole surrounded by a thick accretion disk (accretion torus).

An analytical description of this set-up was first given by [1] in Newtonian gravity and recently by [2] in Kerr geometry[see also 3]. We differ from the latter work only by employing a polytropic equation of state in the form $P = K\rho^{1+1/n}$, but we use the same relation between magnetic pressure $p_{\rm m}$ and the enthalpy w, $p_{\rm m} = K_{\rm m}\mathcal{L}^{\mu-1}w^{\mu}$ with $\mathcal{L} = g_{t\phi}^2 - g_{tt}g_{\phi\phi}$ and μ being a parameter ($\mu = 2$ in our work). We assume that the angular momentum is constant over the whole volume of the torus. The spatial profiles of the fluid density, pressure and the local magnetic field strength measured in a local comoving reference frame can be expressed using Lane-Emden functions $f(r,\theta)$ as

$$\rho = \rho_0 f^n, \quad p = p_0 f^{n+1} \quad B = B_0\left(\frac{\mathcal{L}}{\mathcal{L}_0}\right)^{\mu-1}\left(\frac{1+nc_{s0}^2 f}{1+nc_{s0}^2}\right)^{\mu} f^{\mu n},$$

X-ray Polarimetry: A New Window in Astrophysics, eds. R. Bellazzini, E. Costa, G. Matt and G. Tagliaferri. Published by Cambridge University Press.

where c_s is speed of sound and all quantities with subscript zero are evaluated at the torus center $[r_0, \pi/2]$. At a given point the Lane-Emden function is a solution of the equation

$$\mathcal{U} - \mathcal{U}_{\rm in} + \ln\left(1 + nc_{\rm s0}^2 f\right) + \frac{\mu nc_{\rm s0}^2 [1 + nc_{\rm s0}^2 f]^{\mu-1}}{(\mu - 1)(n+1)\beta_{\rm p}[1 + nxc_{\rm s0}^2 f]^{\mu}} \left(\frac{\mathcal{L}}{\mathcal{L}_0}\right)^{\mu-1} = 0.$$

where $\mathcal{U} = \ln u_t$ is the relativistic effective potential, $\mathcal{U}_{\rm in}$ is its value at the inner edge of the torus and $\beta_p = p_0/p_{\rm m0}$ is the plasma beta-parameter ($\beta = 1$ in our work).

27.2 Polarization from the stationary torus

Because the fluid is highly magnetized, the dominating emission process in our model is synchrotron radiation that produces polarized light. In the local co-moving reference frame, the direction of polarization is perpendicular to the projection of magnetic field onto the polarization plane (i.e. the plane perpendicular to the direction of emitted photons). Therefore, in a polarization basis $\{\mathbf{X}, \mathbf{Y}\}$, where the $\mathbf{Y}$ vector makes an angle φ with that projection, the frequency-integrated local Stokes emissivities are given by

$$J_I = J_0 \left(\frac{\rho}{\rho_0}\right)\left(\frac{B_\perp}{B_0}\right)^{1+\alpha}, \quad J_Q = pJ_I \cos(2\phi), \quad J_U = -pJ_I \sin(2\phi),$$

where J_0 is a constant, $B_\perp$ is a normalized projection of the magnetic field onto the polarization plane basis and $p = (\alpha + 1)/(\alpha + 5/3)$ for the case of the power-law distribution of radiating electrons with spectral index α.

Due to the global orientation of the magnetic field, the synchrotron emission from the torus shows a certain degree of polarization which depends on the orientation of the torus with respect to a distant observer. When the observer sees the torus from above (along its symmetry axis) the total degree of polarization is zero (or negligible) due to axial symmetry. With increasing inclination the net degree of polarization increases and is the largest for observers looking edge-on (Figure 27.1).

27.3 Polarization changes due to torus oscillations

Accretion tori suffer from various perturbations induced by non-steady accretion of mass. Numerical simulations show that perturbations lead to a development of a variety of oscillating modes in the torus body. In the context of polarization it is interesting to examine what effect these internal oscillations have on produced synchrotron radiation, because tracing temporal changes in polarization can be a possible way to probe the geometry of accretion flows.

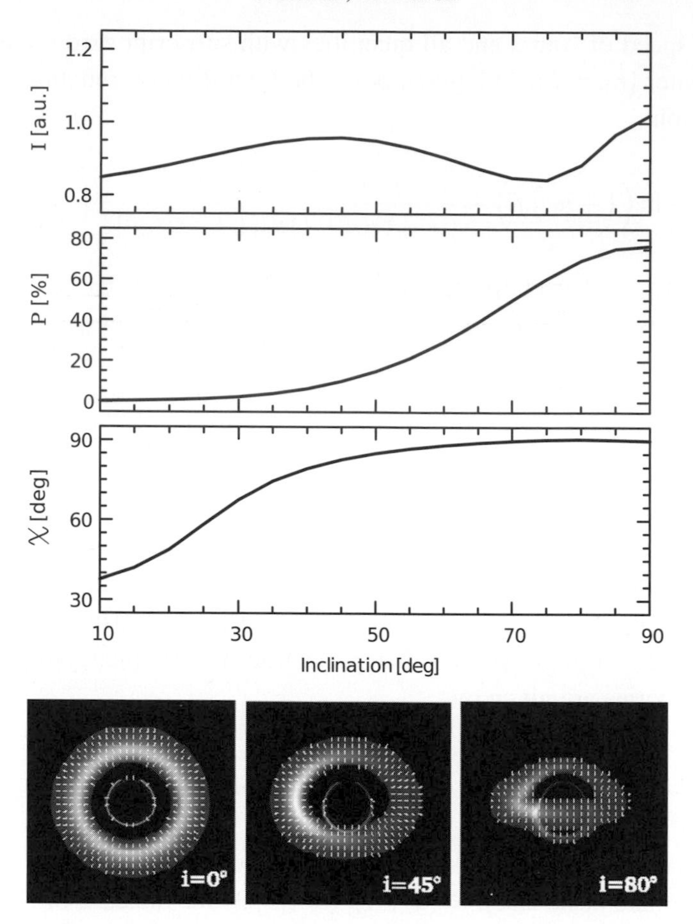

Figure 27.1 The dependence of the total luminosity and of the degree and the angle of polarization on the location of the observer (top). The bottom part of the figure shows three views of the magnetized torus for inclinations 0°, 45° and 80°. On top of the grayscale color map of the emission intensity, there are line segments representing the polarization properties of radiation at a given place: length of lines means the degree, and orientation indicates the angle of polarization.

The equations governing linear perturbations of the torus can be solved analytically in the limit of slender torus, when $c_{s0} \to 0$. In particular, the torus admits two major global modes, radial and vertical epicyclic oscillations, whose poloidal velocity fields are nearly uniform on the torus cross-section, and whose eigenfunctions (in terms of Eulerian velocity perturbation) can be expressed as

$$\delta u^{\alpha} = \mathcal{A}_{\mathrm{r}} \exp[-i(\omega t - m\phi)]\, \delta^{\alpha}_{r}, \quad \delta u^{\alpha} = \mathcal{A}_{\mathrm{v}} \exp[-i(\omega t - m\phi)]\, \delta^{\alpha}_{\theta}, \qquad (27.1)$$

where m is the azimuthal wavenumber. Frequencies of these oscillations are given by $\omega = \omega_{r,\theta} + m\Omega_0$, where Ω_0 and $\omega_{r,\theta}$ are the angular frequency and radial and

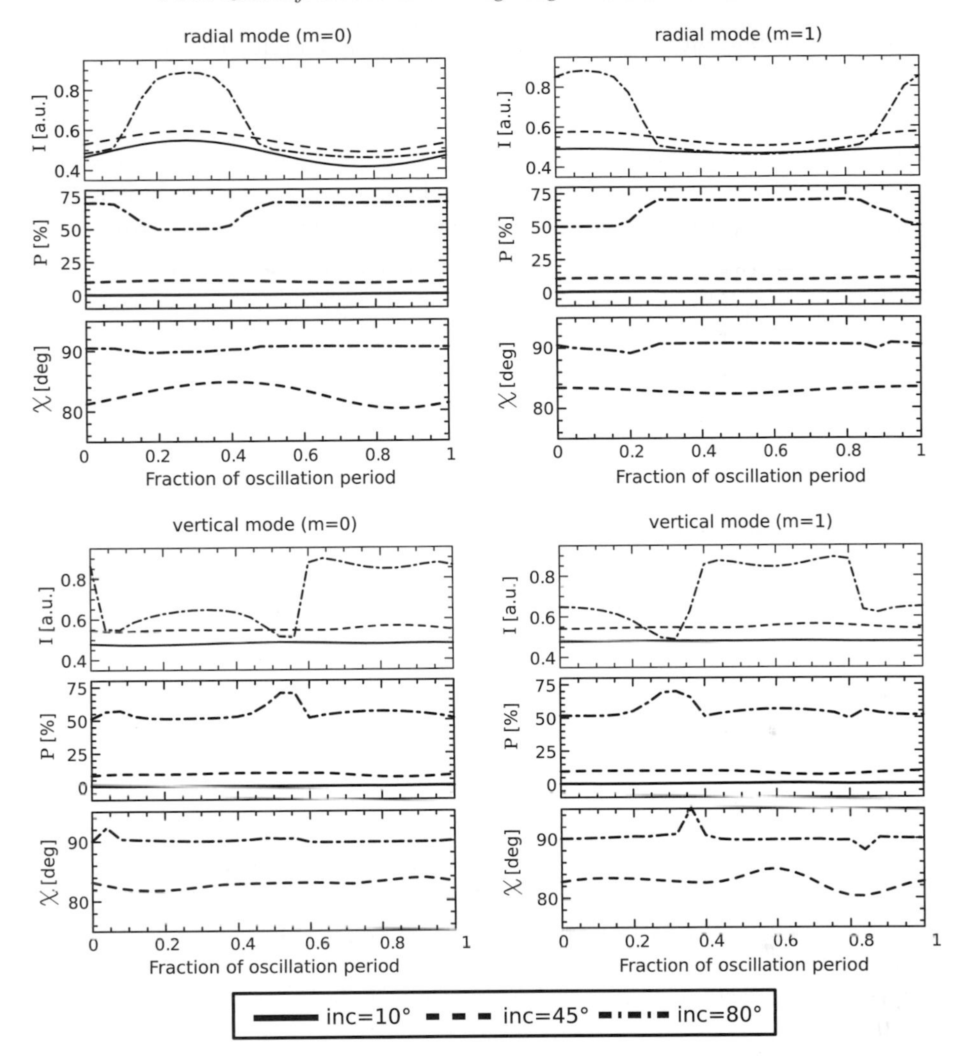

Figure 27.2 The dependence of the total luminosity and of the degree and the angle of polarization on the time phase of some particular oscillation modes: radial and vertical epicyclic modes with wavenumbers $m=0$ and $m=1$. Lightcurves for three inclinations are drawn: 10°, 45° and 80°. Horizontal axes show relative phase of each oscillation – their absolute periods differ. The degree of polarization for 10° inclination is close to zero, hence the corresponding polarization angle curve is not plotted.

vertical epicyclic frequencies at the center of the torus, respectively. Recently, the radial epicyclic mode has been found also in numerical simulations by [3].

Using a numerical ray-tracing technique, we calculate the transport of local Stokes parameters and construct torus light curves as it would appear to a distant observer. We select four distinct oscillation modes, for which we compute time profiles of total intensity, polarization degree and angle. The four modes are radial and vertical epicyclic modes with wavenumbers $m = 0$ and $m = 1$. Resulting lightcurves are summarized in Figure 27.2. The most dramatic changes happen to lightcurves when the source is seen under large inclinations (edge-on). That is because Doppler boosting, gravitational light bending and other relativistic effects are most pronounced in side view. We find that there is a visible difference in temporal evolution of polarization lightcurves between radial and vertical modes. There is, however, little difference between $m = 0$ and $m = 1$ modes (beside some phase shift) of both radial and vertical oscillations. However, these modes can be distinguished by their different oscillation frequencies.

27.4 Conclusions

In our example, one can discriminate between radial and/or vertical oscillation modes easily. On the other hand differences between axisymmetric and non-axisymmetric modes are less apparent – the two cases mainly differ in phases but their phase profiles are quite similar.

Notes

The authors have received support from Czech MSMT grant LC06014 and from project AV0Z10030501.

References

[1] Okada, R., Fukue, J. & Matsumoto, R. (1989). *PASJ* **41**, 133.
[2] Komissarov, S. S. (2006). *MNRAS* **368**, 993.
[3] Montero, P. J., Zanotti, O., Font, J. A., & Rezzolla, L. (2007). *MNRAS* **378**, 1101.

28

X-ray polarization from accreting white dwarfs and associated systems

K. Wu

University College London

A. McNamara

University of Sydney

Z. Kuncic

University of Sydney

We present our results of Monte-Carlo simulations of polarized Compton X-rays from magnetic cataclysmic variables, with realistic density, temperature and velocity structures in the accretion flow. Our study has shown that the X-ray linear polarization may reach about 8% for systems with high accretion rates viewed at a high viewing inclination angle. This value is roughly twice the maximum value obtained by previous studies which assumed a cold, static emission region with a uniform density. We also investigate the X-ray polarization properties of ultra-compact double-degenerate binaries for the unipolar-inductor and direct-impact accretor models. Our study has shown negligible X-ray polarization for the unipolar-induction model. However, the direct-impact accretor model may give X-ray polarization levels similar to that predicted for the magnetic cataclysmic variables.

28.1 Introduction

Magnetic cataclysmic variables (mCVs) and Ultra-compact double degenerate binaries (UCDs) are potential X-ray polarization sources. The mCVs contain a magnetic white dwarf accreting material from a low-mass, Roche-lobe filling companion star. There are two major types: (i) the AM Herculis binaries (AM Hers, also known as polars) and (ii) the intermediate polars (IPs) (see [13]). In AM Hers, the white-dwarf magnetic field ($B \sim 10^7 - 10^8$ G) is strong enough to lock the whole

X-ray Polarimetry: A New Window in Astrophysics, eds. R. Bellazzini, E. Costa, G. Matt and G. Tagliaferri. Published by Cambridge University Press.

system into synchronous rotation. It also prohibits the formation of an accretion disk, and the accretion flow is channelled by the magnetic field into the magnetic polar regions of the white dwarf. The white dwarf in an IP has a weaker magnetic field ($B \sim 10^6$ G). The white-dwarf magnetosphere truncates the inner part of the accretion disk, and the material flow is channelled by the magnetic field from the inner disk rim to the white-dwarf surface. For both AM Her and IP, the supersonic accretion flow becomes subsonic abruptly near the white-dwarf surface, thereby forming an accretion shock. The shock heats up the accreting matter to keV temperature, and X-rays and optical/IR radiation are emitted from the shock-heated gas as it cools and settles onto the white-dwarf surface (see e.g.[16]).

A UCD consists of two white dwarfs revolving around each other in a very tight orbit (which has a linear size of Jupiter). There are debates on what mechanisms generate the X-rays in UCDs. The leading models are the unipolar-inductor (UI) model[19] and the direct-impact accretor (DIA) model[6]. In the UI model, the UCD orbital dynamics are determined jointly by magnetic interaction and gravitational radiation. Magnetic induction similar to that in Jupiter and Io sets up an electric current circuit across the binary. The electrical dissipation at the foot-points of magnetic field lines connecting the two white dwarfs heats up the white-dwarf atmosphere. Because of the convergent field configuration, small hot spots are formed at the surface of the magnetic white dwarf which emit X-rays. In the DIA model, the X-rays are accretion powered as in conventional binary X-ray sources. The close proximity between the two white dwarfs creates a strong tidal interaction and this prohibits the formation of an accretion disk. Mass transfer between the two stars is instead facilitated by a dense material stream. X-rays are emitted from a hot spot where the accretion stream impacts onto the white-dwarf surface. The accretion configuration of the DIA model is somewhat similar to that of AM Hers.

The optical/IR emission from AM Hers is strongly polarized. The circular polarization can be as be as high as several tens percent[17]. The optical/IR polarization is generated by a thermal cyclotron process[1], where energetic electrons in the shock-heated plasma gyrate around the magnetic field. Optical/IR circular polarization has been observed in several IPs (see [11]), but it is much weaker than that of the AM Hers. The origin of the polarization is believed to be the same as that of AM Hers. In the UI model, UCDs are predicted to be strong sources of electron-cyclotron masers[14; 15]. The masers are generated through a loss-cone or a shell instability (see [9]) developed in the electron population when the charged particles stream along the converging magnetic field lines near the polar regions of the magnetic white dwarf. The masers are characterized by a high brightness temperature and almost 100% circular polarization.

Polarization in the low-energy emission of mCVs is produced by processes involving gyration of electrons around a magnetic field. Their X-ray polarization is, however, unrelated to magnetic plasma processes. Instead it is due to scattering, where unpolarized thermal X-ray photons emitted from the shock-heated region are scattered by the electrons that precipitate onto the white dwarf. In this paper we show the results of our calculations of X-ray polarization in mCVs and discuss briefly the astrophysical implications of our findings. We also show some preliminary results from our study of X-ray polarization in ultra-compact double-degenerate binaries.

28.2 Results and discussion

28.2.1 X-ray polarization in magnetic cataclysmic variables

The geometrical setup in our calculations is shown in Figure 28.1. The density, temperature and velocity structure in the post-shock emission region is determined by the hydrodynamics model given in [18; 20]. Unless otherwise stated, thermal free-free emission is the dominant cooling process. We consider a Monte-Carlo approach to simulate the Compton scattering events and the transport of the polarized X-ray photons in the accretion column. A non-linear algorithm based on [2; 3] is used to determine the photon mean-free path and the scattering probability. The scattering

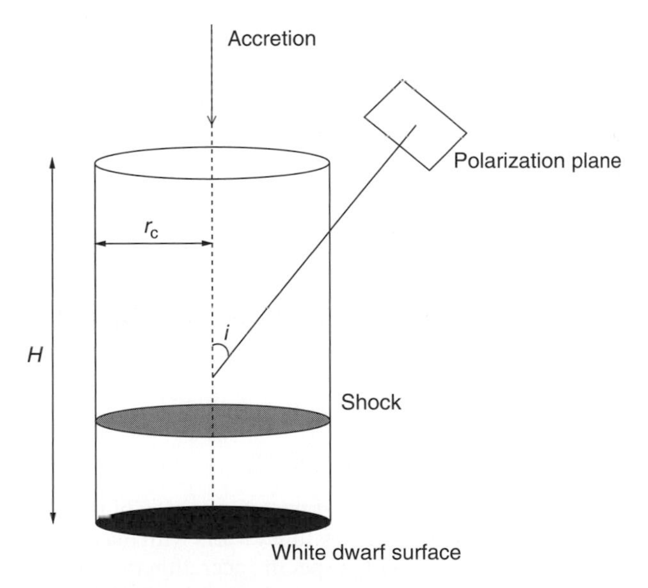

Figure 28.1 An illustration of the emission region in mCVs where the polarized X-rays are generated. The accretion column has a circular cross-section, with a radius r_c. Compton scatterings occur within a height H from the white-dwarf surface.

variables are determined following the prescriptions described in [12]. The polarizations are calculated using the Klein-Nishina cross section and the formulation for photon-electron scattering given in [4; 5]. The polarized photons that escape from the accretion column are binned according to their energy-momentum and summed to give the spectral polarization at specific intervals of viewing inclination angles. The formulation, computational algorithms and numerical simulation procedures are presented in detail in [8].

Figures 28.2 and 28.3 show the results of two example simulations. For the case with a 1.0-$M_\odot$ white dwarf, the maximum value of the linear polarization P_L is

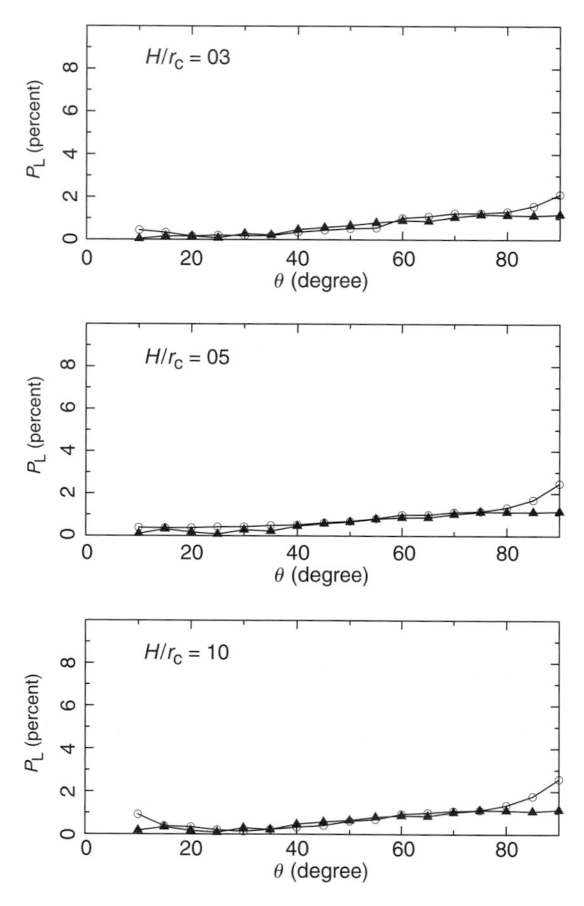

Figure 28.2 Linear polarization P_L of X-rays as a function of the viewing inclination angle θ for accretion onto magnetic white dwarfs with masses of 1 $M_\odot$. Filled triangles correspond to the cases with a specific accretion rate $\dot{m} = 1$ g cm^{-2} s^{-1}; and open circles correspond to the case with ten times higher specific accretion rates. The Thomson scattering optical depths across the accretion shock are $\tau = 0.04$ and 0.3 respectively. The H/r_c ratios are 3, 5 and 10 (panels from top to bottom).

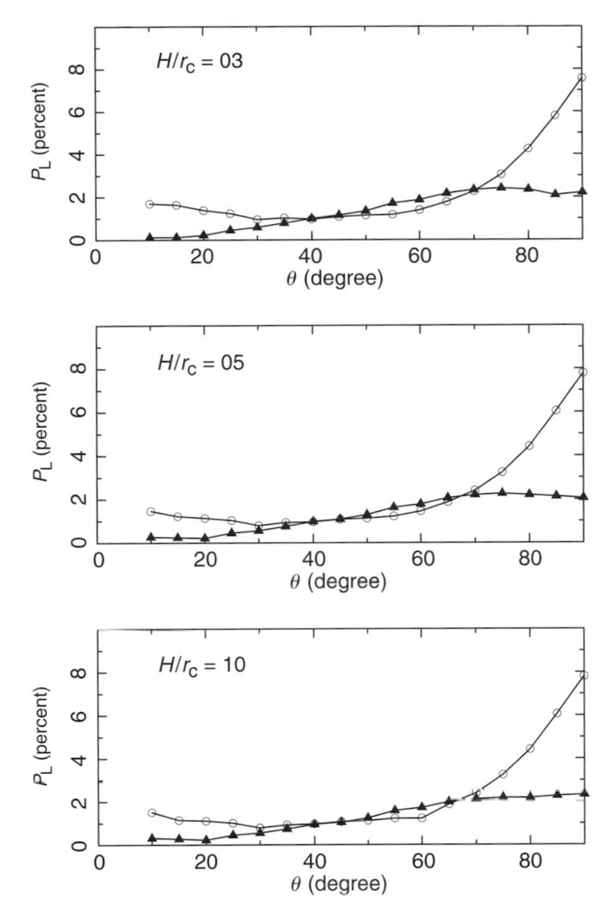

Figure 28.3 Same as Figure 28.2 for white dwarfs with masses of 0.5 $M_\odot$. The Thomson scattering optical depths across the accretion shock are $\tau = 0.1$ and 1.0 for the cases with low and high specific accretion rates respectively.

about 1–2%. For the case with a 0.5-$M_\odot$ white dwarf, the maximum P_L may reach about 8% for viewing angle $\theta \approx 90°$. The polarization increases slightly with θ for low specific accretion rates ($\dot{m} \sim 1$ g cm^{-2} s^{-1}). However, for sufficiently high $\dot{m}$ (~ 10 g cm^{-2} s^{-1}), the polarization could increase substantially at large θ.

Figure 28.4 shows a comparison of the polarization from a structured accretion flow and a cold, static and uniform-density emission region. For low accretion rates, the cold, static, uniform-density emission region underpredicts the linear polarization, especially at large θ. For high accretion rates, it overpredicts the polarization for angles below about 70–80°. The main difference between the two models is that the structured flow model always has a highly dense base region, where most of the scatterings occur, but the static model has a uniform density throughout the

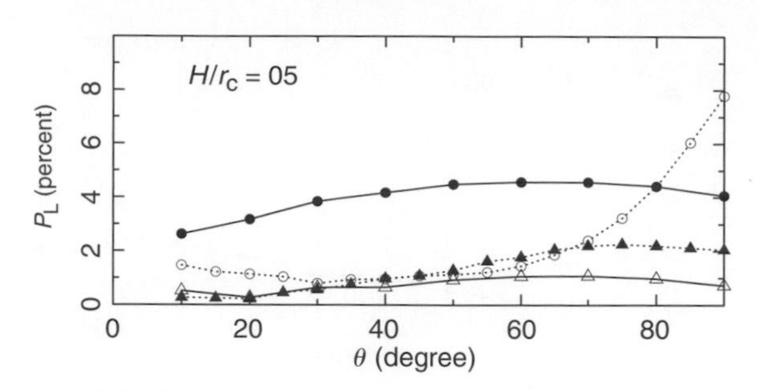

Figure 28.4 Comparison between the X-ray polarizations predicted by a static, cold emission region with a uniform density (solid lines) and by an emission region with structures determined by the hydrodynamic model of [18] (dotted lines). The parameters of the models are the same as those of the case in the middle panel of Figure 28.3. Triangles correspond to cases with $\dot{m} = 1$ g cm^{-2} s^{-1}; and circles correspond to cases with $\dot{m} = 10$ g cm^{-2} s^{-1}.

scattering region. The underprediction of the polarization by the uniform model at low $\dot{m}$ is due to the lack of a dense base layer which gives a substantial scattering optical depth. The overprediction of the polarization by the uniform model at high $\dot{m}$ is caused by a uniformly high scattering optical depth at all heights above the white-dwarf surface, in contrast to the density drop off with height in the stratified accretion flow.

The high density at the base and rapid density drop off with height are also the reasons why the angle dependence of the polarization is insensitive to the accretion-column aspect ratio in the cases with stratified accretion flows. The situation is different for a static, uniform-density scattering region (see [7]), where the effective scattering optical depth depends strongly on the viewing inclination and the aspect ratio of the accretion column. Note the polarization is insensitive to the cyclotron cooling process for the same reason, as scatterings occur mainly in the dense base not the less dense region immediately below the shock (Figure 28.5). Figure 28.6 shows the predicted polarization from a system with the same parameters as those derived for the IP GK Per.

28.2.2 X-ray polarization in ultra-compact double degenerates

The accretion geometry of the DIA UCD is similar to that of the AM Hers, except that an accretion shock might not be formed. In terms of Compton scattering, the model setup would be approximately the same as that of the case with a static, uniform-density accretion column. Although the gas has a bulk motion in the accretion column, the relatively slow speed implies that Compton recoil is the dominant effect in the scattering event. As a first approximation the consequence is not much different from that for the case with cold electrons. The polarization is therefore of the order of 1–4% (see [7]). Thus, as a rough estimate, a polarization of a few

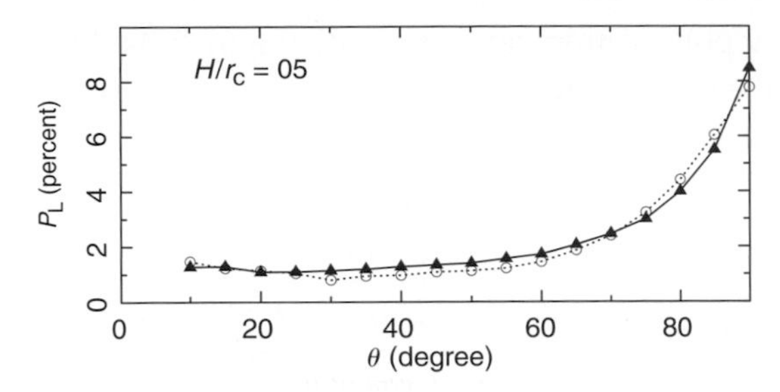

Figure 28.5 Comparison between X-ray polarizations from mCVs with accretion flows dominated by thermal free-free cooling and by cyclotron cooling. The white-dwarf mass is 0.5 $M_{\odot}$, and the specific accretion rate is 10 g $cm^{-2} \times 14\ s^{-1}$. The ratios of efficiencies of cyclotron to thermal free-free cooling at the shock $\epsilon_s = 0$ (open circles) and 10 (filled triangles).

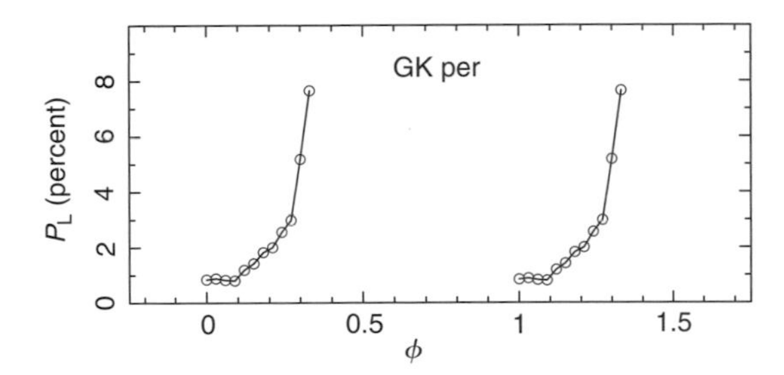

Figure 28.6 Predicted X-ray linear polarization in the mCV GK Per as a function of the white-dwarf spin phase ϕ. In the simulations the white-dwarf mass is set to be 0.63 $M_{\odot}$[10], and the specific accretion rate is assumed to be 10 g $cm^{-2}\ s^{-1}$.

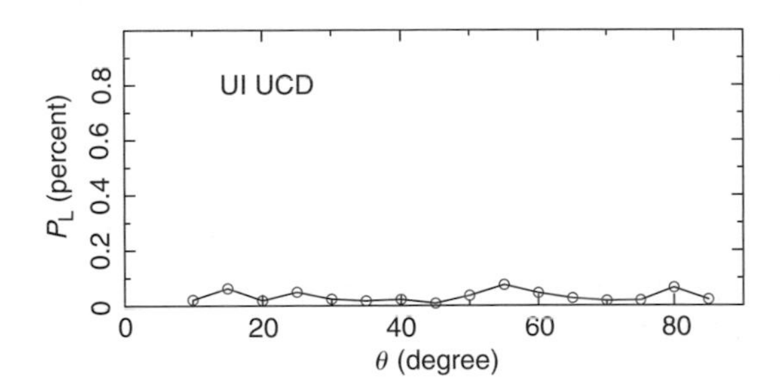

Figure 28.7 Linear polarization as a function of viewing inclination angle for a UI UCD model. In the model, the bulk Lorentz parameter of the fast streaming electrons $\Gamma = 10^3$, the effective Thomson scattering depth across the column $\tau = 10^{-3}$. The H/r_c ratio is 100.

percent would be expected from the DIA UCD model. The UI UCD does not have a dense accretion column. Instead there is a stream of relativistic electrons. The dominant effect is Doppler shift instead of recoil. Moreover, almost all events are head-on as seen by the relativistic electrons. Figure 28.7 shows our simulation for a model UI UCD. As the scattering optical depth is small, there is no substantial polarization. X-ray polarization can clearly distinguish the DIA and the UI model for UCD, despite the fact that the two models could have very similar X-ray spectral and timing properties.

References

[1] Chanmugam, G. & Dulk, D. A. (1981). *ApJ* **244**, 569–578.
[2] Cullen, J. G. (2001). PhD Thesis, University of Sydney.
[3] Cullen, J. G. (2001). *JCoPh* **173**, 175–186.
[4] Heitler, W. (1936). *Quantum Theory of Radiation*. Oxford, Oxford University Press.
[5] Jauch, J. M. & Rohrlich, F. (1980). *The Theory of Photons and Electrons*. Berlin, Springer-Verlag.
[6] Marsh, T. & Steeghs, D. (2002). *MNRAS* **331**, L7–L11.
[7] Matt, G. (2004). *MNRAS* **423**, 495–500.
[8] McNamara, A. L., Kuncic, Z. & Wu, K. (2008). *MNRAS* **386**, 2167–2172.
[9] Melrose, D. A. & Dulk, D. A. (1982). *ApJ* **259**, 844–858.
[10] Morales-Reuda, L., Still, M. D., Roche, P., Wood, J. H. & Lockley, J. J. (2002). *MNRAS* **329**, 597–604.
[11] Piirola, V., Hakala, P. & Coyne, G. V. (1993). *ApJ* **410**, L107–L110.
[12] Pozdnyakov, L. A., Sobol, I. M., Sunyaev, R. A. (1983). *Astrophys. Space Phys. Rev.* **2**, 189–331.
[13] Warner, B. (1995). *Cataclysmic Variables*. Cambridge, Cambridge University Press.
[14] Willes, A. J. & Wu, K. (2004). *MNRAS* **348**, 285–296.
[15] Willes, A. J., Wu, K. & Kuncic, Z. (2004). *PASA* **21**, 248–251.
[16] Wu, K. (2000). *Space Sci. Rev.* **93**, 611–649.
[17] Wu, K. & Wickramasinghe, D. T. (1990). *MNRAS* **246**, 686–698.
[18] Wu, K., Chanmugam, G. & Shaviv, G. (1994). *ApJ* **426**, 664–668.
[19] Wu, K., Cropper, M., Ramsay, G. & Sekiguchi, K. (2002). *MNRAS* **331**, 221–227.
[20] Wu, K., Cropper, M., Ramsay, G., Saxton, C. & Bridge, C. (2003). *ChJAS* **3**, 235–244.

29

Polarization of pulsar wind nebulae

N. Bucciantini

Astronomy Department, University of California at Berkeley

Pulsar wind nebulae (PWNe) were the first objects where non-thermal polarized synchrotron emission was detected. They are one of the best astrophysical *labs* available for the study of high-energy processes like particle acceleration, properties of relativistic outflows, and non-thermal emission. Their broad-band spectrum makes them a suitable target for many instruments, and to date they are the only objects for which there is clear and undisputed evidence for high-energy X-ray polarized emission. In recent years a canonical model has been established which has proved incredibly successful in explaining many of the observed features. All of this makes PWNe a prime candidate for any future X-ray polarimetry study. I will review here the current MHD model, what we know from polarization in the optical and radio band, and what we might learn from next-generation polarimetry.

29.1 Introduction

Pulsar wind nebulae (PWNe) are bubbles of relativistic particles and magnetic field created when the ultra-relativistic wind from a pulsar interacts with the ambient medium, either SNR or ISM. The prototype, and the best studied of this entire class of objects, is the Crab Nebula. The canonical model of PWNe was first presented by Rees & Gunn[23], developed by Kennel & Coroniti[20; 21], and is based on a relativistic MHD description. The pulsar wind is confined inside the SNR, and slowed down to non-relativistic speeds in a strong termination shock (TS). At the shock the toroidal magnetic field of the wind is compressed, the plasma is heated and particles are accelerated to high energies; they are then advected in the post-shock flow as it expands toward the edge of the nebula.

X-ray Polarimetry: A New Window in Astrophysics, eds. R. Bellazzini, E. Costa, G. Matt and G. Tagliaferri. Published by Cambridge University Press.

Despite its simplicity the MHD model can explain many of the observed properties of PWNe (see [2] for a more detailed review). Acceleration at the TS accounts for the continuous, non-thermal, very broad-band spectrum, extending from radio frequencies to X-rays, modeled as synchrotron emission. The under-luminous region, centered on the location of the pulsar, is interpreted as the ultra-relativistic unshocked wind. The emission is highly polarized and the nebular magnetic field is mostly toroidal, as expected from the compression of the pulsar wind. High-energy X-ray-emitting particles have a short lifetime for synchrotron losses, and they are present only in the vicinity of the TS; in contrast radio particles fill the entire volume: this increase of the size of the nebula at smaller frequencies is observed in the Crab Nebula.

The details of the flow structure in the PWN and its emission properties depend critically on the boundary conditions with the SNR. At the beginning the PWN expands inside the cold SN ejecta: this phase, referred to as the *free-expansion phase*, lasts for about 1000–3000 yr, and the pulsar luminosity is high and almost constant. This is the present phase of the Crab Nebula, and PWNe in this phase are expected to shine in high-energy X-rays. Given the relative ratio of typical pulsar kick velocities and the expansion rate of the PWN, one can assume, in modeling this phase, the pulsar to be centrally located.

29.2 The MHD model

Recent optical and X-rays images from HST, CHANDRA, and XMM-Newton have shown that the inner region of PWNe is characterized by a complex axisymmetric structure, generally referred to as *jet-torus structure* (Figure 29.1). This was first observed in Crab, and has subsequently been detected in many other PWNe: a main emission torus, corresponding to the equatorial plane of the pulsar rotation, multiple

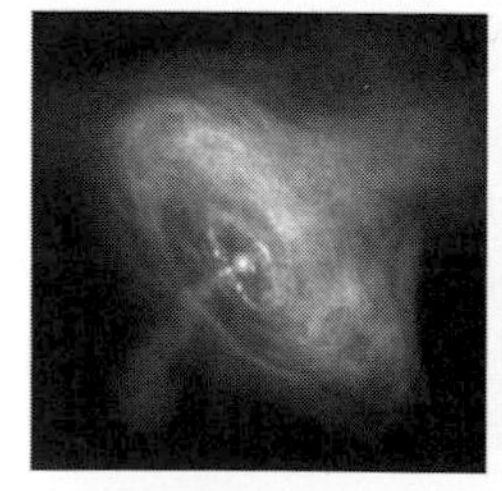   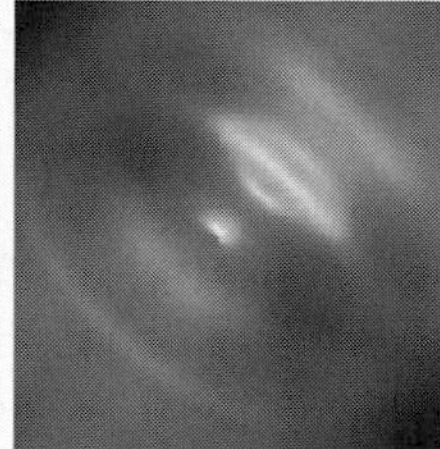

Figure 29.1 From left to right: CHANDRA image of the Crab Nebula, simulated synchrotron emission map in the CHANDRA band based on multidimensional MHD simulations; optical HST image of the central region (wisps, knot and torus) of the Crab Nebula[13]; simulated synchrotron emission map in optical based on multidimensional MHD simulations.

arcs or rings together with a central knot, located in the vicinity of the pulsar, and one or two opposite jets along the polar axis.

Such structure poses several problems for the simple 1D radial model: the difference in brightness between the front and back sides of the torus in Crab Nebula, requires a post-shock flow velocity ~0.3–0.4 c, much higher than what is expected for subsonic expanding flows. Moreover the existence of an inner ring, separated from the torus, is in contrast with the assumption of a smooth flow, and finally the knot seems to be located inside the wind region. The most interesting feature is, however, the jet, because pulsar-wind models[1; 11; 18; 3] show no presence of collimated energetic outflows.

Without going into many details of the current multidimensional MHD models (see [17; 6; 7]) let us summarize the key features. If one assumes the correct force-free latitudinal dependence for the pulsar wind ($L \propto \sin^2\theta, B \propto \sin\theta$[11; 3; 18; 26]), an oblate TS naturally forms, because of the higher equatorial energy flux, with a cusp in the polar region. Given the obliquity of the TS at higher latitudes, the post-shock flow in the nebula can have speeds ~0.3–0.5 c. Hoop-stresses are more efficient in the mildly relativistic nebular flow, and the collimation of a jet occurs in the post-shock region[22; 16]. It is also possible in numerical models to take into account modification of the wind properties associated with oblique rotators by assuming that the magnetic field in the equatorial region of the striped wind is dissipated. Indeed from detailed modeling it seems that this unmagnetized equatorial region is required to explain the observational aspect of Crab Nebula.

Figure 29.1 shows a comparison between observation and simulated emission maps based on the result of MHD nebular models. The emission is computed by assuming that particles are accelerated at the TS, and then evolved as they flow in the nebula to account for adiabatic and synchrotron losses. Results show that the MHD model can recover all of the observed axisymmetric features both in optical and X-ray. It can also recover the X-ray spectral index maps as well as the integrated spectrum. The Doppler boosted torus and wisps, the knot, the jet, and several other features are correctly reproduced in size, location, and brightness.

It is known that PWNe show a short time variability at high energies. The best examples are the wisps in the Crab Nebula which show variability on a time-scale of months, much shorter than typical sound crossing times, in the form of an outgoing wave pattern. MHD nebular models recover such variability: the shear between flow channels causes the formation of eddies on scales typical of the TS, that are subsequently advected away from the TS. These eddies in turn change the shape of the TS, and the Doppler boosting responsible for the arches and rings. The typical duty-cycle of about one year and the outgoing wave pattern are recovered.

29.3 Polarization

Synchrotron emission in the strong nebular field makes these objects the ones with the highest degree of linear polarization observed. Radio maps[27; 8; 19] clearly show the toroidal structure of the field to high accuracy (interestingly being a pair plasma, there are no dispersive effects like Faraday rotation associated with the nebular field), but the polarized fraction is far from the maximal value. This is because the radio emitting particles sample the entire nebular volume. While it is expected that the internal flow should preserve the toroidal geometry of the field, it is unlikely that the interaction with the SNR will act similarly. On the contrary, the role of large-scale instability at the boundary of the nebula, or internal instability of the flow that grows as fluid moves away from the TS, are thought to be responsible for the considerable degree of depolarization that is observed at low frequency.

Optical resolved spectroscopy might be of great help. Unfortunately the only object with bright optical emission is the Crab Nebula. Polarized optical maps of Crab have been recently published by Hester[14], using HST. There are interesting results that confirm previous observations[24]: the polarized map shows a filamentary structure, quite distinct from the unpolarized one, which is interpreted as a signature of magnetic toroidal flux-tubes. Polarized intensity is maximal in the wisps-knot region reaching values of $\sim$50% in the knot and $\sim$30–20% in the wisps, with lower values in the torus ($\sim$10%). The base of the jet appears to be completely depolarized, despite being bright, and in general the brightest regions do not coincide with those with higher polarization. Interestingly, these results also outline the importance of high resolution: low-resolution polarimetry results in a polarized structure that has nothing to do with the nebula, and only by careful investigation of small resolved features, and foreground subtraction, it is possible to derive the local magnetic field structure.

Recently Słowikowska et al.[25], have presented phase-resolved optical spectroscopy of the Crab Pulsar. The data show the presence of a continuous contribution, with a very high degree of polarization ($\sim$30%), and polarization angle aligned with the nebular axis within few degrees, that confirms previous results[24]. It is more likely that such emission comes from the knot (which is not resolved from the pulsar), but one cannot exclude a magnetospheric or wind contribution.

Results at high energy are not of the same quality. Recent Integral (IBIS/SPI[10; 5]) phase-resolved data in the 100 keV–1 MeV range confirm the presence of a continuous polarized component, with high degree of polarization $\sim$50–80%, and polarization angle aligned with the nebular axis, fully consistent with the optical data. Given that particles responsible for emission in the INTEGRAL band have very short synchrotron lifetime ($<$1 yr), it is likely that, despite the low angular resolution of the instrument, the observations are sampling a small region downstream of the TS, possibly coincident with the wisps and/or knot. The nebular interpretation for the optical and hard X-ray polarization is consistent with the observations. Interestingly, if both optical and hard X-ray polarization have the same origin (the fact that they trace the symmetry axis of the nebula confirms this point), then it is possible to constrain quantum gravity theory[15; 9] to higher accuracy than from existing optical GRBs measures.

The only data in the Chandra band $\sim$1–10 keV, dates back to Weisskopf et al.[28]. In contrast with other measures, they show a lower degree of polarization $\sim$20% and a different angle of about 30°, not consistent with the nebular axis, but consistent with the low-resolution optical data by Hester[14]. Similar results have been recently obtained by the PHENEX experiment[12]. The possible interpretation of this discrepancy lies in the fact that none of these observations is able to resolve the nebula. The low degree of polarization, consistent with optical measures in the torus, might be due to the depolarizing effect at the edge of the nebula, which still contributes to the emission at those energies, while, as seen in Figure 29.2 the lack of resolution is most likely responsible for the inferred polarization angle.

Bucciantini et al.[4] were the first to present a study of the polarization properties derived using the MHD nebular model of PWNe. The high degree of polarization that is predicted on axis is an artifact of the perfect axisymmetry of the model. However, there are key properties of the polarization that properly match the observations. Polarized fraction is very high $\sim$50–60% in the knot and wisps region, and there is a depolarized region between the wisps and the torus, as seen in optical. Unfortunately it is not possible to test the polarization angle swing expected from the high-speed wisps because of the lack of sufficient resolution in soft X-ray.

The decreasing polarization of the emission coming from larger distances from the central PSR, as seen in optical, and the lower soft X-ray polarization, can also

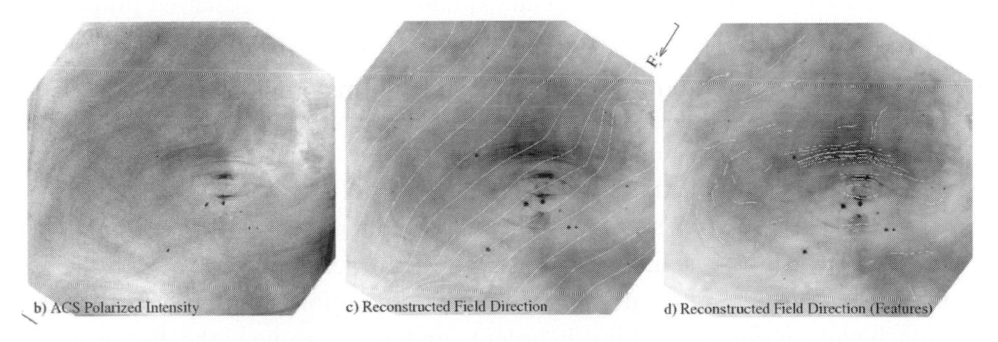

Figure 29.2 Optical polarization in the Crab Nebula[14]. From left to right: polarized intensity; total intensity with polarization angle (magnetic field direction) at low resolution; polarization angle (magnetic field direction) of selected features after background and foreground subtraction[14].

be explained if one considers that any instability that might destroy the ordered structure of the magnetic field might grow only toward the edge of the nebula, leaving the central region less affected. The depolarization of the bright base of the jet, also fits the model: the base of the jet, where nebular flow tends to converge, is quite dynamic and the MHD model suggests that dissipation might occur in that region.

Variability is also important, and perhaps it might explain some of the discrepancies between similar observations in the hard X-ray band and the optical, which were conducted at different epochs.

29.4 Conclusion

Pulsar wind nebulae are perhaps the best sources where linear polarization at high energies can be studied. Unfortunately the lack of specifically designed instruments, and in particular the lack of high angular resolution, makes data interpretation more difficult. Recent successes in the optical band clearly show the possibility of high angular resolution spectroscopy. In particular, it has been realized in recent years that the synchrotron luminosity is not a tool well-suited to the investigation of the properties of magnetic fields, because it mostly depends on the internal velocity structure. There are several indications that suggest that instabilities might destroy the ordered magnetic field at the base of the jet, or in the torus, that only polarization studies can address. While it is unlikely that any X-ray polarimeter could achieve CHANDRA resolution in the near future, it is still possible to derive useful information, and to study particular features (both axisymmetric and not) using the more modest resolution of IXO (5 arcsec) or perhaps even the lower values expected for POLARIS (15 arcsec).

An advantage in studying PWNe comes from the fact that we have a well-established and reliable model that can explain many of the observed properties. Theoretical models can in this sense guide the observation and the interpretation of the results much better than for other objects. We also need to consider that PWNe are persistent objects (unlike GRBs) and there is always the possibility of improving the data quality with long-term observations. The fact that these system are also broad-band allows one to conduct polarization study over a large energy range, to test for consistency. Perhaps a combination of low-resolution X-ray polarimetry with high-resolution X-ray imaging (in order to constrain the brightness contribution of different features at the time of observation) might provide the best approach with current technology, in order to understand some of the discrepancies of recent results.

The current MHD model of PWNe makes a strong prediction: that the fluid properties of any feature (i.e. magnetic field structure), should be the same, no matter at what energy the feature is observed. If different polarization properties were to be seen in X-ray with respect to other bands, this would challenge the current model, and force new theoretical inquiries into the properties of PSR-PWN systems.

Acknowledgments

NB was supported by NASA through Hubble Fellowship grant HST-HF-01193.01-A, awarded by the Space Telescope Science Institute, which is operated by the Association of Universities for Research in Astronomy, Inc., for NASA, under contract NAS 5-26555.

References

[1] Bogovalov, S. V. (2001). *A&A* **371**, 1155.
[2] Bucciantini, N. (2008). *40 Years of Pulsars: Millisecond Pulsars, Magnetars and More, AIP Conf. Proc.*, **983**, 186.
[3] Bucciantini, N., et al. (2006a), *MNRAS* **368**, 1717.
[4] Bucciantini, N., Del Zanna, L., Amato, E., & Volpi, D. (2005b). *A&A* **443**, 519.
[5] Dean, A. J., et al. (2008). *Science* **321**, 1183.
[6] Del Zanna, L., Amato, E., & Bucciantini, N. (2004). *A&A* **421**, 1063.
[7] Del Zanna, L., Volpi, D., Amato, E., & Bucciantini, N. (2006). *A&A* **453**, 621.
[8] Dodson, R., Lewis, D., McConnell, D., & Deshpande, A. A. (2003). *MNRAS* **343**, 116.
[9] Fan, Y.-Z., Wei, D.-M., & Xu, D. (2007). *MNRAS* **376**, 1857.
[10] Forot, M., et al. (2008). *ApJL* **688**, L29.
[11] Gruzinov, A. (2005). *Phys. Rev. Let.* **94**, 021101.
[12] Gunji, S., et al. (2007). *SPIE* 6686.
[13] Hester, J. J., et al. (1995). *ApJ* **448**, 240.
[14] Hester, J. J. (2008). *ARA&A* **46**, 127.
[15] Kaaret, P. (2004). *Nature* **427**, 287.
[16] Khangoulian, D. V. & Bogovalov, S. V. (2003). *Astron. Let.* **29**, 495.
[17] Komissarov, S. S. & Lyubarsky, Y. E. (2004). *MNRAS* **349**, 779.
[18] Komissarov, S. S. (2006). *MNRAS* **367**, 19.
[19] Kothes, R., Reich, W., & Uyanıker, B. (2006). *ApJ* **638**, 225.
[20] Kennel, C. F. & Coroniti, F. V. (1984a). *ApJ* **283**, 710.
[21] Kennel, C. F. & Coroniti, F. V. (1984b). *ApJ* **283**, 694.
[22] Lyubarsky, Y. E. (2002). *MNRAS* **329**, L34.
[23] Rees, M. J. & Gunn, J. E. (1974). *MNRAS* **167**, 1.
[24] Schmidt, G. D., Angel, J. R. P., & Beaver, E. A. (1979). *ApJ* **227**, 106.
[25] Słowikowska, A., Kanbach, G., Kramer, M., & Stefanescu, A. (2009). [arXiv:0901.4559].
[26] Spitkovsky, A. (2006). *ApJL* **648**, L51.
[27] Velusamy, T. (1985). *MNRAS* **212**, 359.
[28] Weisskopf, M. C., et al. (1978). *ApJL* **220**, L117.

30

X-ray polarization of gamma-ray bursts

D. Lazzati

Department of Physics, North Carolina State University, Raleigh

The degree and the temporal evolution of linear polarization in the prompt and afterglow emission of gamma-ray bursts is a very robust diagnostic of some key features of gamma-ray-burst jets and their micro and macro physics. In this contribution, I review the current status of the theory of polarized emission from GRB jets during the prompt, optical flash, and afterglow emission. I compare the theoretical predictions to the available observations and discuss the future prospects from both the theoretical and observational standpoints.

30.1 Introduction

Gamma-ray bursts (GRBs) are the brightest explosions in the present day Universe. Unfortunately, our understanding of their physics is still incomplete, probably due to the fact that they are short-lived, point-like sources.

Polarization is a formidable tool to improve our understanding of GRB jets: their geometry, magnetization, and radiation mechanism could in principle be pinned down with a comprehensive and time-resolved analysis of linear polarization. Observationally speaking, however, polarization is not easy to measure. So far, only the optical afterglow has robust polarization measurements[4; 9] but the diverse features and the sensitivity of the models to detail has made their interpretation, at best, controversial.

In this review, I describe the theory underlying the production of polarized radiation in GRBs in their three main phases. I will focus on X-ray polarization but the discussion will be general, since the frequency dependence of GRB polarization is very weak, especially at frequencies where Faraday rotation is not relevant.

X-ray Polarimetry: A New Window in Astrophysics, eds. R. Bellazzini, E. Costa, G. Matt and G. Tagliaferri. Published by Cambridge University Press.

30.2 Prompt emission

The interest in the properties of linear polarization in the prompt emission of GRBs increased dramatically with the claim that the prompt emission of GRB 021206 had a linear polarization fraction of ~80%[3]. Even though the claim was subsequently put into doubt[20; 24], it generated a suite of models against which any subsequent polarization observation will be compared. In this section we present these models and compare them to observations.

- **Toroidal magnetic field model –** Gamma-ray burst jets are commonly believed to be produced by the effects of strong magnetization (either of a neutron star[2] or of a massive accretion disc onto a black hole[17]) combined with fast spinning. In such conditions, sufficiently far from the jet engine, the magnetic field is expected to be predominantly toroidal.

  Synchrotron from a toroidal field configuration does not produce polarized radiation in normal conditions. However, due to relativistic aberration, only a small section of the whole toroidal structure produces the radiation that is detected by an observer at infinity. As a consequence, the radiation observed appears to come from a region of uniform magnetic field and is maximally polarized (see Figure 30.1). Due to additional aberration effects, the polarization angle direction is distorted in the edges of the visible zone, slightly reducing the maximum detectable polarization[16].

  Independently of details, this model predicts that almost all GRBs are strongly polarized during their prompt emission. The polarization position angle does not change with

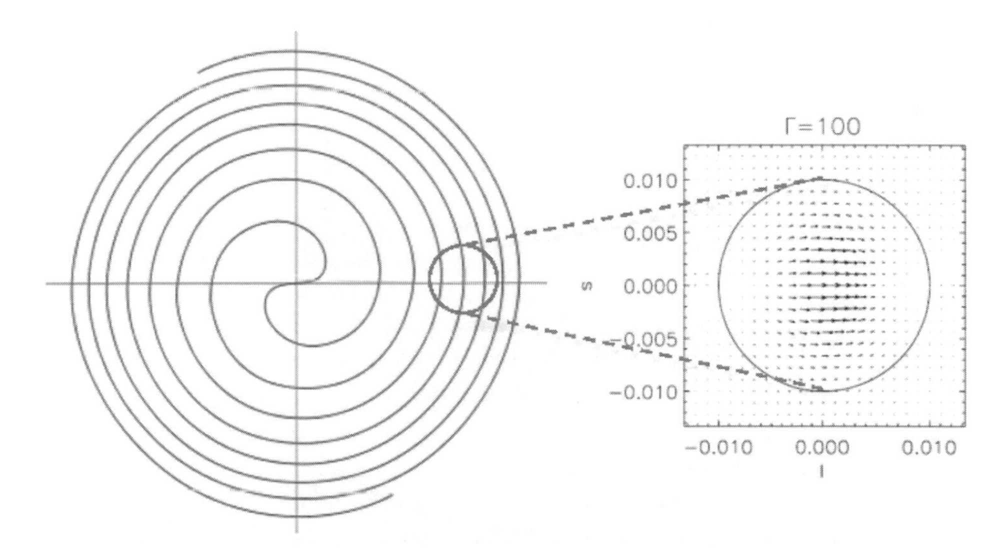

Figure 30.1 Front view of the toroidal magnetic field in a fireball. The circle highlights the fact that, due to relativistic aberration, only a small fraction of the fireball is visible to the observer at infinity. Additional relativistic aberration effects reduce the maximum polarization, as shown in the right sub-panel[16].

time, since the electric field vector always points towards the pole of the field, i.e. the jet axis. Only a very small fraction of bursts, those seen within an angle $1/\Gamma$ from the jet axis, should display little to no polarization.

- $1/\Gamma$ **viewing angle effects –** The main reason why most models for GRB polarization predict small values of polarization (few to ten percent) is due to the fact that typically an observer collects radiation from different regions with different polarization orientations and the net signal is small. If, however, the fireball configuration is such that only one emission zone is observed, high polarization can be detected by the observer at infinity[23]. Consider a fireball with an opening angle $\theta_j \sim 1/\Gamma$ observed at an angle $\theta_o = 1/\Gamma$ from its edge. Due to relativistic aberration, in the co-moving frame the fireball velocity and the line of sight are at a right angles. Both synchrotron radiation from a planar magnetic field[6; 21; 8] and bulk inverse Compton radiation[14] are maximally polarized in that configuration.

  This model can in principle account for polarization up to 100 percent. Differently from the toroidal field model, only a small fraction of GRBs should be polarized, due to the low probability for the particular viewing configuration to be attained. As discussed for the toroidal model the electric vector points towards the jet axis and the polarization angle is therefore constant throughout the prompt emission evolution.

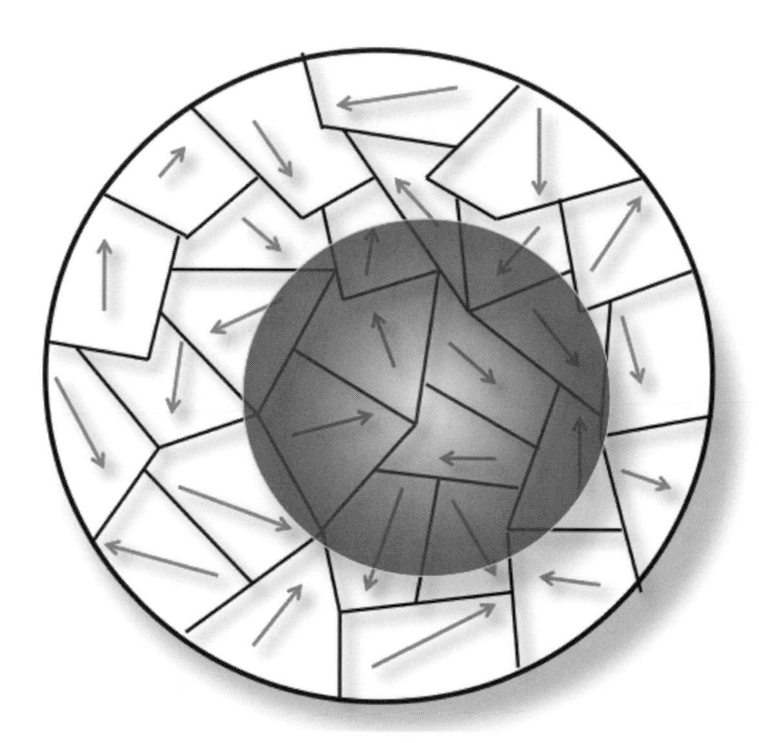

Figure 30.2 Cartoon of the front view of a fireball with magnetic domains. The arrows show the direction of the field in the domains. The shaded circle emphasizes the fact that only some of the domains are visible by the observer.

- **Magnetic domains** – If the magnetic field generated by a relativistic collisionless shock can reorganize into a uniform configuration, the fireball surface would be covered with magnetic patches, each with a different field orientation, but with a uniform field within[10]. As a result of the speed of the field re-organization and of relativistic aberration, the observer at infinity sees radiation from approximately $N \sim 100$ domains. The resulting net polarization is therefore reduced by a factor $\sqrt{N} \sim 10$. This model predicts that all GRBs should be mildly polarized (in the 10 percent range), with rapid fluctuations of the polarization angle. The model cannot account for very high polarized fractions, like those possibly observed in the prompt emission of GRBs.
- **Fragmented fireballs** – The main weakness of the "$1/\Gamma$ effects" model is that it requires a very unlikely viewing configuration. Such limitation would not be present if the fireball is fragmented in shotguns[11], cannonballs[5], or mini-jets[25]. If we model a fireball as a series of identical fragments, each producing radiation with the same efficiency and moving at the same speed, polarization and intensity from a fragment are strictly correlated[15]. The brightest light observed comes from the fragment that is exactly pointing at the observer. Due to cylindrical symmetry, the radiation is unpolarized. At the $1/\Gamma$

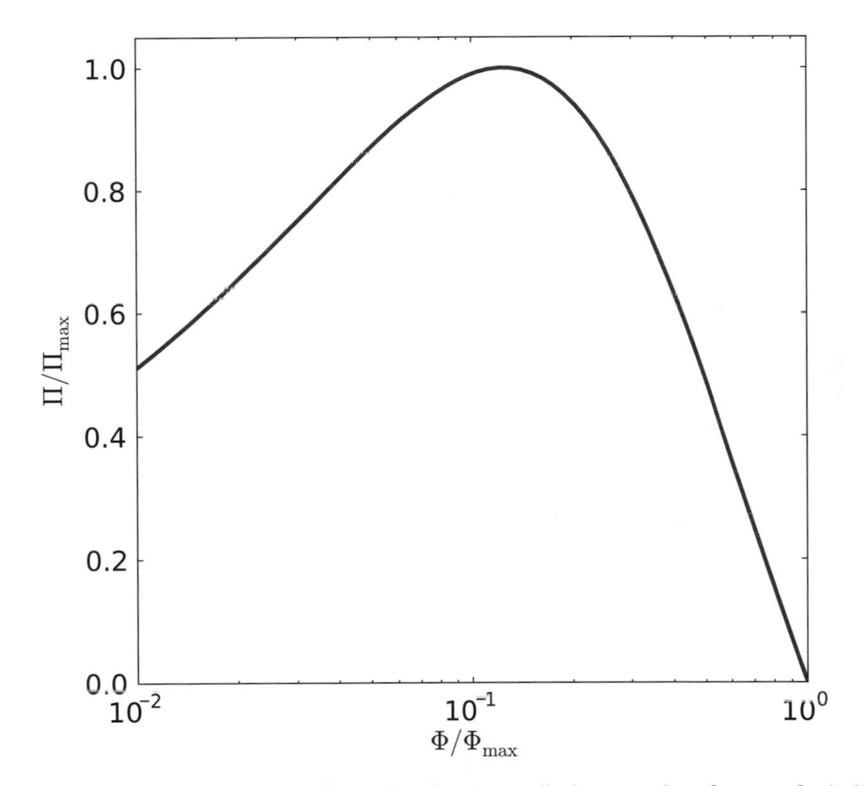

Figure 30.3 Polarization vs. intensity for the radiation coming from a fireball made by a large number of identical fragments with negligible opening angle ($\theta_{jet} \ll 1/\Gamma$).

configuration, the radiation intensity is decreased by a factor ~ 10, and the polarization is maximum. For viewing angles $\theta_o > 1/\Gamma$, both the intensity and the polarization decrease (see Figure 30.3). Most bursts from fragmented fireballs are highly polarized if a time-resolved analysis is performed, but they are weakly polarized if the whole prompt emission is considered. That is because the electric vector points towards each fragment, and so the position angle fluctuates randomly from a pulse to the next.

30.3 Afterglow

The polarization of afterglow radiation has been observed with robust results, but the comparison of observational data with models is difficult. Afterglow radiation is known to be produced by synchrotron from relativistic electrons gyrating into a shock-generated magnetic field. Detailed calculations show that the polarization from a uniform fireball is intimately connected to the evolution of the light curve[6; 21] and is very weakly dependent on the frequency of photons (at least above optical frequencies[19]). Initially, the polarization is vanishingly small. At times before the jet break, a small polarization of a few percent is observed, with a position angle perpendicular to the direction towards the jet axis. At a time approximately coincident with the jet break time, the polarization vanishes again. Subsequently, it reappears, rotated by 90°, reaches a maximum of ~ 10 percent and eventually vanishes again (see Figure 30.4).

The characteristic behavior of a 90° rotation of the polarization angle at a time roughly coincident with the jet break time is in principle a formidable prediction and was actively looked for in observations, with no success[12; 13]. It was

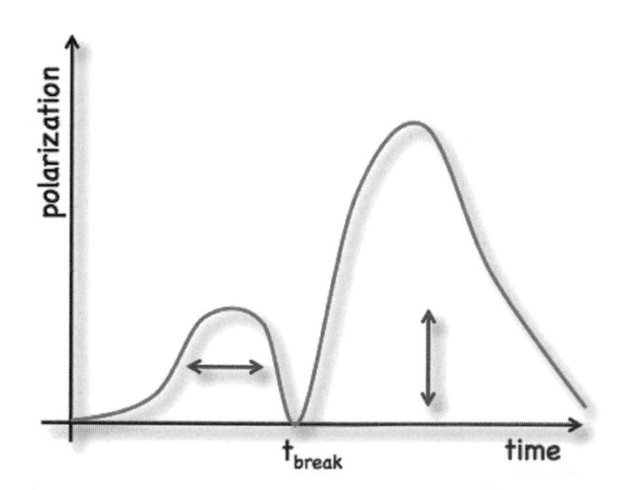

Figure 30.4 Polarization of afterglow radiation from a uniform fireball. The arrows indicate the direction of polarization.

subsequently realized that the polarization curve is very sensitive to the brightness profile of the fireball, and that fireballs with a bright core and less energetic wings produce a completely different polarization curve, with maximum polarization around the break time and a constant position angle[19]. Even more complicated is the case of a fireball with bright spots randomly distributed on the emitting surface. The polarization in that case would be virtually unpredictable.

30.4 Early afterglow

The polarization of the prompt emission is in principle full of important information to understand the physics of GRB jets. However, polarization in the X-ray and γ-ray regimes is hard to observe. Optical polarization is relatively easy to observe. However, models are too sensitive to details and we have not been able to obtain many robust clues from optical polarization measurements. A potential game changer is polarization of the early optical afterglow, also known as the optical flash. The optical flash is believed to be due to electrons in the fireball energized by the reverse shock[22]. If that is the case (see [1] for alternative models) the optical flash should have the same polarization characteristic as the prompt emission (and therefore carry a lot of insight) combined with the same ease of observation as the afterglow polarization[13].

30.5 Discussion

After discussing the polarization of the various stages of GRB emission in detail, we here compare them with each other and with observations and focus more on the X-ray aspects and future perspectives. Prompt emission polarization is certainly the most appealing from the theoretical point of view. Models are able to deliver univocal interpretation for the various observational scenarios. The few available observations are, however, inconclusive and contradictory. Early observations claimed a high polarization for the overall burst[3]. More recent observations find, instead, that the polarization is indeed large, but the position angle varies from pulse to pulse[7]. The observations are different and so too are the implications. Constant position angle and high polarization point to a toroidal magnetic field model, while variable angle is indicative of a fragmented fireball scenario. The perspective of an early afterglow polarization measurement is exciting, but the optical flash has no emission in the X-rays, and its theoretical interpretation is still a matter of open debate. While a positive measurement of large polarization would be interesting, a no-polarization result, as the one for GRB 060418[18], would be open to very many interpretations. Afterglow observations are plagued by the model sensitivity to details, with the notable exception of polarization of the X-ray flares (Fan, this volume) which are supposed to be due to engine activity and could therefore be polarized in the same way as the prompt emission.

At the end of the day, what would be the best choice for an X-ray polarimeter? Disregarding technical challenges, a theoretician would try to observe prompt emission first, X-ray flashes second, and the afterglow emission only as a last resort.

References

[1] Beloborodov, A. M. (2002). *ApJ* **565**, 808–828.
[2] Bucciantini, N., Quataert, E., Arons, J., Metzger, B. D. & Thompson, T. A. (2008). *MNRAS* **383**, L25–L29.
[3] Coburn, W. & Boggs, S. E. (2003). *Nature* **423**, 415–417.
[4] Covino, S. et al. (1999). *A&A* **348**, L1–L4.
[5] Dado, S. & Dar, A. (2009). [arXiv:0901.4260].
[6] Ghisellini, G. & Lazzati, D. (1999). *MNRAS* **309**, L7–L11.
[7] Götz, D., Laurent, P., Lebrun, F., Daigne, F., & Bošnjak, Ž. (2009). *ApJ* **695**, L208–L212.
[8] Granot, J. (2003). *ApJ* **596**, L17–L21.
[9] Greiner, J. et al. (2003). *Nature* **426**, 157–159.
[10] Gruzinov, A. & Waxman, E. (1999). *ApJ* **511**, 852–861.
[11] Heinz, S. & Begelman, M. C. (1999). *ApJ* **527**, L35–L38.
[12] Lazzati, D. et al. (2003). *A&A* **410**, 823–831.
[13] Lazzati, D. et al. (2004). *A&A* **422**, 121–128.
[14] Lazzati, D., Rossi, E. M., Ghisellini, G., & Rees, M. J. (2004). *MNRAS* **347**, L1–L5.
[15] Lazzati, D. & Begelman, M. C. (2009). *Submitted to ApJL.*
[16] Lyutikov, M., Pariev, V. I., & Blandford, R. D. (2003). *ApJ*, **597**, 998–1009.
[17] MacFadyen, A. I. & Woosley, S. E. (1999). *ApJ* **524**, 262–289.
[18] Mundell, C. G. et al. (2007). *Science* **315**, 1822.
[19] Rossi, E. M., Lazzati, D., Salmonson, J., & Ghisellini, G. (2004). *MNRAS* **354**, 86–100.
[20] Rutledge, R. E. & Fox, D. B. (2004). *MNRAS* **350**, 1288–1300.
[21] Sari, R. (1999). *ApJ* **524**, L43–L46.
[22] Sari, R. & Piran, T. (1999). *ApJ* **520**, 641–649.
[23] Waxman, E. (2003). *Nature* **423**, 388–389.
[24] Wigger, C., Hajdas, W., Arzner, K., Güdel, M. & Zehnder, A. (2004). *ApJ* **613**, 1088–1100.
[25] Yamazaki, R., Ioka, K., Nakamura, T., & Toma, K. (2006). *Advances in Space Research* **38**, 1299–1302.

31

Central engine afterglow from GRBs and the polarization signature

Y. Z. Fan

Niels Bohr International Academy, University of Copenhagen, Denmark

There are two kinds of gamma-ray burst (GRB) afterglows. One is the *fireball afterglow* that is the radiation of the external shock(s) driven by the GRB remnant. The other is the emission from the late ejecta launched by the prolonged activity of the central engine, i.e. the *central engine afterglow*. The former seems to be only weakly polarized and thus is not suitable for the upcoming X-ray polarimetry. For the latter, the polarization property is less clear. Some central engine afterglows, such as energetic flares and the plateau followed by a sharp drop, might be highly polarized because the outflows powering these behaviors may be Poynting-flux dominated. Furthermore, the breakdown of the symmetry of the visible emitting region may be hiding in some X-ray data and will give rise to interesting polarization signatures. For example, in the high latitude emission model for the sharp X-ray decline strong polarization evolution is possible. An *XRT*-like detector but with polarization capability on board a *Swift*-like satellite would be suitable to detect these possible signals.

31.1 The central engine afterglow

In the context of the standard fireball model of GRBs, the prompt γ-rays and the afterglow emission are powered by internal shocks and external shocks, respectively (see [17] for a review). Before 2004, most of the afterglow data were collected hours after the prompt γ-ray emission and were found to be consistent with the external forward shock model, though at times energy injection, a wind medium profile, or a structured/patchy jet were needed. We call the emission powered by external shocks the *"fireball afterglow"* or "afterglow". An alternative possibility

X-ray Polarimetry: A New Window in Astrophysics, eds. R. Bellazzini, E. Costa, G. Matt and G. Tagliaferri. Published by Cambridge University Press.

for the production of the afterglow is the continued activity of the central engine via either the "late internal shocks" or "late magnetic dissipation". This idea has been put forward by Katz, Piran & Sari already in 1998 to interpret the afterglow of GRB 970228[14]. However, the agreement of the predictions of the external shock afterglow model[20] with most subsequent multi-wavelength afterglow observations strongly disfavors the *central engine afterglow* model. The central engine afterglow model is short of predictive power. The fireball afterglow model, instead, has predicted smooth light curves and in particular the intrinsic relation between the flux in different bands (e.g., [20]) as well as between the spectral slopes and the temporal decay. In particular, for fireball afterglow, there are *two very general constraints*: (1) The ratio between the variability timescale of the emission δt and the occurrence time t has to be ≥ 1[16]. (2) The decline of the fireball afterglow emission cannot be steeper than $t^{-(2+\beta)}$ (where β is the spectral index) unless the edge of the GRB ejecta is visible. This is because the GRB outflow is curving and emission from high latitude (relative to the line of sight) will reach us at later times and give rise to a decline shallower than $t^{-(2+\beta)}$[15]. These limitations do not apply, of course, to a central engine afterglow (see Figure 31.1). For example, for the X-ray flares, the duration ($\sim \delta t$) is determined by the re-activity process of the central engine and can be much shorter than the occurrence time. On the other hand, since the ejection time ($\sim t_{\rm eje}$) of the last main pulse of the flare is close to its peak time $t_{\rm p}$, the net flux of the high latitude emission of the pulses can be approximated by $(t_{\rm p} - t_{\rm eje})^{-(2+\beta)}$, which can be much steeper than $t^{-(2+\beta)}$ (see also [10]).

There are several kinds of central engine afterglows.

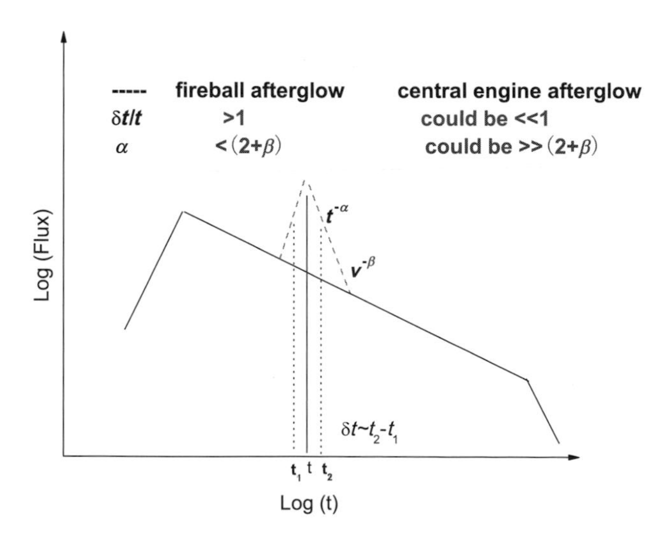

Figure 31.1 Two constraints that help us to distinguish between the *central engine afterglow* and the *fireball afterglow* (from [5]).

Very early rapid X-ray decline. If the central engine does not turn off abruptly, its weaker and weaker activity will give rise to rapidly decaying emission[7]. This model can account for some very early rapid X-ray declines identified by the *Swift* satellite[1], in particular those decaying with time more slowly than $t^{-(2+\beta)}$, for which the high latitude emission interpretation[22] is invalid.

X-ray/optical flares. In 2005, Piro et al. [18] published an analysis of the X-ray data of GRB 011121, in which two X-ray flares after the prompt γ-ray emission are evident. They interpreted the X-ray flares, in particular the first one, as the onset of the forward shock emission. Fan & Wei [7] applied the *decline argument* to these flares and suggested a central engine origin. Many other possibilities like a reverse shock, a density jump, a patchy jet, energy injection and a refreshed shock have been convincingly ruled out[22]. The XRT on board *Swift* confirmed Piro *et al.*'s discovery[3]. By now X-ray flares have been well detected in $\sim$40% of *Swift* GRBs and most violate the two constraints of the fireball afterglow. Though the physical processes powering these delayed X-ray bursts are not clear yet, it is most likely that they are related to a re-activity of the central engine.

Power-law decaying X-rays. For GRB 060218 the inconsistency of the X-ray afterglow flux with the radio afterglow flux and the very steep XRT spectra support here the central engine afterglow hypothesis[6]. The X-ray afterglow of GRBs 060607A, 070110, 060413, 060522, 060607A and 080330[21; 23] is distinguished by a very sharp X-ray drop, challenging the fireball afterglow interpretation. The luminosity and the temporal behavior of these X-ray flat segments before the sudden drop are roughly consistent with those of the dipole radiation of a millisecond magnetar. So these afterglow photons may be from the magnetic dissipation of a millisecond magnetar wind, as suggested by Gao & Fan [11].

31.2 The polarization signatures

31.2.1 Significantly polarized central engine afterglow?

The radiation mechanism of the central engine afterglow is likely to be the same as that of the prompt emission, which unfortunately remains unclear. The physical composition of the late outflow launched by the prolonged activity of the central engine has not been well understood, either. As already mentioned, some X-ray afterglow flat segments followed by a sudden drop may be powered by the magnetic dissipation of the millisecond magnetar wind[11; 21; 23]. If correct, these X-ray photons are expected to be significantly magnetized.

The X-ray flares, in particular those following short GRBs, may be highly polarized[9]. The physical reason is that in the double-neutron-star merger scenario, the accretion of the fallback material onto the nascent black hole is very limited. For an accretion rate $\dot{M} \sim 10^{-3}\ M_{\odot}/s$, the neutrino mechanism seems to be too inefficient to successfully launch energetic outflow that can power the observed flares.

Alternatively, a relativistic jet could be launched from a black-hole-torus system through MHD processes. For example, the spin energy of the black hole might be tapped by magnetic fields through the Blandford-Znajek mechanism[2]. The jet luminosity could be estimated as $L_{\rm BZ} \approx 2.5 \times 10^{47}(a/0.5)^2(B/10^{14}{\rm G})^2$ erg s^{-1}, where B is the magnetic field at the central engine and a is the spin parameter of the central BH. This power is also adequate to power the X-ray flares as long as the black hole spin energy is essentially not tapped during the prompt emission phase. In such a case, the jet is also Poynting-flux dominated.

In a Poynting-flux-dominated flow, the observed X-ray flare emission could be due to dissipation of the magnetic fields. Because of the ordered magnetic field and because of the narrow visible-emitting region, the synchrotron emission has a preferred polarization orientation (i.e. perpendicular to the direction of the toroidal field and the line of sight). Consequently, the linear polarization of the synchrotron emission of each electron could not be effectively averaged out and the net emission should be highly polarized.

31.2.2 Strong polarization evolution accompanying the sharp decline of the prompt X-ray emission and the flares?

If the sharp X-ray decline is the high latitude emission of a main internal shock pulse, there should be very interesting polarization signals. As is known, GRB outflow is likely to be jetted. Following [12], we assume that the half-opening angle of the ejecta is $\theta_{\rm j}$, and that the line of sight (L.o.S.) makes an angle $\theta_{\rm v}$ with respect to the jet's central axis (C.A., see Figure 31.2a for illustration). The probability of observing the ejecta along its C.A. is vanishingly small since it corresponds a very small solid angle. For typical bright GRBs, the L.o.S. is likely to be within the cone, i.e. $\theta_{\rm v} < \theta_{\rm j}$. At early time, the high latitude emission is from zones around the L.o.S. satisfying $\theta \leq \theta_{\rm j} - \theta_{\rm v}$, so the net polarization of the detected photons vanishes because of the symmetry; but at later time, the high latitude emission is from $\theta > \theta_{\rm j} - \theta_{\rm v}$, the symmetry is broken, and the observed net polarization is not zero any longer.

In this model, the photons arriving at t were emitted from $\theta \approx (2ct/R_0)^{1/2}$, where $R_0 \sim R_{\rm prompt}$ is the radius of the main internal shock pulse and c is the speed of light. Below we adopt the simplified calculation of [12]. The notations used in Figure 31.2b are related to R_0, $\theta_{\rm j}$ and $\theta_{\rm v}$ as $R \approx R_0 \sin\theta_{\rm j}$ and $R_1 \approx R_0 \sin(\theta_{\rm j} - \theta_{\rm v})$. When $\rho > R_1$, the ring is not a complete circle. The missing parts are in $(0,\ \psi_1)$ and $(2\pi - \psi_1,\ 2\pi)$ as shown in Figure 31.2b, where

$$\psi_1 \approx \begin{cases} \frac{\pi}{2} - \arcsin[\frac{2R_1R - R_1^2 - \rho^2}{2\rho(R-R_1)}], & \text{for } \rho > R_1; \\ 0, & \text{for } \rho < R_1, \end{cases} \tag{31.1}$$

where $\rho \approx R_0 \sin\theta$.

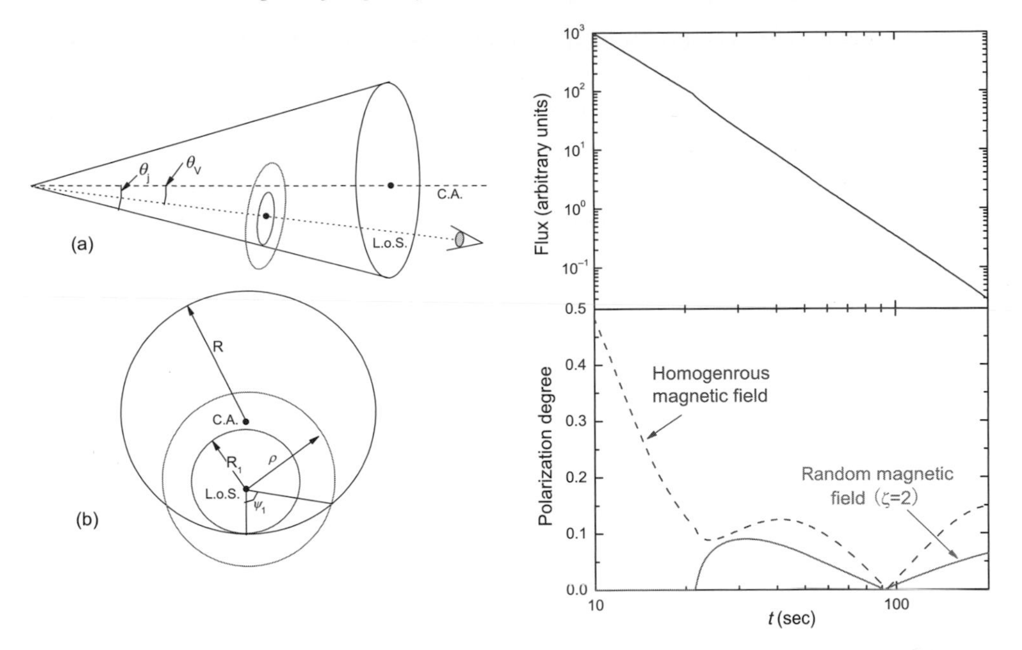

Figure 31.2 Left: the high latitude emission model for the X-ray decline. The larger the angle with respect to the line of sight (L.o.S), the later the emission arrives with us because of the geometry effect. (b) Sketch of the geometrical set-up used to compute the polarization signal (see also [12]). Right: the sharp X-ray decline (upper panel) and the corresponding polarization light curves (lower panel) after a time-shift correction, in the case of the high latitude emission model. For this figure we assumed $\Pi_0 = 60\%$, $\Gamma = 50$, $\theta_j = 0.1$, $\theta_v = 0.07$, $R_0 = 10^{15}$ cm and $\beta_X = 1.15$ (from[8]).

In principle, the GRB outflow could be either baryon-rich or Poynting-flux dominated. For the former the magnetic field is generated in the shock front and is likely to be random. For the latter the magnetic field is coherent on a large scale. The resulting polarization light curves are expected to be different in these two cases, as shown in the right panel of Figure 31.2.

31.3 Discussion

For the GRB fireball afterglow, polarimetry has been made in the optical band and the polarization degree is found to be just a few percent[4; 13]. In the X-ray band, no polarimetry has been performed. *In principle the fireball X-ray afterglow will be only a little more polarized than in the optical band. Such a weak signal seems unlikely to be measured in the foreseeable future. For the central engine afterglow, the polarization properties may be very different.* As shown in Section 31.2, some peculiar behaviors in the X-ray afterglows of *Swift* GRBs, such as energetic flares and the plateau followed by a sharp drop, might be highly linearly polarized because the outflows powering these behaviors may be Poynting-flux dominated[9]. Furthermore, the breakdown of the symmetry of the visible emitting region may be hiding in these X-ray data and will give rise to interesting polarization signatures. For example, we find strong polarization evolution accompanying the sharp X-ray declines, particularly in the high latitude emission model[8]. An *XRT*-like detector but with polarization capability on board a *Swift*-like satellite would be suitable to test our predictions.

References

[1] Tagliaferri, G., et al. (2005). *Nature* **436**, 985–988.
[2] Blandford, R. D. & Znajek, R. L. (1977). *MNRAS* **179**, 433–456.
[3] Burrows, D. N., et al. (2005). *Science* **309**, 1833–1835.
[4] Covino, S. et al. (1999). *A & A* **348**, L1–L4.
[5] Fan, Y. Z., Piran, T. & Wei, D. M. (2008). *AIPC* **968**, 32–35.
[6] Fan, Y. Z., Piran, T., Xu, D. (2006). *JCAP* **0609**, 013.
[7] Fan, Y. Z. & Wei, D. M. (2005). *MNRAS* **364**, L42–L46.
[8] Fan, Y. Z. Xu, D. & Wei, D. M. (2008). *MNRAS* **387**, 92–96.
[9] Fan, Y. Z., Zhang, B. & Proga, D. (2005). *ApJ* **635**, L129–L132.
[10] Gao, W. H. (2009). *ApJ* **697**, 1044–1047.
[11] Gao, W. H. & Fan, Y. Z. (2006). *Chin. J. Astron. Astrophys.* **6**, 513–516.
[12] Ghisellini, G. & Lazzati, D. (1999). *MNRAS* **309**, L7–L11.
[13] Greiner, J., et al. (2003). *Nature* **426**, 157–159.
[14] Katz, J. I., Piran, T. & Sari, R. (1998). *Phys. Rev. Lett.* **80**, 1580–1582.
[15] Kumar, P. & Panaitescu, A. (2000). *ApJ* **541**, L51–L54.
[16] Nakar, E. & Piran, T. (2003). *ApJ* **598**, 400–410.
[17] Piran, T. (2004). *Rev. Mod. Phys.* **76**, 1143–1210.
[18] Piro, L., et al. (2005). *ApJ* **623**, 314–324.
[19] Proga, D., MacFadyen, A. I., Armitage, P. J. & Begelman, M. C. (2003). *ApJ* **599**, L5–L8.
[20] Sari, R., Piran, T. & Narayan, R. (1998). *ApJ* **497**, L17–L20.
[21] Troja, E., et al. (2007). *ApJ* **665**, 599–607.
[22] Zhang, B., et al. (2006). *ApJ* **642**, 354–370.
[23] Zhang, X. H. (2009). *Res. Astron. Astrophys.* **9**, 213–219.

32

GRB afterglow polarimetry past, present and future

S. Covino

INAF/Brera Astronomical Observatory, Merate (LC), Italy

Gamma-ray bursts and their afterglows are thought to be produced by an ultrarelativistic jet. One of the most important open questions is the outflow composition: the energy may be carried out from the central source either as kinetic energy (of baryons and/or pairs), or in electromagnetic form (Poynting flux). While the total observable flux may be indistinguishable in both cases, its polarization properties are expected to differ markedly. The later time evolution of afterglow polarization is also a powerful diagnostic of the jet geometry. Again, with subtle and hardly detectable differences in the output flux, we have distinct polarization predictions.

32.1 Introduction

Polarimetry is a powerful diagnostic tool to study spatially unresolved sources at cosmological distances, such as gamma-ray burst (GRB) afterglows. Radiation mechanisms that produce similar spectra can be disentangled by means of their polarization signatures. Also, polarization provides unique insights into the geometry of the source, which remains hidden in the integrated light.

Historically, essentially all interpretative studies about GRB afterglow polarimetry have been based on the cosmological fireball model[24; 31], which we will also use as a reference for our discussion. Afterglow polarization studies have indeed the advantage that different models are often almost indistinguishable in terms of radiation output in the optical, but produce markedly distinct predictions about polarization.

X-ray Polarimetry: A New Window in Astrophysics, eds. R. Bellazzini, E. Costa, G. Matt and G. Tagliaferri. Published by Cambridge University Press.

In this proceeding, we will briefly review in Section 32.2 what we have derived by optical afterglow polarimetric observations and discuss the most recent development in the field in Secttion 32.3. For a deeper discussion about the physical ingredients generating a polarized flux in GRB afterglow radiation one can refer to other proceedings in this volume[6; 14; 16].

32.2 What have we learnt so far?

We report below what we consider the three most important achievements obtained by afterglow polarimetric observation in GRB research. Generally speaking, two general families of models have been developed to explain why GRB afterglows can be polarized and the time evolution of polarization. One possibility is that the emission originates in causally disconnected regions of highly ordered magnetic field, each producing polarization almost at the maximum degree. Reference [13] predicted a $\sim$10% polarization. If the regions have a statistical distribution of energies, the position angle can be different at various wavelengths. This value is greater than that observed in many GRB afterglows[4] as most of the positive detections so far derived are below $\sim$3%. In an alternative scenario[10] the magnetic field is ordered in the plane of the shock. In a spherical fireball, such a field configuration would give null polarization, but if a collimated fireball is observed off-axis (as it is most probable), a small degree of polarization would be predicted, with a well-defined temporal evolution. Here the ultrarelativistic motion toward the observer and the physical beaming of the outflow are fundamental ingredients.

32.2.1 GRB afterglows polarization

After a few unfruitful attempts, and not by chance as soon as the first unit of the ESO-VLT become operational, a low although highly significant polarization for the afterglow of GRB 990510 (Figure 32.1) was successfully detected for the first time[2; 29]. This simple observational finding carried already a lot of information. First of all, the detection of polarized flux from a GRB afterglow can be and has been considered a clear signature for synchrotron emission, although various alternative explanations indeed exist. In general, the detected polarization ($1.7\% \pm 0.2\%$) would require an emission processes involving particle acceleration. In the external shock phenomenon we have particle acceleration at the shock front and, once we consider the ultrarelativistic motion toward the observer and the physical beaming of the outflow, some level of polarization in the afterglows is naturally predicted. It is possible to have some degree of polarization adopting other scenarios, but in no case is a polarized flux a natural output of the model, as it is for the cosmological fireball model. To my knowledge, this is still one of the most convincing, although admittedly often unrecorded, observational proofs supporting the standard afterglow model.

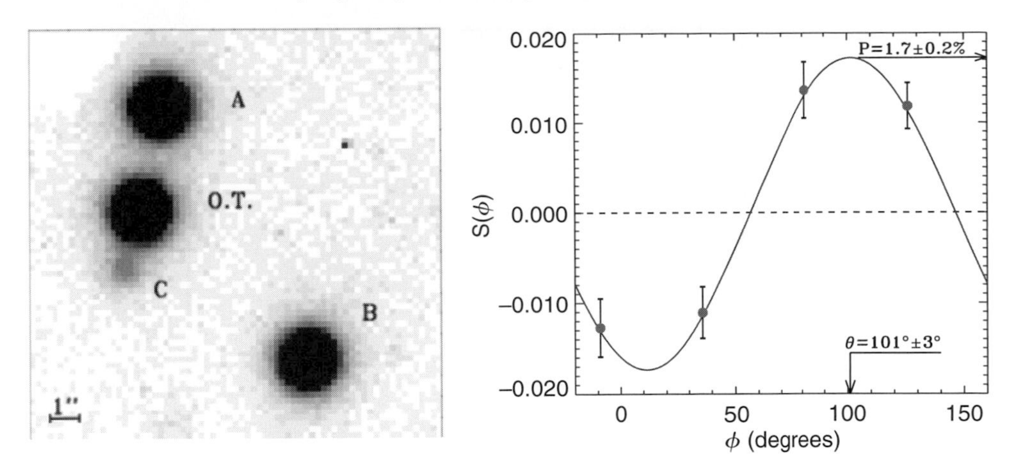

Figure 32.1 The field of GRB 990510 observed by the ESO-VLT equipped with FORS1 in the R band (left). The net polarization of GRB 990510 (right). From [2].

32.2.2 Afterglow polarization variable in time

The detection of varying polarization on time scales comparable to those of the afterglow evolution immediately implies that the observed polarization is intrinsic to the source and not, for instance, due to scattering against material along the line of sight. The first convincing evidence of time-varying polarization was obtained for GRB 020813[1; 18], where a decrease of the polarization degree from ∼3% down to less than 1% (Figure 32.2), with constant position angle, was recorded from a few hours to half a day after the burst. Evolution was also singled out in GRB 021004[17].

The most striking example is, however, GRB 030329. Due to its relatively small distance, a very high-quality dataset was obtained covering more than two weeks[12]. Strong, somewhat erratic, variations of the polarization degree and position angle during the afterglow evolution were singled out. Polarization variations occurred on a time scale comparable to that of the afterglow flux variability, offering a direct link between the two phenomena[11], although in the case of GRB 030329 the late-time rise of the supernova (associated with the GRB) component played an important role.

32.2.3 Afterglow polarization and the geometry of GRB jets

Since GRB sources are unresolved, any model for producing polarization requires some kind of anisotropy in the emitting fluid. The simplest configuration envisages emission from homogenous jets observed off-axis. In this case, as shown by [10; 26], the polarization time evolution presents two maxima, reaching a zero level in between, where a flip of the polarization angle by 90° also occurs. A clear prediction which in principle is easy to test with observations. More complex jet

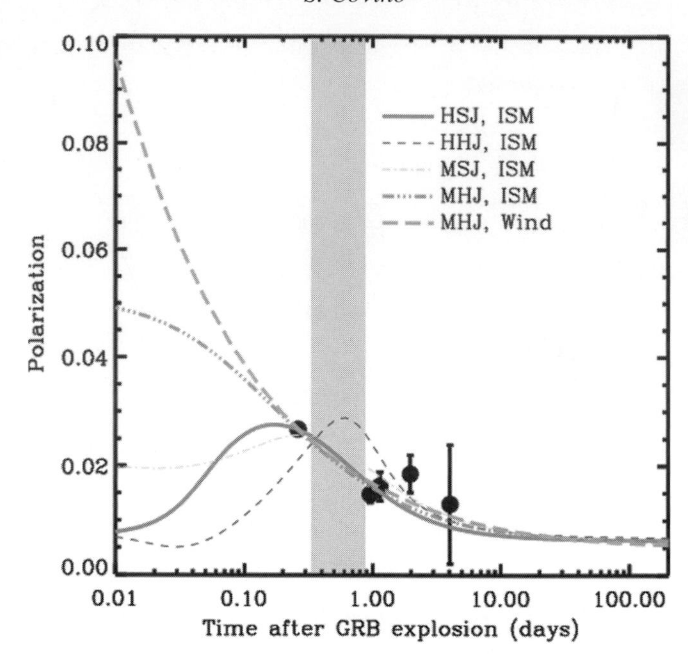

Figure 32.2 Polarimetric curve of GRB 080213. The observations are compared to predictions for several families of models as discussed in [18]. The shaded area shows where a break, possibly a jet break, was observed during the optical afterglow evolution[3].

structures have also been proposed, i.e. in which the energy content per solid angle is decreasing towards the wings of the jet. Such configurations may allow for a unified view of GRBs, in which differences among events arise only (or mostly) from the orientation of the observer with respect to the jet core. The polarization behaviour is in this case markedly different, with a single broad maximum[25; 18].

Up to now, a full set of (late-time) observations of polarization evolution could effectively be compared with models only in the case of GRB 020813. The main result of accurate modeling[18] is that homogeneous-jet-model predictions with shock-generated magnetic fields are in clear disagreement with the observations. This is one of the strongest direct observational evidences against the homogeneous-jet scenario so far obtained (see also the case of GRB 030328, [21]).

32.3 *Swift* and the early afterglow

Time-resolved polarimetric observations of late-time (later than about 1 hour) afterglow are extremely demanding, even for 8-m-class telescopes, due to the low polarization detected and the rapid fading of the afterglows. Moreover, the complexity of afterglow behaviors compared to the theoretical predictions have made it difficult to apply further observational tests and derive unambiguous answers.

However, after the launch of the *Swift* satellite[9], early afterglow observations became feasible thanks also to the network of ground-based robotic telescopes devoted to GRB follow-up. Early afterglow observations can provide powerful diagnostics for many physical ingredients of GRB models, and again polarimetry can help to solve one of the hottest issues of GRB research.

Within the cosmological fireball model a hot fireball[24] expands driven by internal energy. An alternative scenario which attracted great theoretical interest has also been developed[28; 27; 20; 19; 30], the "electromagnetic outflow", where most of the energy is carried to large distances from the central source in electromagnetic form (Poynting flux). Although dramatically different physics are involved, these two scenarios may result in a similar radiation output.

Things are different if polarization is considered. In the early phases optical emission can be generated by the forward shock, i.e. the afterglow, or by the reverse shock responsible for the optical flash. Reference [11] showed that the optical flash, if due to the reverse shock, shares the same magnetic field configuration as the prompt emission, and therefore the same level of linear polarization. If the fireball is electromagnetically dominated, the first tens of minutes of the afterglow may be $>40\%$ polarized[15]. Optical flash polarization properties probe the magnetic field structure within the original outflow, while the afterglow emission probes the magnetic field structure in the shocked external medium, as well as the jet angular structure. Producing strong polarization in the optical flash requires a large-scale-ordered magnetic field possibly advected from the inner engine, while if the magnetic field is shock generated no polarization is expected.

32.3.1 Early afterglow observations

To date, the only early polarimetric measurement was performed by [23] deriving an 8% upper limit just after the onset of the afterglow of GRB 060418 (Figure 32.3). This result could strongly limit the possible role of magnetic fields in driving the outflow dynamics. However, in this case the optical emission is likely due to the forward shock only[22; 8], and the predicted polarization level depends on still poorly known details of the transfer of magnetic energy from the outflow to the shocked circumburst medium[11; 30; 7; 5], so that low or null polarization is still compatible with the theoretical expectations and these measurements are not yet conclusive.

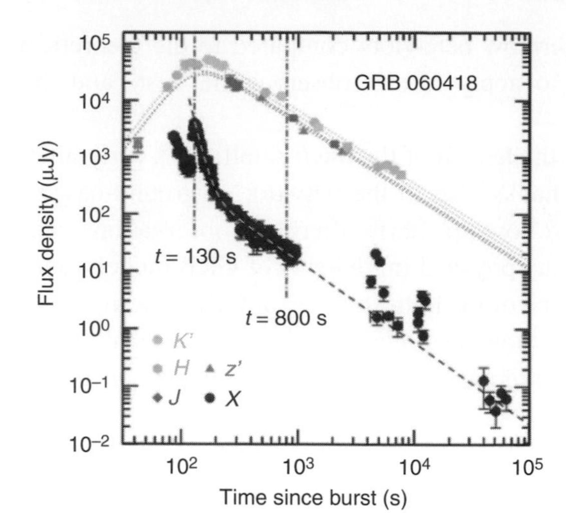

Figure 32.3 Early afterglow near-infrared and X-ray light curves obtained by the REM telescope and by *Swift*[22]. The Liverpool telescope polarimetric measurement[23] was carried out about three minutes after the high-energy event, just after the peak of the near-infrared light curve which was interpreted as the afterglow onset.

References

[1] Barth, A. J., et al. (2003). *ApJ* **584**, L47.
[2] Covino, S., et al. (1999). *A&A* **348**, L1.
[3] Covino, S., et al. (2003). *A&A* **404**, 5.
[4] Covino, S., et al. (2004). *ASPC* **312**, 169.
[5] Covino, S., (2007). *Science* **315**, 1798.
[6] Fan, Y. Z., (2009). These proceedings, Chapter 31, pp. 209–214.
[7] Fan, Y. Z. & Piran, T. (2006). *MNRAS* **369**, 197.
[8] Jin, Z. P. & Fan, Y. Z. (2007). *MNRAS* **378**, 1043.
[9] Gehrels, N., et al. (2004). *ApJ* **611** 1005.
[10] Ghisellini, G. & Lazzati, D. (1999). *MNRAS* **309**, L7.
[11] Granot, J. & Königl, A. (2003). *ApJ* **594**, L83.
[12] Greiner, J., et al. (2003). *Nature* **426**, 157.
[13] Gruzinov A. & Waxman E. (1999). *ApJ* **511**, 852.
[14] Laurent P. (2009). These proceedings, Chapter 34, pp. 230–237.
[15] Lazzati, D. (2006). *New Journ. Phys* **8**, 131.
[16] Lazzati, D. (2009). These proceedings, Chapter 30, pp. 202–208.
[17] Lazzati, et al. (2003). *A&A* **410**, 823.
[18] Lazzati, et al. (2004). *A&A* **422**, 121.
[19] Lyutikov, M. (2006) *New Journ. Phys.* **8**, 119.
[20] Lyutikov, M., Pariev, V. I., & Blandford, R. D. (2003). *ApJ* **597**, 998.
[21] Maiorano, E., et al. (2006). *A&A* **455**, 423.
[22] Molinari, E., et al. (2007). *A&A* **469**, 13.
[23] Mundell, C. et al. (2007). *Science* **315**, 1822.
[24] Piran, T. (1999). *Phys. Rep.* **314**, 575.

[25] Rossi, E. et al. (2004). *MNRAS* **354** 86.
[26] Sari, R. (1999). *ApJ* **524**, L43.
[27] Thompson, C. (1994). *MNRAS* **270**, 480.
[28] Usov, V. V. (1992). *Nature* **357**, 472.
[29] Wijers, R. et al. (1999). *ApJ* **523**, L33.
[30] Zhang, B. & Kobayashi, S. (2005). *ApJ* **628**, 315.
[31] Zhang, B. & Mészáros, P. (2003). *Int. J. Mod. Phys. A* **19**, 2385.

33

Gamma-ray polarimetry with SPI

A. J. Dean

School of Physics and Astronomy, The University of Southampton

Young energetic pulsars are capable of accelerating electrons to extremely high energies, which subsequently become visible in the X-ray and low-energy γ-ray domain through synchrotron radiation. The recent polarization measurement of the Crab γ-ray emission provides a powerful investigative tool for the physical conditions and geometry of the magnetic field close to the pulsar. The Crab nebula has been found to emit linearly polarized γ-rays during the off-pulse phase with an efficiency close to the maximum allowable by physics. The close alignment between the electric vector and the spin axis of the neutron star places severe boundaries on theoretical models. The off-pulse γ-radiation is constrained to originate somewhere close to the pulsar, but outside the light cylinder. Three contenders are identified: the striped-wind model; radiation from within the inner jet; Doppler-boosted radiation from knot-like features close to the pulsar.

33.1 Introduction

Observations in the γ-ray regime allow us to scrutinize some of the most energetic emission processes associated with cosmic sources. Unlike other wave bands that mainly show emission due to thermal reprocessing within hot gases, the majority of γ-ray emission is decidedly non-thermal in nature and is generally produced directly by electrons and other elementary particles in magnetic fields. The electrons which produce the γ-rays are extremely energetic and often at the upper limit of the capability of the accelerator that produces them. Coupled with the extreme penetrating power of γ-ray photons, this has the consequence that, through observing γ-rays, we are looking directly at the heart of the engine responsible for their

X-ray Polarimetry: A New Window in Astrophysics, eds. R. Bellazzini, E. Costa, G. Matt and G. Tagliaferri. Published by Cambridge University Press.

acceleration. To produce electrons with Lorentz factors of 10^7–10^8 or more requires an extremely ordered machine of well-defined geometry. Consequently the study of the polarization characteristics of γ-ray emission provides an ideal means for the detailed understanding of the physics, geometry and inner working of the most powerful cosmic particle accelerators.

Imaging and spectroscopy provide fundamental precursors to the identification of source-model scenarios. However, ambiguities often remain, and the precise physical and geometrical conditions at the site of the emission are left to informed guesswork. The linear polarization of the emission provides a powerful diagnostic tool to constrain models of emission. Electrons are accelerated on ordered magnetic field lines and release γ-rays. The emission will be polarized, with the direction and percentage of polarization related to the emission mechanism and the geometry of the fields. Although not designed as polarimeters the two main INTEGRAL instruments, SPI and IBIS, are capable of studying the polarization characteristics of the stronger sources, and the first measurements are beginning to emerge for GRB 041219a[21], and the Crab nebula[4; 9]. Here the Crab results are discussed.

33.2 The polarization analysis technique

Measuring polarization at γ-ray energies in the 100 keV–10 MeV range is usually accomplished using Compton scattering. The probability that the photon will scatter is given by the Klein-Nishina differential cross-section. In the case of a linearly polarized photon, this is given by

$$\frac{d\sigma}{d\Omega} = \frac{r_0^2}{2}\left(\frac{E'}{E}\right)^2 \left[\frac{E'}{E} + \frac{E}{E'} - 2\sin^2\Theta\cos^2\phi\right], \tag{33.1}$$

where r_0 is the classical electron radius, E'/E is the ratio between the scattered photon and the incident photon energies and ϕ is the azimuthal scattering angle, defined as the angle to the polarization electric vector[18], and Θ is the elevation angle. The secondary photon will preferentially scatter 90° to the direction of electric vector. Measurement of the amplitude and azimuthal distribution of the scattered photons thus enables the degree of polarization and vector direction of the polarized beam to be evaluated. The scattering takes place between detector pixels, and both the SPI and IBIS instruments on INTEGRAL can be used in this way. In practice there are a number of systematic effects in the background noise that generate non-symmetrical azimuthal distributions and which can greatly compromise the polarization sensitivity; e.g. for the case of the data set used for the SPI measurement[4], on a statistical basis the polarization of the Crab was measurable to a precision of 0.03%[32], in practice systematic errors limited the accuracy to $\pm 10\%$. Fortunately the advent of modern fast computing clusters has now made large scale simulation of an instrument's response possible. Combining the instrument data with results obtained from detailed Monte-Carlo mass-model (MM) simulations[5], it is now possible to use the MM to suppress the systematic contributions for all sources. This is where the real power of the MM is felt. The MM also suppresses the negative effects due to dead pixels and non-uniform background, making sensitive polarization measurements possible with imperfect instrumentation.

33.3 The Crab polarized γ-ray emission

Recent measurements of the Crab by INTEGRAL-SPI[4] have shown the emission to be highly polarized in the γ-ray domain, a measurement which has been independently confirmed using the entirely separate instrument INTEGRAL-IBIS[9]. Both sets of measurements were made at photon energies close to the electron rest mass, the most suitable for polarization studies using Compton scatters. The SPI measurements were made in the off-pulse period (phase 0.5–0.8) and over the energy range 0.1–1.0 MeV, whereas the IBIS measurements covered all the pulsar's phases and energies above 0.2 MeV. The electric vector of off-pulse polarization is found by SPI to be $123° \pm 11°$ (measured from north, anticlockwise on the sky), and by IBIS to be $122.0° \pm 7.7°$. There is no noticeable change in polarization angle with phase. The percentage of polarization has been found to be very high, in the case of SPI, this is reported to be $46 \pm 10\%$, and the IBIS team report a value greater than 72% at the 95% confidence level. The IBIS team found no significant polarization in the pulsed emission. As in the X-rays[33] both of the γ-ray observations encompass the entire nebula and pulsar, due to the angular resolution of the instruments involved. Since the two instruments suffer from entirely different systematic errors, the very good agreement between the two sets of results greatly relieves any doubts that either result is purely a systematic effect.

33.4 The source of the γ-ray emission within the Crab nebula

The Crab is the only PWN system for which we have a γ-ray polarization measurement currently available. The production of polarized emissions provides an important diagnostic of the underlying physical system, requiring a highly ordered machine, but from where precisely do they originate? The soft γ-ray images show that the centroid of the IBIS emission lies close to the pulsar, with an accuracy of 20 arcsec between 18 and 60 keV[2], a zone encompassing the pulsar and surrounding jet/torus structure. We know that the lifetime of the 500 TeV synchrotron electrons radiating in the average 160 G nebular field[1] responsible for the soft γ-rays will be short (<1 yr) and that they do not travel far from their source, so all the imagery tells us is that they must originate somewhere close to the pulsar. Likewise an attempt to match the 2.23 power law spectral index measured by INTEGRAL with the morphological spectral maps of Chandra[23] leads to the same conclusion. The polarization data have the potential to help better locate the origin of the soft γ-rays.

The best fit values for the angle of polarization for the non-pulsed soft γ-ray emission show a remarkable degree of alignment with the $124° \pm 0.1°$ direction of the spin axis of the neutron star projected onto the sky[24], and this is unlikely to be accidental. The extremely high degree of linear polarization, amounting to at least 60% of the maximum allowed by physics[18], dictates that the γ-ray emission is generated in an exceptionally ordered machine. Such a high degree of linear polarization is not compatible with any significant emission from the tangled magnetic fields associated with the Crab plerion. An inspection of radio polarization measurements, e.g. [30], of this outer region demonstrates that a high degree of order does not exist there. The alignment with the spin axis and the high degree of polarization also implies that any significant production of highly energetic electrons does not take place within the nebula and must be generated close to the pulsar. It is also difficult to envisage how such a high net degree of polarization could result from the torus. The bare polarization measurements demand that the soft γ-rays are generated somewhere inside the torus region or within the jet structure.

Polarization measurements of the Crab optical radiation have been reported by [12; 27]. The key measurement parameters for the non-pulsed optical emission are: a mean polarization angle of $\sim$123° and a 33% degree of polarization. The angular resolution possible in the optical band locates the origin of the optical emission to within about an arcsec or so of the pulsar. The optical and γ-ray polarization characteristics are remarkably consistent, strongly suggesting that, for the non-pulsed radiation at least, both originate from the same region close to the pulsar. The high energy requirement of the parent electrons and their short lifetimes implies that the polarized γ-ray emission comes from a region well within the wind termination shock. The predicted polarization characteristics for mechanisms within the light cylinder, including the polar-cap model, the outer-gap model and the two-pole-caustic model have been extensively discussed in the literature[7; 29]. Whilst the different models have varying degrees of success in describing individual features, such as pulse profiles and pulse separations, none of these models can well reproduce the measured optical and γ-ray polarization characteristics with the generally accepted values for the magnetic dipole offset angle ($\sim$60°) and view angle ($\sim$61°)[24]. The inclusion of the off-pulse soft γ-ray polarization characteristics leads us to conclude that the soft γ-rays must be generated somewhere outside the pulsar's light cylinder and within the X-ray ring that marks the termination shock. Three realistic alternatives remain:

Curvature radiation within the inner jet structure. Chandra images of the Crab[31] indicate that the jet is curved, so energetic electrons streaming along the jet will radiate by curvature radiation with the electric vector parallel to the jet, as seen by INTEGRAL. In the relativistic limit, the emission frequency is given by $\nu_c = (3c/4\pi R)\gamma^3$ where γ is the Lorentz factor of the radiating electrons and R is the radius of curvature of the field lines. Taking an estimate of $R \sim 2.5 \cdot 10^{18}$ cm for the Crab jet, in order to generate the few hundred keV photons detected by INTEGRAL, electrons with energies of typically 10^{15} eV are required, which is entirely plausible in the context of the Crab pulsar. Analysis of radio and optical polarization studies of jets in radio galaxies[17; 25] indicates that the apparent magnetic field vector is parallel to the jet axis for the inner reaches of the jet. For radiation from within the inner jet, we would expect the direction of the electric vector associated with the polarized radiation to lie approximately parallel to the spin axis of the neutron star, as observed by INTEGRAL.

Synchrotron radiation within a jet hotspot. A number of bright knot-like and other features within the Crab inner jet region have been reported from the study optical and X-ray images[10; 11]. Some of the features move along the jet with speeds of up to 0.4c. Some of the knots appear to be persistent over periods of several years or more. Rapidly moving knot-like features are a common characteristic seen in a number of the γ-ray-emitting PWN systems. An MHD model[16] identifies the $0.65''$ southeast knot as due to relativistic beaming from high-velocity plasma flowing just above the termination shock at high altitudes, where its velocity vector points directly at the observer. Detailed polarization maps computed for the Doppler-boosted features[6] show a high degree of linear polarization close to the axis of symmetry, in accordance with the INTEGRAL results. The radio and optical emissions from AGN knots are often polarized[25] but the polarization vectors show no preference for the direction of the parent jet axis, making this interpretation an unlikely option.

Striped-wind models. The very good correlation between the pulsar and PWN X-ray luminosities found by Kargaltsev and Pavlov [13] is not to be expected unless the pulsed radiation comes not from the magnetosphere, but from the wind [14]. The idea of a striped pulsar wind was originally introduced by Coroniti [3], and has been extensively discussed in the literature (e.g. [15]). This model places the site for the production of the pulsed and unpulsed radiation outside the light cylinder and within the termination shock. The energy transport is dominated by a Poynting flux with the pulsar driving a low-frequency wave in the wind, consisting of stripes of toroidal magnetic field of alternating polarity propagating around the equatorial plane. The wind expands until the ram pressure is balanced by the surrounding nebula at the termination shock and the wave is dissipated. Relativistic beaming is responsible for the phase coherence of the synchrotron radiation, and is also responsible for the X-ray dark zone 'observed' between the pulsar and the termination shock (Weisskopf et al. [31]). The striped-wind radiation thus appears to originate from an object of diminished angular size with a correspondingly higher surface brightness. This model predicts that the magnetic field configuration for synchrotron radiation automatically aligns the polarization vector with the spin axis of the pulsar, as observed in γ-rays. This property is inherent to the striped-wind model. The polarization characteristics expected from the striped-wind model have been discussed in [26; 15].

33.5 Future γ-ray polarization measurements

The existing γ-ray polarization measurements have been carried out with satellite-borne instrumentation not explicitly designed for polarization studies. Consequently they have unavoidably low polarization sensitivities. Any balloon-borne polarization-specific payload will suffer shorter exposure periods. Thus until a dedicated instrument comes along, only the brightest γ-ray emitters may be studied with any hope of success. However, on the plus side, γ-rays with energies greater than 100 keV are generally the products of extremely high-energy electrons which were created within a powerful accelerator and so it is likely that the emission will often be polarized, thus allowing this powerful diagnostic tool to be used to study different classes of high-energy emitting objects.

Pulsar wind nebulae. The near-future scientific goals associated with the polarization studies of PWN in the context of currently available instrumentation are quite straightforward: to accumulate enough good quality on-source data to identify the underlying mechanism for the production of the γ-rays. Of the ten PWN systems detected by INTEGRAL, the Vela and MSH 15-52 systems provide the best opportunities for such tests. The first step will be to see if the γ-ray polarization vector habitually aligns with the axis of rotation of the neutron star, such that the Crab value is not purely coincidental. We have seen, above, that there are three contending scenarios, all of which are considered capable of explaining the alignment of the γ-ray polarization vector with the pulsar spin axis. The subsequent challenge will be to identify which model best explains the detailed measurements.

Astrophysical jets. The measurement of hard X-ray/γ-ray polarization from bright microquasar sources would also be of great interest. In both hard and soft states these systems should have steep power-law tails out to at least 1 MeV[20], which must be associated with non-thermal particle populations and, by implication, with the jets in these systems (see [8] for a review). In particular, the γ-ray and in some cases also X-ray emission (e.g. [19]) may be optically thin synchrotron emission, and a measure of the linear polarization degree and angle would provide information on the ordering and orientation of the magnetic field in the jet very close to the black hole. Among the best candidates are Cygnus X-1 and GRS 1915+105, both powerful jet sources where the jet orientation on the sky is well measured (e.g. [28; 22]), allowing for a direct comparison with any γ-ray polarization angle. Furthermore, in GRS 1915+105 the system often oscillates between states in which the jet is on and then off; compiling data in each state should reveal differences which would be clearly attributable to the presence of a jet, perhaps even revealing how the large-scale magnetic field near the black hole re-orders as the jet reforms. The strong sources, if strongly polarized, should yield measurements of the vector directions at the $\sim 10^\circ$ level.

33.6 Conclusions

The brightest persistent source in the low-energy γ-ray sky, the Crab nebula, has been found to emit linearly polarized γ-rays during the off-pulse phase with an efficiency close to the maximum allowable by physics. The close alignment between the electric vector of the polarized γ-ray emission and the spin axis of the neutron star places severe boundaries on theoretical models. The off-pulse γ-radiation is constrained to originate through magneto-bremsstrahlung processes somewhere close to the pulsar, but outside the light cylinder. Three contenders for the production of the highly polarized γ-radiation are identified: the striped-wind model; radiation from within the inner jet; Doppler-boosted radiation from knot-like features close to the pulsar. Further γ-ray polarization measurements are required on pulsars other than the Crab in order to ascertain whether the spin axis/electric vector alignment is a general feature for pulsars, or whether the apparent alignment of the Crab is coincidental.

References

[1] Aharonian, F. et al. (2004). *ApJ* **614**, 897.
[2] Bird, A. et al. (2007). *ApJ* supplement, **170**, 175.
[3] Coroniti, F. V (1990). *ApJ* **349**, 538.
[4] Dean, A. J. et al. (2008). *Science* **321**, 1183.
[5] Dean, A. J. et al. (2003). *SSRv* **105**, 285.
[6] Del Zanna, l. et al. (2006). *A&A* **453**, 621.
[7] Dykes, J., Harding, A. K. & Rudak, K. (2004). *ApJ* **606**, 1125.
[8] Fender, R. P. M., Belloni, T. & Gallo, E. (2004). *MNRAS* **355**, 1105.
[9] Forot, M., Laurent, P., Grenier, I. A., Gouiffes, C. & Lebrun, F. (2008). *ApJ* **688**, L29.
[10] Hester, J. J. et al. (1995). *ApJ* **448**, 240.
[11] Hester, J. J. et al. (2002). *ApJ* **577**, L49.
[12] Kanbach, G. et al. (2005). *Am.Inst. Phys. Conf. Proc.* **801**, 306.
[13] Kargaltsev, O. & Pavlov, G. G. (2008). [astro-ph/0801.2602].
[14] Kirk, J. G., Skjeraasen, O. & Gallant, Y. A. (2002). *A&A* **388**, L29.
[15] Kirk, J. G., Lyubarsky, Y. & Petri, J. (2007). [astro-ph/0703116].
[16] Komissarov, S. & Lyubarsky, Y. (2004). *Ap&SS* **293**, 107.
[17] Laing, R. A. & Bridle, A. H. (2002). **336**, 328.
[18] Lei, F., Dean, A. J. & Hills, G. L. (1997). *SSRv.* **82**, 309L.
[19] Markoff, S., Nowak, M. A. & Wilms, J. (2005). *ApJ* **635**, 1203.
[20] McConnell, M. L. et al. (2000). *ApJ* **543**, 928.

[21] McGlynn, S. et al. (2007). *A&A* 466: 895–904.
[22] Mirabel, I. F. & Rodriguez, L. F. (1994). *Nature* **371**, 46.
[23] Mori, K. et al. (2004). *ApJ* **609**, 186.
[24] Ng, C.-Y. & Romani, R. W. (2004). *ApJ* **601**, 479.
[25] Perlman, E. S. et al. (2006). *ApJ* **651**, 735.
[26] Petri, J. & Kirk, J. G. (2005). *ApJ* **627**, L37.
[27] Sowikowska, A., Kanbach, G. & Stefanescu, A. (2007). The First GLAST Symposium, AIP Conference Proceedings, **921**, 419.
[28] Stirling, A. M. et al. (2001). *MNRAS* **327**, 1273.
[29] Takata, J., Chang, H. K. & Cheng, K.S. (2007). *ApJ* **656**, 1044.
[30] Velusamy, T. (1985). *MNRAS* **212**, 359.
[31] Weisskopf, M. C. et al. (2000). *ApJ* **536**, L81.
[32] Weisskopf, M. C. et al. (2006). [astro-ph/061148].
[33] Weisskopf, M. C. et al. (1978). *ApJ* **220**, L117.

34

INTEGRAL/IBIS observations of the Crab nebula and GRB 041219A polarization

P. Laurent
CEA/DSM/IRFU/APC

M. Forot & D. Gotz
CEA/DSM/IRFU/AIM

F. Lebrun
CEA/DSM/IRFU/APC

Neutron stars generate powerful winds of relativistic particles that form bright synchrotron nebulae around them. Polarimetry provides a unique insight into the geometry and magnetic configuration of the wind, but high-energy measurements have failed until recently. The INTEGRAL-IBIS telescope[1] has been used in its Compton mode to search for linearly polarized emission for energies above 200 keV from the Crab nebula. The asymmetries in the instrument response are small and we obtain evidence for a strongly polarized signal at an angle parallel to the pulsar rotation axis. This result confirms the detection recently reported by [1], and extends the polarization measure for all the pulsar's phases. We also report the recent observation of a variable polarization signal from the long GRB 041219A. The achieved sensitivity opens a new window for polarimetric studies at energies above 200 keV.

34.1 Introduction

Gamma-ray polarimetry has been possible with Compton telescopes since the 1970s. Photons that are Compton scattered between two detectors follow an azimuthal distribution around the source direction that allows quantification of the degree and direction of linear polarization because the photon is preferentially scattered in a plane at right angles to its incident electric vector. It was unsuccessful until recently because of intrinsic asymmetries in the detector response and

X-ray Polarimetry: A New Window in Astrophysics, eds. R. Bellazzini, E. Costa, G. Matt and G. Tagliaferri. Published by Cambridge University Press.

nonuniformities in the large background signals. They induce pseudo-polarimetric signals, even from an unpolarized source, that limit the sensitivity to any detection. For instance, attempts at detecting polarization in bright γ-ray bursts have been met with varying success and need confirmation (see [10] and references therein). With its double layer of finely pixellated detectors[14], the IBIS telescope on board the INTEGRAL satellite is well suited for polarimetry studies, between 200 keV and 5 MeV. The background is efficiently subtracted by deconvolving the coded-mask shadowgram on the primary detector. The angular resolution is energy independent and the two detector planes are close enough to detect events scattered at large angles with a rather uniform response in azimuth.

We first describe the polarimetry method we have developed, and check the level of asymmetries in the instrument response using different unpolarized sources. We then present evidence for the detection of polarization from the Crab nebula and GRB 041219A, at energies above 200 keV.

34.2 Polarimetry with the INTEGRAL/IBIS Compton mode

Photons entering IBIS are Compton-scattered in the first detector plane, ISGRI[6], at a polar angle θ from their incident direction and at an azimuth ψ from their incident electric vector. They are then absorbed in the second detector, PiCsIT[7]. The azimuthal profile $N(\psi)$, in Compton counts recorded per azimuth bin, follows:

$$N(\psi) = S[1 + a_0 cos(2\psi - 2\psi_0)] \qquad (34.1)$$

for a source polarized at an angle PA = $\psi_0 - \pi/2 + n\pi$ and with a polarization fraction PF = a_0/a_{100}. The a_{100} amplitude is expected for a 100% polarized source. Unfortunately, the IBIS polarimetric capacities have not been calibrated on the ground, due to the tight planning of space missions. We have then evaluated a_{100} to be 0.30 ± 0.02 for a Crab-like $E^{-2.2}$ spectrum between 200 and 800 keV, using GEANT3 Monte-Carlo simulations of IBIS and its detailed mass model[5], using the GLEPS package for polarization.[2] This value agrees with the early estimate of 0.3 obtained in the 200–500 keV band by [8].

Events recorded in ISGRI and PiCsIT within the same time window of 3.8 μs are tagged as 'Compton' events, but do not all result from Compton scattering. Chance coincidences can occur between ISGRI and PiCsIT events independently coming from the source, the sky, or the instrumental background. These coincidences are generally called spurious events. Most of the 'Compton'-tagged events are due to background events that will be removed by the shadowgram deconvolution; 5% are due to a small fraction of spurious coincidences that must be removed with high accuracy because they induce a false source detection in the sky image. (Indeed, among the spurious events, there are some for which the ISGRI or PiCsIT events are really coming from the source: for instance, an ISGRI low-energy Crab event could be associated by chance with a PiCsIT background event; these events coming from the source are then not removed by the deconvolution process.) 2% come from true Compton events from the source.

To measure $N(\psi)$, we first optimize the true-Compton to spurious ratio by applying angular cuts on the difference between the true source direction and the reconstructed one (from the events geometry and energies, assuming Compton scattering). The remaining spurious counts described above are estimated by randomizing in time all the ISGRI and PiCsIT events recorded during the observation[2]. After subtraction of these spurious counts, the ISGRI shadowgram is deconvolved to remove the background and get source counts. The whole process is applied for events in regularly spaced bins in azimuth to derive $N(\psi)$. To improve the polarimetric sensitivity, we keep only fully-coded observations, at off-axis angles $< 5°$.

Confidence intervals on a_0 and ψ_0 are not given by the $N(\psi)$ fit to the data since the variables are not independent. They have been derived from the probability density distribution of measuring a and ψ from N_{pt} independent data points in $N(\psi)$ over a π period, based on Gaussian distributions for the orthogonal Stokes components[15; 16; 18]:

$$dP(a,\psi) = \frac{N_{pt}\ S^2}{\pi\ \sigma_S^2} exp[-\frac{N_{pt}\ S^2}{2\ \sigma_S^2}[a^2 + a_0^2 - 2aa_0cos(2\psi - 2\psi_0)]]\ a\ da\ d\psi, \tag{34.2}$$

where σ_S notes the error on the profile mean S. The errors on each a or ψ dimension are obtained by integrating $dP(a,\psi)$ over the other dimension.

34.3 Calibration and performances of the IBIS/Compton polarimeter

To avoid a false polarimetric claim, we have looked for residual modulation of instrumental origin in $N(\psi)$ for a series of unpolarized sources, using on-axis and off-axis radioactive calibration sources, empty fields and spurious samples.

The non-axisymmetric geometry of the detectors (grids, corners, square mask pattern) can induce a small modulation. Its amplitude has been measured from the data of bright on-axis calibration sources at 392 and 662 keV. Source counts dominate over spurious and background ones. We find modulation fractions $PF = 0.066 \pm 0.013$ and $0.049^{+0.016}_{-0.013}$, respectively, indicating that the detector's square geometry induces a significant, but small modulation at an angle near 40°, primarily because of the sensitivity difference between two adjacent PiCsIT modules. In these studies, the reference frame has been oriented to allow a direct angle comparison with the Crab data below.

In order to check for systematics due, for instance, to the background or the analysis process, we have computed the polarimetric pattern of a 'pseudo source' located 1.5° away from the Crab, that is out of the Crab point spread function but still in the IBIS field of view. We used the same set of observations and analysis software as the ones used for the Crab polarimetric measurements. We concluded that the fluctuations were consistent with statistical variations, so that the possible systematic fluctuations can be neglected.

Despite the efficient subtraction of spurious events, a small residual contamination may also modify the shape of $N(\psi)$. We have analyzed a sample of in-flight spurious events recorded between 200 and 800 keV during Crab observations. The profile shows a small and well-defined modulation with $\mathrm{PF} = 0.15^{+0.04}_{-0.03}$ at an angle of $175.10° \pm 7.80°$ due to the detector's segmentation.

34.4 Crab observations

The Crab nebula has been repeatedly observed by INTEGRAL between 2003 and 2007 and a total of 1.2 Ms of fully-coded observations can be used for polarimetry. The pulsed lightcurve in the 200–800 keV band has been constructed with the Jodrell Bank ephemerides of the pulsar[9]. We have considered four phase intervals for polarimetry studies (see Table 34.1): the two main peaks, the off-pulse interval and the bridge interval, dominated by the nebular emission. The interval boundaries were taken from [3] and were not adapted to enhance a possible signal. Table 34.2 and Figure 34.1 show the results and illustrate the contrast between the modulation obtained for the nebular emission versus the flat profile of the pulsed emission.

There is no significant indication of polarization in the pulsed peaks. The chance probability of a random fluctuation reaches 33.5% and the signal shows no modulation at the 95% confidence level over all angles. This behaviour is consistent with the radio and optical data where PF drops below 10% as the angle largely flips within each peak[12; 13]. Conversely, the chance is low that the modulation seen in the off-pulse emission above 200 keV could be of random origin. The $P(a,\psi)$ probability density yields a probability of 2.6×10^{-3} that a random fluctuation produces an amplitude a_0 larger than the recorded one. Adding the bridge and off-pulse data strengthens the signal and gives a chance probability of $10\ ^{3}$ that an unpolarized source could produce this modulation. The observed modulation strongly differs from that of the spurious events recorded in the same Crab observations and from the instrumental asymmetries. So, the evidence for a strongly polarized signal holds against both statistical and systematic uncertainties. The polarization of the

Table 34.1 *Crab neutron star phase intervals.*

Name	Phase band
P_1	$0.88 < \phi < 0.14$
B	$0.14 < \phi < 0.25$
P_2	$0.25 < \phi < 0.52$
OP	$0.52 < \phi < 0.88$

Table 34.2 *Polarization angle and fraction with respect to the Crab pulsar phase ϕ.*

Phase interval	Polarization angle	Polarization fraction	Chance probability
P_1 and P_2	$70° \pm 20°$	$0.42^{+0.30}_{-0.16}$	33.5%
OP	$120.6° \pm 8.5°$	$> 0.72^a$	0.26%
OP and B	$122.0° \pm 7.7°$	$> 0.88^a$	0.10%
All	$100° \pm 11°$	$0.47^{+0.19}_{-0.13}$	2.8%

[a]: The lower limits for the polarization fraction are given at the 95% confidence level.

DC emission appears to be strong enough to still yield a marginal signal in the total Crab emission (see Table 34.2).

The off-pulse-and-bridge emission is polarized at an angle of $122.0° \pm 7.7°$ which is fully consistent with the north-to-east angle $\psi = 124° \pm 0.1°$ of the pulsar rotation axis projected on the skyplane[11]. The large data sample was taken in different observing modes, none of which was particularly aligned with the pulsar axis in the sky. At the 95% confidence level, we find fractions $PF > 72\%$ and $PF > 88\%$ in the off-pulse and off-pulse-and-bridge emission, respectively, for any values of the polarization angles. These values, which also do not take into account the uncertainties in the a_{100} computation, are both consistent, at the 95% level, with a 77% polarized signal along the pulsar rotation axis, which is the maximum polarization fraction allowed for synchrotron radiation in a uniform magnetic field and from a power-law distribution of electrons with the spectral index $p = 3.454 \pm 0.026$ recorded at these energies[3].

Our off-pulse measurement confirms the recently observed polarization in hard X-ray using INTEGRAL/SPI data[1]. Indeed, they measure in the off-pulse region, between 100 keV and 1 MeV, an angle of $123.0° \pm 11°$, very close to our own values (see Table 34.2). However, their polarization fraction $46 \pm 10\%$ seems marginally

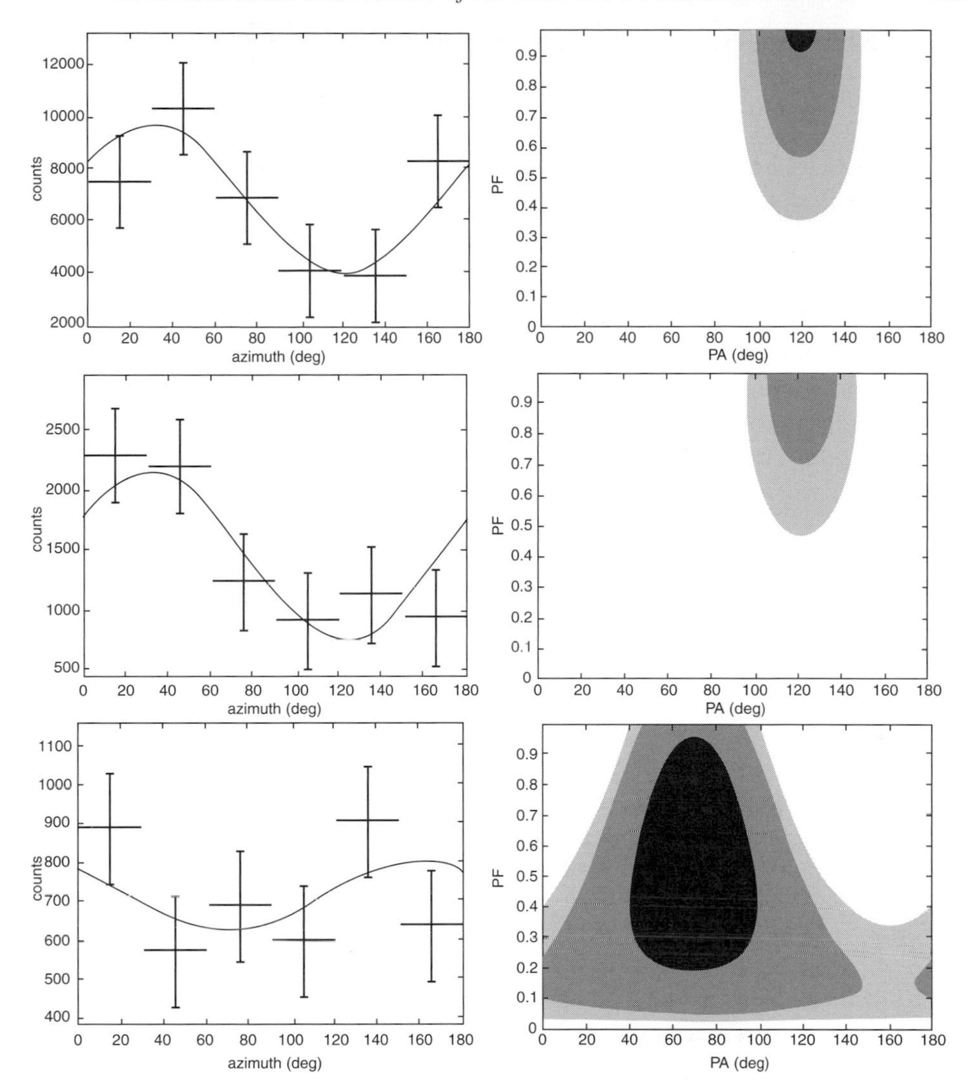

Figure 34.1 Azimuthal profile, modulation angle, PA, and fraction, PF = a_0/a_{100}, measured for the Crab data between 200 and 800 keV, in the off-pulse (top), off-pulse and bridge (middle) and two-peak (bottom) phase intervals. The error bars for the profile are at one sigma. The 68%, 95% and 99% confidence regions are shaded from dark to light gray. The SPI result[1] is indicated in the top figure by a cross.

consistent with our lower limits. However, our lower limit, as noticed previously, is computed for any value of the polarization angle; the SPI and IBIS results are consistent at the 95% level, if we fix the angle at the SPI measured best fit value, as it can be seen on Figure 34.1.

34.5 Observation of variable polarization from GRB 041219A

We also used the IBIS telescope on board the INTEGRAL satellite to measure the polarization of the prompt γ-ray emission of the long and bright γ-ray burst GRB 041219A in the 200–800 keV energy band[4]. We find a variable degree of polarization ranging from less than 4% over the first peak to $43 \pm 25\%$ for the whole second peak. Time-resolved analysis of both peaks indicates a high degree of polarization, and the null average polarization in the first peak can be explained by the rapid variations observed in the polarization angle and degree. The azimuthal distributions of the GRB flux for the different time intervals are reported in Figure 34.2.

Our results are consistent with different models for the prompt emission of GRB at these energies, but they favour synchrotron radiation from a relativistic outflow with a magnetic field which is coherent on an angular size comparable with that of the emitting region ($\approx 1/\Gamma$). Indeed this model has the best capabilities to maintain a high polarization level, and to produce the observed variability.

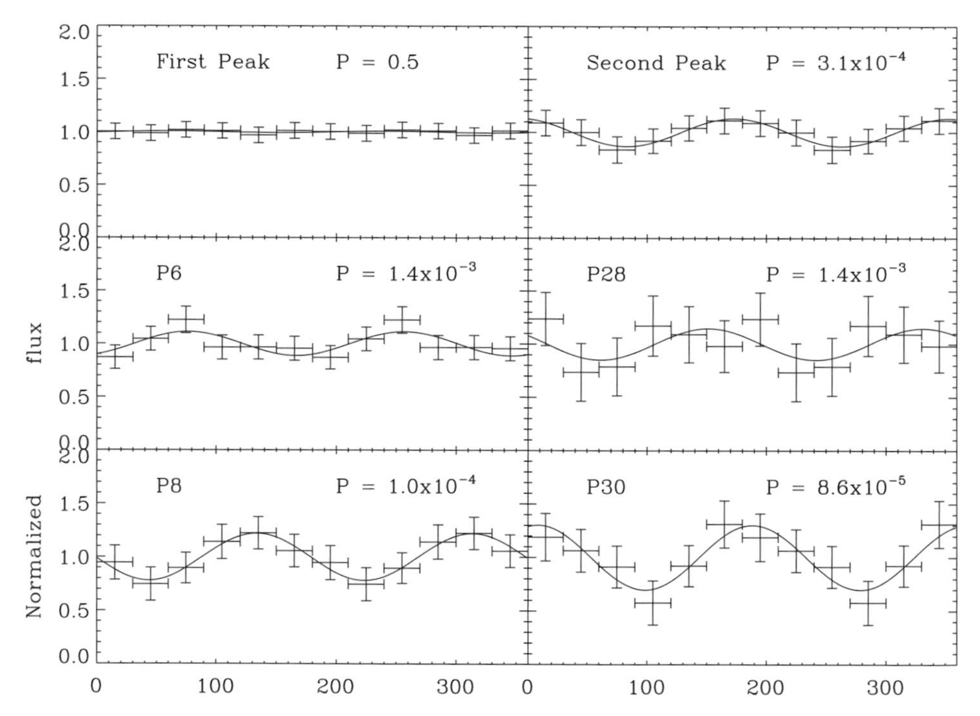

Figure 34.2 Polarigrams of the different time intervals that have been analyzed. For comparison purposes, the curves have been normalized to their average flux level. The crosses represent the data points (replicated once for clarity) and the continuous line the fit done using Equation (34.1). The chance probability of a nonpolarized signal is reported in each panel.

Notes

1. INTEGRAL is an ESA project with instruments and science data centre funded by ESA member states with the participation of Russia and USA.
2. GLEPS is a package for handling polarization in Geant 3 developed by Dr. Mark McConnell at University of New Hampshire, USA

References

[1] Dean, A.J., et al. (2008). Science **321**, 1183.
[2] Forot, M., Laurent P., Lebrun F., & Limousin O. (2007). *ApJ* **668**, 1259.
[3] Kuiper, L., Hermsen, W., Cusumano, G., Diehl, R., Schvnfelder, V., Strong, A., Bennett, K., & McConnell, M. L. (2001). *A&A* **378**, 918.
[4] Gotz, D., Laurent, P., Lebrun, F., Daigne, F., & Bosnjak, Z. (2009). *ApJ* **695**, L208.
[5] Laurent, P., et al. (2003). *A&A* **411**, L185.
[6] Lebrun, F., et al. (2003). *A&A* **411**, L141.
[7] Labanti, C., et al. (2003). *A&A* **411**, L149.
[8] Lei, F., Dean, A. J., & Hills, G. L. (1997). *Space Science Reviews* **82**, 309.
[9] Lyne, A. G., Pritchard, R. S. & Graham-Smith, F. (1993). *MNRAS* **265**, L1003. `http://www.jb.man.ac.uk/ pulsar/crab.html`
[10] McGlynn, S. et al. (2007). *A&A* **466**, 895.
[11] Ng, C.-Y., & Romani, R. W. (2004). *ApJ*, **601**, 479.
[12] Slowikowska, A., Kanbach, G., & Stefanescu, A. (2006). *Proc. IAU General Assembly*, Prague.
[13] Slowikowska, A., Kanbach, G., Kramer, M., & Stefanescu, A. (2008). Proc. of high time resolution astrophysics: The universe at sub-second timescales. *AIP Conference Proceedings* **984**, 51.
[14] Ubertini, P., et al. (2003). *A&A* **411**, L131.
[15] Vaillancourt, J. E., (2006). *PASP* **118**, 1340.
[16] Vinokur, M. (1965). *Annales d'Astrophysique* **28**, 412.
[17] Weisskopf, M. C., Silver, E. H., Kestenbaum, H. L., Long, K. S., & Novick, R. (1978). *ApJ* **220**, L117.
[18] Weisskopf, M. C. (2006). *Proc. of Neutron Stars and Pulsars*, Bad Honnef, Germany.

35

Fermi results on the origin of high-energy emission in pulsars

S. McGlynn, M. Axelsson & F. Ryde

Oskar Klein Centre, AlbaNova University Centre, Stockholm

on behalf of the *Fermi* LAT collaboration

The *Fermi* Gamma-ray Space Telescope has detected approximately 50 pulsars since launch in June 2008. We present results from recent observations of three pulsars (J2021+3651, Vela and J0030+451) and discuss their implications for the origin of the high-energy emission. The unprecedented sensitivity of *Fermi* has led to many new constraints on theoretical models. These results may give new insight into the physics of X-ray production and thereby contribute to our understanding of the polarization.

35.1 Introduction

The Large Area Telescope, LAT[1], is the primary instrument on *Fermi*. It covers an energy range of 20 MeV to more than 300 GeV, with an angular effective area of 8000 cm^2 on-axis at 1 GeV. Event times can be measured with a precision of better than 1 μs.

35.2 J2021+3651

J2021+3651 was previously observed by EGRET[2] as a young and energetic pulsar. It was also observed by AGILE in the 100–1500 MeV band[3]. Two pulses are clearly distinguished in the histogram (Figure 35.1). The first peak has a lag with respect to the radio of 0.162 ± 0.004 (statistical) $\pm$ 0.01 (systematic) in phase. The spectrum was fitted with an exponential cut-off power law with index $\Gamma = 1.5 \pm 0.1 \pm 0.1$ and cut-off energy at $2.4 \pm 0.3 \pm 0.5$ GeV (Figure 35.1). The rejection of a super-exponential cut-off fit to the spectrum favours an outer-magnetosphere

X-ray Polarimetry: A New Window in Astrophysics, eds. R. Bellazzini, E. Costa, G. Matt and G. Tagliaferri. Published by Cambridge University Press.

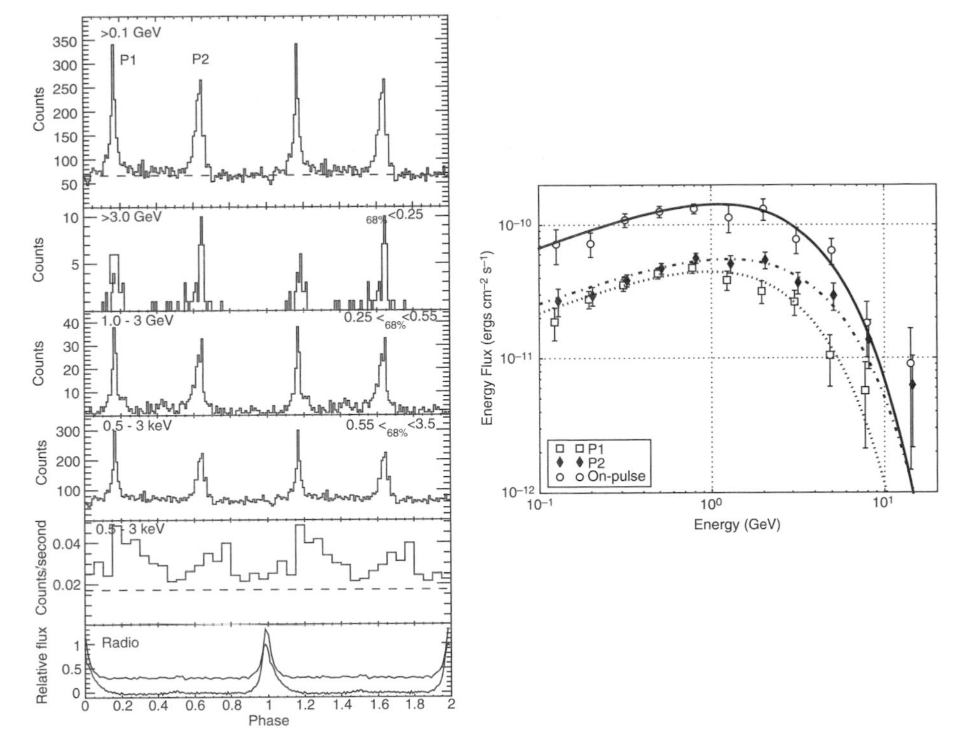

Figure 35.1 Left: Phase histograms of J2021+3651 with energy ranges indicated in the top left corner of each plot (two phases are shown). Bottom panel: Radio lightcurve at 1950 MHz. Right: Spectral energy distribution for Pulse 1 (P1), Pulse 2 (P2) and summed pulses (Total). The error bars are statistical. The pulses are offset by 2.5% from the summed pulses for clarity.

origin[4]. Narrow gamma-ray pulses also arise naturally in outer-magnetosphere models, where the peaks are created by pile-up of events from different altitudes in the magnetosphere[5].

35.3 Vela pulsar

Vela is one of the brightest persistent sources in the gamma-ray sky and is an excellent target for GeV observatories. The pulses are extremely well defined with 32 400 photons above 0.03 GeV (Figure 35.2) and a clear evolution of the pulses with energy. P1 shrinks in relation to P2 with increasing energy, while a third pulse becomes obvious at $E > 1$ GeV. The best fit to the spectrum is an exponential cut-off power law with index $\Gamma = 1.51\,^{+0.05}_{-0.04}$ and cut-off energy at $2.86 \pm 0.09 \pm 0.17$ GeV (Figure 35.2). The EGRET data from [6] is shown for comparison. The fast rise and slow fall of P1 implies a leading caustic, while the slow rise and abrupt fall of P2 indicates a trailing caustic[7]. The polar-cap model is therefore disfavoured for Vela.

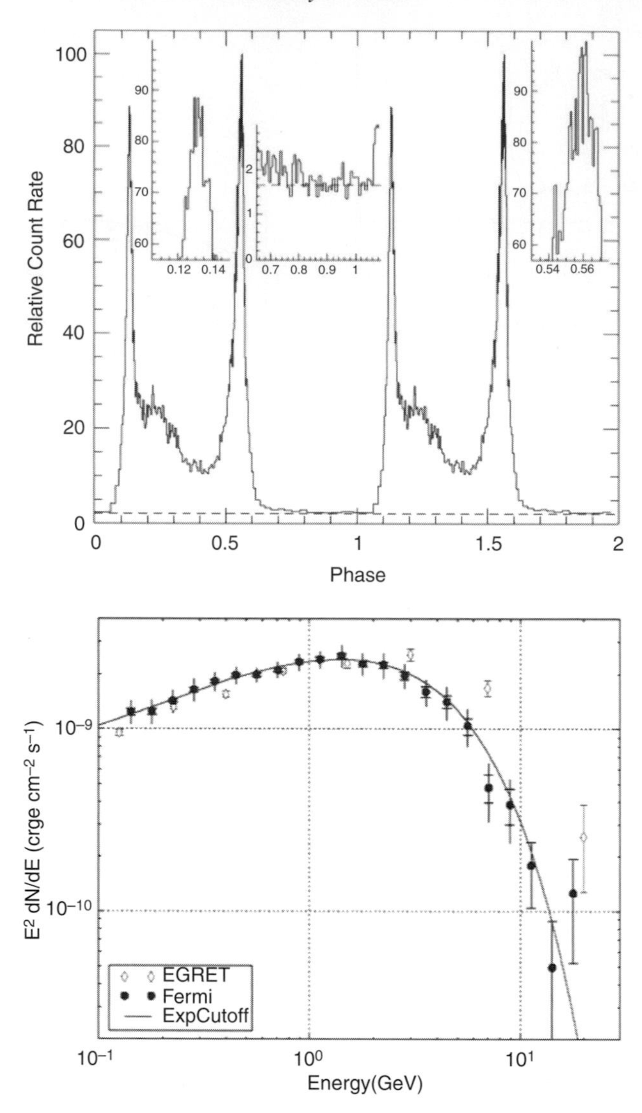

Figure 35.2 Top: 0.1–10 GeV phase histogram of the Vela pulsar with variable width bins (200 counts/bin). The pulse profile is normalized to 100 at the pulse peak. The insets show the pulse structure near each peak (top and bottom) and the average background (central inset). Bottom: Phase-averaged SED. *Fermi*–LAT data are shown in black; EGRET data points (white circles) are shown for comparison. Statistical and systematic errors are shown.

35.4 J0030+451

J0030+451 is the second millisecond pulsar ($P = 4.87$ ms) discovered in gamma-rays. These pulsars are generally old stars in binary systems, with a pulse shape similar to normal pulsars. There is no evidence for pulsed emission between 20 and 100 MeV. The first pulse, P1, appears to be harder than Pulse 2. The radio and X-ray

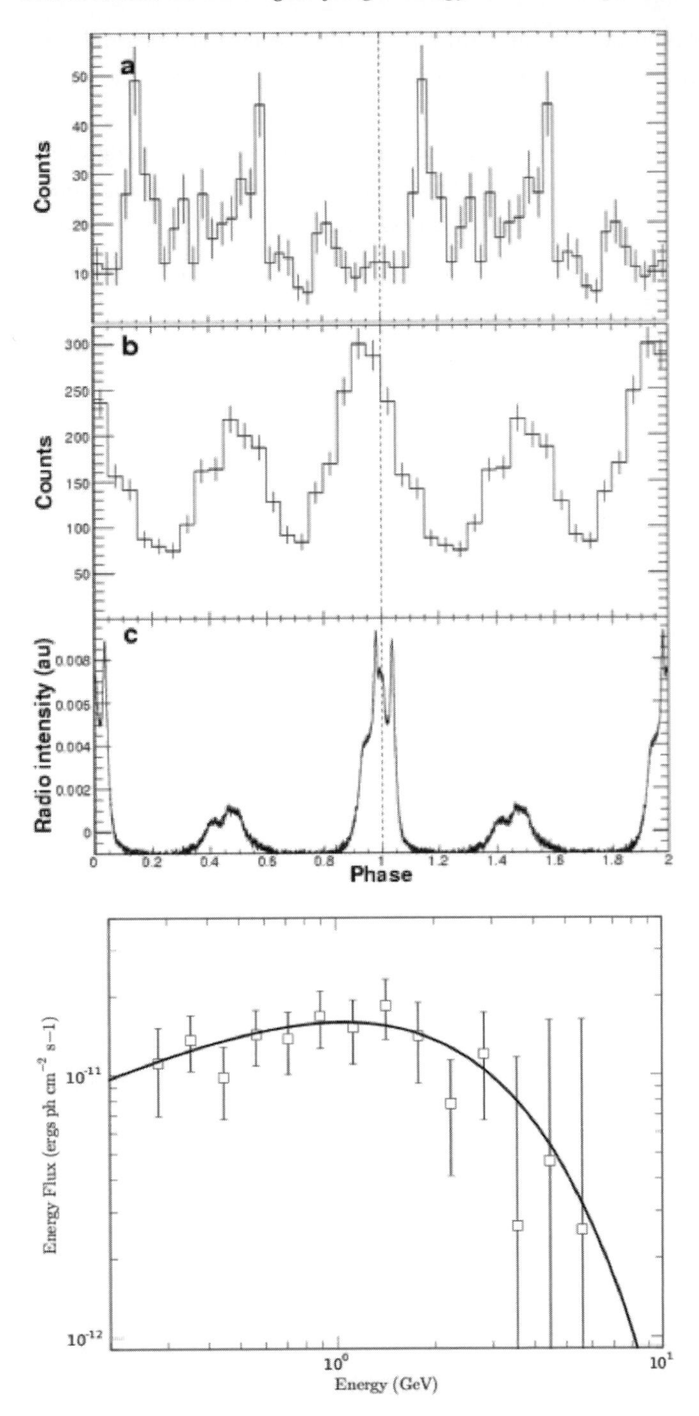

Figure 35.3 Top: Multiwavelength phase histograms of J0030+45: (a) gamma-ray emission with energy $E > 100$ MeV. (b) XMM-Newton histogram in the energy range 0.3–2.5 keV. (c) Radio profile at 1.4 GHz. Bottom: Phase-averaged SED fit with an exponential cut-off power law. The error bars are statistical.

emission are aligned, but the gamma-ray emission is offset from both (Figure 35.3). The spectrum was fitted with an exponential cut-off power law with cut-off energy at 1.7 ± 0.4 ± 0.5 GeV (Figure 35.3). The X-ray and radio emission are phase aligned and thus seem to have a common origin in the magnetosphere, but the gamma-rays may be produced in a different region[8]. This may have implications for polarization models, which often assume that the X-ray and gamma-ray emission are correlated.

35.5 Conclusions

The lag between the radio and gamma-ray pulses and a spectrum with an exponential cut off close to 2 GeV are common features of all the pulsars discussed here. Emission from the outer magnetosphere seems to be favoured over other emission mechanisms, particularly the polar-cap model. The different models have varying polarization characteristics, and future polarization measurements could discriminate between the emission regions. This may have implications for simulating polarization signals from pulsars, where the X-ray emission is extrapolated to gamma-ray energies and assumed to have a similar form.

References

[1] Atwood, W. B., Abdo, A. A., Ackermann, M. et al. (2009). *ApJ* **697**, 1071–1102.
[2] Roberts, M. S. E., Hessels, J. W. T., Ransom, S. M., et al. (2009). *ApJL* **577**, L19–L22.
[3] Halpern, J. P., Camilo, F., Giuliani, A., et al. (2008). *ApJ* **688**, L33–L36.
[4] Abdo et al. (2009). *ApJL* **695**, L72–L77.
[5] Dyks, J. & Rudak, B. (2003). *ApJ* **598**, 1201–1206.
[6] Kanbach, G., Arzoumanian, Z., Bertsch, D. L., et al. (1994). *A & A* **289**, 855–867.
[7] Abdo et al. (2009). *ApJ* **696**, 1084–1093.
[8] Abdo et al. (2009). Submitted to *ApJ*.

36

Diagnostics of the evolution of spiral galaxies in a cluster environment

M. Weżgowiec
Obserwatorium Astronomiczne UJ

M. Ehle
ESA, European Space Astronomy Centre

M. Urbanik, K.T. Chyży & M. Soida
Obserwatorium Astronomiczne UJ

R. Beck
Max-Planck-Institut für Radioastronomie

B. Vollmer
CDS, Obs. Astronomique de Strasbourg

We present X-ray and radio polarimetric observations of selected Virgo Cluster spiral galaxies. The X-ray extended emission traces hot-gas filaments in galactic halos and is sensitive to the environmental effects exerted by interactions inside the cluster, like ram-pressure stripping. The radio polarization studies provide clues about alignment, distortion, compression and strength of detected magnetic fields. When used together, the two types of observations constitute an excellent tool for examining disturbances in galactic disks and halos caused by interactions of galaxies with the intracluster medium or between the galaxies themselves. The coming of age of X-ray polarimetry could provide us with unprecedented tools to explore further the evolution of galaxies in a cluster environment.

36.1 Introduction

There are a number of environmental effects in clusters of galaxies that modify spiral galaxies causing their HI deficiency, Hα spatial truncation and distortions, as well

X-ray Polarimetry: A New Window in Astrophysics, eds. R. Bellazzini, E. Costa, G. Matt and G. Tagliaferri. Published by Cambridge University Press.

as strong dynamical and morphological evolution[1]. They influence also galactic magnetic fields, which can be stretched and compressed while the distribution of the ISM is being changed. The Virgo Cluster is the best system to study such processes due to its proximity and large spiral galaxy content. In this work we investigate galaxy–galaxy and galaxy–ICM interactions in the Virgo Cluster in the radio and in the soft X-ray bands. We make use of the VLA and Effelsberg 100-m radio telescope to obtain high sensitivity for the extended total power and polarized nonthermal radio emission. For the X-ray part of our studies, the XMM-Newton observations assure high sensitivity for the diffuse, extended X-ray emission of hot gas.

In the following we present preliminary results for a couple of Virgo galaxies from our sample. The detailed analysis and interpretation of these and other cluster members will be presented in the PhD thesis by Weżgowiec.

36.2 Virgo spirals in the hot ICM

NGC 4254 is an Sc galaxy situated outside the Virgo Cluster core. It is a rapidly star-forming galaxy with no signs of any gas deficiency. The distribution of the radio emission of NGC 4254 (Figure 36.1, left) is disturbed and shows an unusual, sharp and bright polarized ridge in the southern disk.

This, together with the map of the soft X-ray emission (Figure 36.1, right), which shows no out-of-disk emission or shocks, leads to the conclusion that perturbations

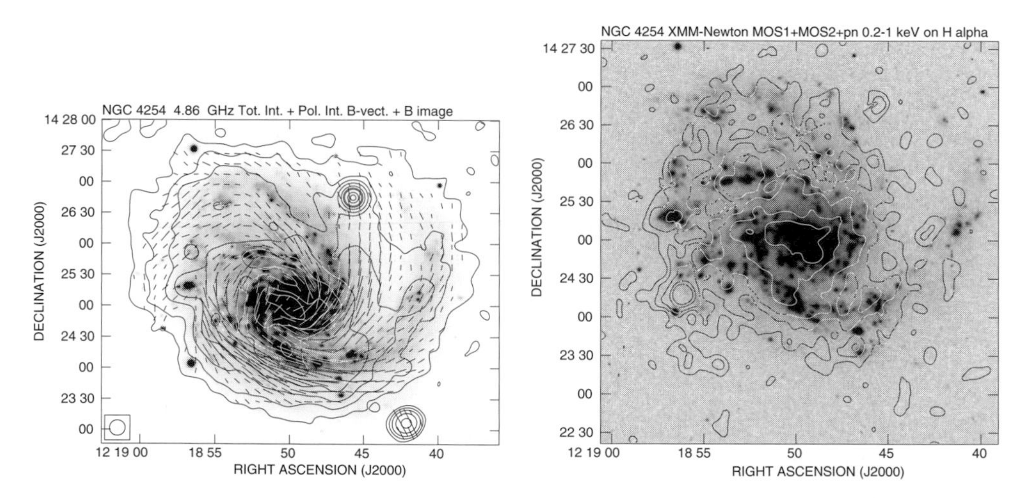

Figure 36.1 Left: Contours of total radio intensity of NGC 4254 at 8.46 GHz with a resolution of 15″ and B-vectors proportional to the polarized intensity overlaid on an optical blue image. The contours are at 30, 80, 160, 250 μJy/beam area (from [3]). Right: X-ray contours of soft 0.2–1 keV emission with 15″ resolution overlaid on an Hα image of NGC 4254.

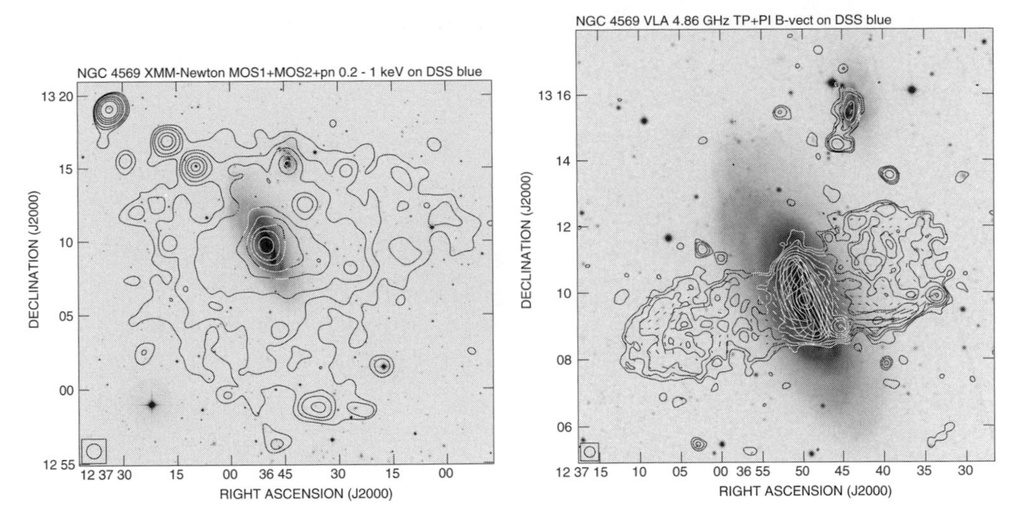

Figure 36.2 Left: X-ray contours of soft emission with 1' resolution overlaid onto the DSS blue image of NGC 4569. The contours are 3, 5,8, 16, 25, 40, 60, 100 × rms. Right: High resolution radio total power map of NGC 4569 at 4.86 GHz with superimposed B-vectors of polarized intensity, overlaid upon the blue image from the DSS. The contour levels are 3, 5, 8, 20, 50, 100, 200, 400 × 8.6 μJy/beam area. The polarization vector of 1' corresponds to a polarized intensity of 0.17 mJy/beam area. The angular resolution is 20".

in NGC 4254 are caused by tidal interactions, possibly with the nearby dark galaxy VIRGOHI 21.

NGC 4569, an SABa galaxy situated quite close to the Virgo Cluster centre, shows extended radio lobes (Figure 36.2, right) uncommon in normal (anaemic) spirals[2]. Such structures might suggest the galaxy's nuclear activity in the past.

Extra-planar emission is visible in a soft X-ray map at high angular resolution as 'filaments' on the NW side of the galactic disk, possibly tracing outflowing hot gas. A lower resolution version of the soft X-ray map (Figure 36.2, left), which is sensitive for more extended structures, actually reveals the presence of a giant hot gas halo elongated roughly in the direction of the extended radio lobes. North of NGC 4569, still within its halo, is the irregular galaxy IC 3583 which is suspected to interact with NGC 4569[6].

The comet-like galaxy NGC 4654 shows polarized emission concentrated on one side of the galaxy (Figure 36.3). Recent simulations of the magnetic field of this galaxy[7; 5] suggest that the best explanation of the radio polarization structure of NGC 4654 are the combined effects of ram pressure and tidal interactions. A comparison with sensitive X-ray data to check for any peculiarities in the distribution of the soft X-ray emission is still lacking.

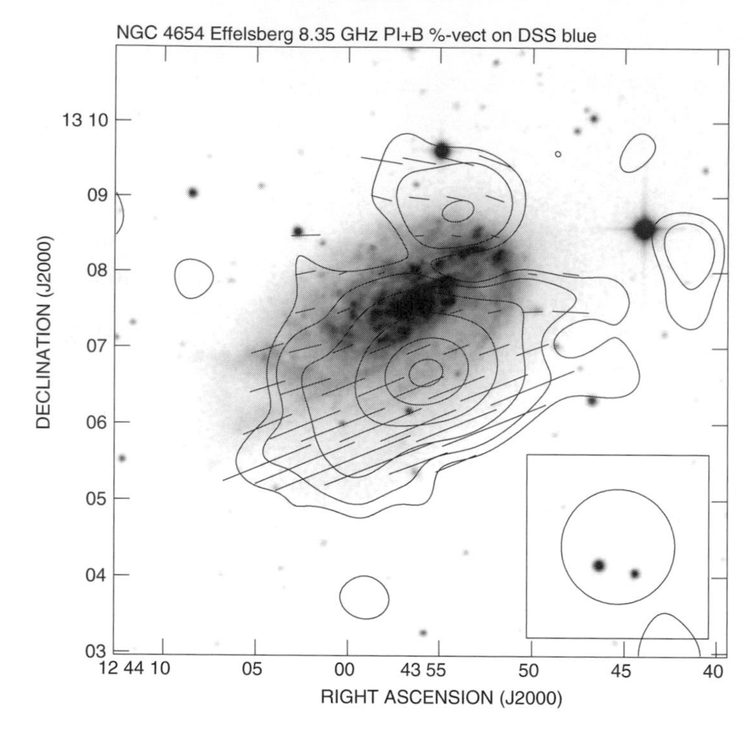

Figure 36.3 The map of polarized intensity of NGC 4654 at 8.45 GHz with apparent B-vectors of polarization degree overlaid on to the image from the DSS (from [8]). The contours are 3.5, 5, 8, 13, 18, 21 $\times$ 0.03 mJy/beam area and a vector of $1'$ length corresponds to the polarization degree of 12.5%. The map resolution is $1.5'$.

36.3 Summary and outlook

Virgo Cluster galaxies show evidence of a variety of effects occurring in a cluster environment, which can be investigated in various spectral domains. Especially, our polarized radio intensity together with X-ray studies can help to trace magnetic field distortions as well as hot-medium spatial and spectral distribution. Observations of a large sample of Virgo Cluster objects will provide valuable information about the whole system and characteristic features of its environment.

It will be highly interesting to see if the planned generation of X-ray polarimeters will be suited to measure any polarization of the X-ray photons around Virgo Cluster members at the interface between the ISM and ICM.

Although galactic magnetic fields are likely too weak to polarize X-ray photons, it will be interesting to observe if outflows and interaction processes are leaving their mark in the predicted polarization of resonance X-ray lines from clusters of galaxies (see e.g. [4]).

Notes

This work is supported by a grant from the Polish Ministry of Science and High Education (no. 3965/B/H03/2008/34) and the ESAC Faculty Visiting Scientist Programme.

References

[1] Boselli, A. & Gavazzi, G. (2006). *PASP* **118**, 517.
[2] Chyży, K. T., Soida, M., Bomans, D. S. et al. (2006). *A&A* **447**, 465.
[3] Chyży, K. T., Ehle, M., & Beck, R. (2007). *A&A* **474**, 415.
[4] Sazonov, S. Yu, Churazov, E. M., & Sunyaev, R. A. (2002). *MNRAS* **333**, 191.
[5] Soida, M., Otmianowska-Mazur, K., Chyży, K. T., & Vollmer, B. (2006). *A&A* **458**, 727.
[6] Tschöke, D., Bomans, D. J. et al. (2001). *A&A* **380**, 40.
[7] Vollmer, B. (2003). *A&A* **398**, 525.
[8] Weżgowiec, M., Urbanik, M., Vollmer, B. et al. (2007). *A&A* **471**, 93.

Part III

Future missions

Part III

37

Gravity and Extreme Magnetism SMEX (GEMS)

J. Swank, T. Kallman & K. Jahoda
GSFC

K. Black
GSFC/Rock Creek Scientific

P. Deines-Jones
GSFC

P. Kaaret
University of Iowa

The prime scientific objectives of the NASA Small Explorer mission, Gravity and Extreme Magnetism SMEX, or "GEMS", are to determine the effects of the spin of black holes, the configurations of the magnetic fields of magnetars, and the structure of the supernova shocks which accelerate cosmic rays. In the cases of both stellar black holes and supermassive black holes, sensitivity to 1% polarization is needed to make diagnostic measurements of the net polarizations predicted for probable disk and corona models. GEMS can reach this goal for several Seyferts and quasars and measure the polarizations of representatives of a variety of other classes of X-ray sources, such as rotation-powered and accretion-powered pulsars. GEMS uses foil mirrors to maximize the collecting area achievable within the SMEX constraints. The polarimeters at the mirror foci are time projection chambers which use the photoelectic effect to measure the polarization of the incident photon. We have built laboratory models with good efficiency and modulation in the 2–10 keV range. An attached small student experiment would add 0.5 keV sensitivity for bright, soft sources. The instrument has a point spread function which allows measurement of structures in the brighter nearby supernova remnants. GEMS' Orbital Sciences spacecraft will rotate at a rate of 0.1 revolutions per minute during observations, so that systematic errors due to the detector can be detected and corrected. A program of 35 sources can be observed in 9 months. GEMS is designed for a two-year lifetime which will allow a General Observer program that would more than double the number of sources measured. For subsets of black holes, neutron stars

X-ray Polarimetry: A New Window in Astrophysics, eds. R. Bellazzini, E. Costa, G. Matt and G. Tagliaferri. Published by Cambridge University Press.

and supernova remnants, GEMS will measure the polarization of several sources, solving important questions while establishing the sensitivity required for future missions.

37.1 Introduction

The Gravity and Extreme Magnetism SMEX (GEMS) was proposed to NASA's 2007 SMEX AO in January 2008. It was one of six proposed missions selected for phase A studies. A study report was submitted in December 2008. The review continued during the spring of 2009 with presentations to the reviewers and selection officials to address questions. After this meeting on the Coming of Age of X-Ray Polarimetry, GEMS was one of two SMEX missions selected to have an opportunity for flight. In this paper, we describe the particular science objectives of GEMS that have defined the requirements for the mission. A program has been proposed, which will be reviewed before launch in consideration of what is then known. There are many other targets for which GEMS could be used. We give a brief overview of the instrument and the mission. The instrument depends on a polarimeter which uses the photoelectric effect. The polarimeter is a time projection chamber. We briefly describe the polarimeter design and our demonstrations of its performance.

37.2 GEMS scientific objectives

37.2.1 A sensitive survey

At this point in the history of efforts to measure the polarization of the X-ray flux from astrophysical sources, still the only definitive measurements have been for the Crab nebula[1; 2]. INTEGRAL has obtained provocative results for the Crab pulsar above 10 keV[3]. Theoretical predictions between a few and tens of percent polarization have been been made for several types of black holes, neutron stars, and other supernova remnants. We are interested in answering questions about these sources. The GEMS sensitivity is shown in Figure 37.1. This will open the frontier of X-ray polarimetry of the main emission from these sources, which is in the energy range 2–10 keV.

The predicted polarizations depend on several parameters in general. Several targets in a class should be observed to sample these parameters. We selected 15 black holes, 11 pulsars and neutron stars, and six supernovae remnants. (Pulsars in two supernova remnants and the Galactic center source Sgr B2 bring the number of targets to 35.) Plans for subsequent polarimetry missions will know which classes of sources turn out to have polarization that is most interesting and should be further explored. GEMS results will be a guide to the importance of polarization in this energy range.

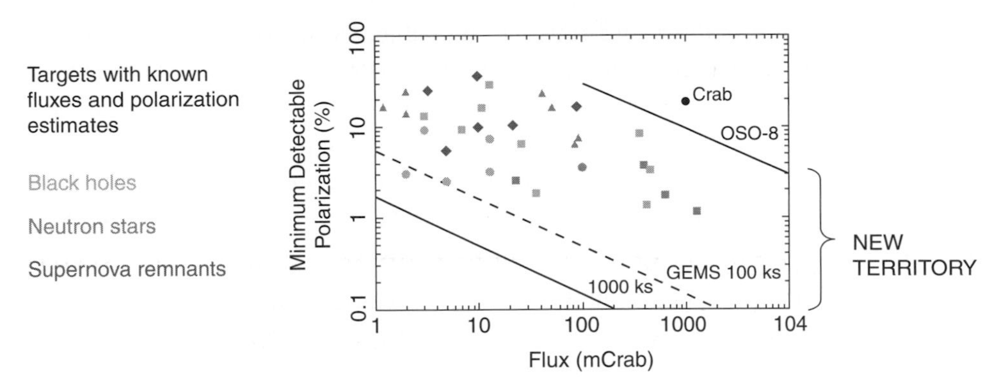

Figure 37.1 GEMS sensitivity and discovery phase space. The minimum detectable polarization (MDP) at 99% confidence is shown for observations of 10^6 and 10^5 seconds. Possible values for polarization of proposed targets, according to some predictions, are shown at the known fluxes of the targets.

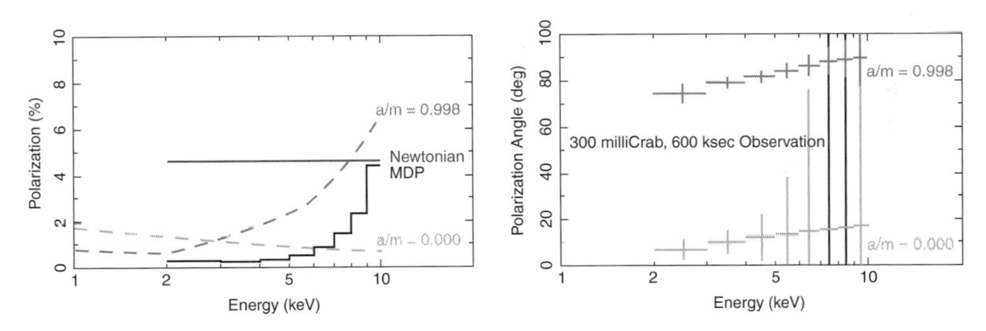

Figure 37.2 General Relativity effects on the polarization. The GR predictions for accretion on a black hole in the high-soft state are distinct from Newtonian predictions and, in GR, high- and low-spin are distinguished.

37.2.2 Black holes

Rees[4], near the beginning of X-ray astronomy, pointed out that X-ray polarization had been predicted by Chandrasekar for accretion disks. In 1980, Connors *et al.*[5] calculated the effects of a black hole on the polarization and its angle. Their results were explored in more detail by Li et al.[6], using the same approach. Schnittman and Krolik[7] used a different approach and it makes a difference. The model of an optically thick disk accreting on to a spinning black hole can be specified in detail. The polarization and its angle determine the black hole spin and the disk inclination. This makes this type of target especially interesting. Figure 37.2 shows simulations of possible results for a high-state black hole of high inclination (75°).

Polarization is a degree of freedom that has not yet been available for the exploration of the geometry of X-ray sources. Modeling the X-ray spectral and timing data for black holes in their hard states is still subject to frustrating uncertainty. Is the accretion in a spherical corona or a sandwich of a dense disk? There has not been consensus that all the data definitively pick out one or the other. Polarization measurements can resolve this ambiguity. Predicted levels of polarization are sensitive to the inclination angles of a geometrically thick, scattering corona. AGN are the most challenging targets we have for GEMS, because most of them are relatively faint X-ray sources and the predicted polarizations are mostly in the few percent range. We propose six Seyferts and quasars as part of our observing program, for observation times which can detect a level of 1% for the polarization averaged over the GEMS energy band.

37.2.3 Neutron stars

The proposed GEMS program includes neutron stars with pulsations arising from different mechanisms. Questions still attend the location of the x-ray emission in rotation-powered pulsars, in accretion-powered pulsars, and in magnetars. Anomalous X-ray pulsars and soft-gamma repeaters are both estimated to have fields on the order of 5×10^{14} gauss. The radiative transfer in the neutron-star crust is expected to leave only the polarization perpendicular to the plane common to the magnetic field and the propagation direction. Van Adelsberg and Lai[8] have calculated the polarizations to be expected from the neutron star surface as it rotates about an axis. Magnetospheric fields are also estimated to be very strong in these sources and would give rise to polarization of the radiation emitted by plasma confined there. Pulse-phased polarization will help define the geometry and determine the location of the emission, and test ideas about these sources.

37.2.4 Supernova remnants

X-ray polarization confirmed the idea that the X-ray emission from the Crab Nebula is synchrotron radiation. Some remnants with clear shell structure and no identified pulsar wind nebula also have synchrotron emission from accelerated electrons, as indicated by power-law spectra and TeV emission. From *Suzaku* and *Chandra* observations, Uchiyama et al.[9] deduced that magnetic fields in the shocks are amplified from the swept-up fields and that the synchrotron interpretation is secure. Questions remain about the way the fields are amplified and whether the fields in the shocks are turbulent. X-ray polarization will be an important diagnostic tool. The GEMS instrument does not image the sky, but the field of view is small compared to the diameters of several supernova remnants and selective pointings can examine the polarization of the shells.

37.3 Overview of the GEMS mission

37.3.1 X-ray polarimeter instrument

The GEMS primary instrument, the X-ray polarimeter instrument, or XPI, uses the photoelectric effect to measure the linear polarization of X-ray flux in the energy range 2–10 keV. It comprises three telescopes which focus source flux into photoelectric polarimeters which employ a time projection chamber (TPC) readout geometry and have been developed at GSFC for this application (Black et al.[10]). The polarimeters record images of the photoelectron tracks produced after the absorption of X-rays. The initial direction of each track is correlated with the photon electric field direction; the distribution of directions from an ensemble of tracks gives the source polarization.

Costa et al.[11] first showed that gas detector development and read out technology had reached the point where photoelectrons in the 2–10 keV range could be tracked and the initial direction determined. This work was followed by development of thin gas pixel detectors (e.g. [12]). The TPC detectors use the same effects, but with a geometry which allows the detector to have higher quantum efficiency.

Within the polarimeter, an X-ray ionizes a gas atom, usually ejecting a K electron, whose initial direction is correlated with the photon electric field, and an isotropically emitted Auger electron. The photoelectron loses energy within the gas, leaving an ionization track which drifts in a uniform field to a micro-pattern gas detector (MPGD) multiplication stage (a gas electron multiplier) and a 1-D array of readout strips. Continuously sampling the output provides the second dimension of the track image. The data from these strips allows an image of the primary-electron track to be reconstructed as indicated in the TPC diagram in Figure 37.3. From this image the initial direction of the ejected electron is deduced.

The mirror focal lengths are 4.5 m. This was the focal length to the XRS instrument on *Suzaku*. The GEMS mirrors will be 33 cm, rather than the XRS value, 40 cm in diameter. The combined area of three telescopes is 510 cm^2 at 6 keV. The point spread function of these telescopes will be 1.5 arc-minutes. The effective field of view is 12 arc-minutes in diameter.

The XPI and the spacecraft bus rotate around the science axis at 0.1 rpm. Submillisecond time tagging of each event relates the track direction (measured in detector coordinates) to sky coordinates. The ensemble of track directions in detector coordinates is a sensitive measure of uncalibrated asymmetries in the detector response (and thus provides input to the calibration model). Since the experiment uniformly samples all sky angles at all detector angles, such asymmetries are effectively averaged out.

The SMEX program encourages the inclusion of student experiments which enhance the baseline mission. GEMS includes a student experiment from the

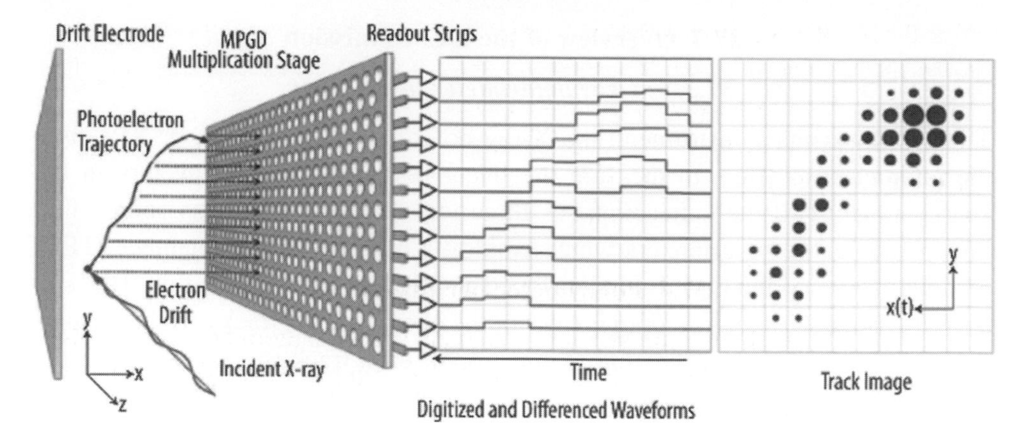

Figure 37.3 Diagram of a TPC and the track reconstruction. The "y" coordinate comes from the strip number. The "x" coordinate from the relative arrival time of the signal. The track is made over a time that is very short compared to the drift time. The velocity of drift is calibrated from other measurements.

University of Iowa. It uses the polarization dependence of Bragg reflection and the rotation of the instrument to provide additional sensitivity for bright, soft sources, in a narrow window near 0.5 keV.

37.3.2 Mission design

The SMEX missions are to be launched on Pegasus class rockets. Accommodation of the 4.5-m focal length requires an in-orbit deployment mechanism. GEMS employs a coilable boom supplied by ATK Space Systems. The mission is designed with a lifetime goal of 2 years, which is accommodated by an altitude of 575 km and an inclination of 28.5°.

The solar arrays are sized such that with the rotation, the pointing can be $90° \pm 30°$ from the sun. This means that a given source on the equator can be observed for as much as 2 months, and higher latitude sources for longer periods. A given pointing position may be occulted by the earth every orbit and passage through the high charged particle regions of the South Atlantic Anomaly will interrupt data for minutes to a half-hour during about half of the 15 spacecraft orbits each day. Taking an average efficiency estimate of 50% good observing time, a nine-month observing program for 35 sources has been developed, which would sample nine types of sources.

Spacecraft operations will be conducted at the multimission operations center of Orbital Sciences Corporation, with data retrieved by a science operations center at Goddard. The data will be archived in HEASARC. Software will be developed within the FTOOLS system that is distributed by HEASARC. Calibration and examination of the data from a new instrument are estimated to take six months, after which data and results will be accessible from HEASARC.

If the proposed science objectives are achieved on schedule, the goal of a two-year lifetime would make a a General Observer program possible.

37.4 Polarimeter performance

The TPCs will each have four MPGDs with a hexagonal pattern and pitch of 140 μm. The anode strips which detect the charge have a pitch of 121 μm. The MPGDs were developed by Riken and SciEnergy (Hayato and Tamagawa, this volume). Figure 37.4 shows the effective area and the modulation for the configuration. The modulation was measured at 2.7, 3.5, and 4.5 keV and estimated at other energies, for 180 torr of Dimethyl Ether (DME).

The MDP shown in Figure 37.1 for observations with a data accumulation time T has been calculated as $MDP = (4.29/(\mu r))\sqrt{(r+b)/T}$, where r is the source rate and b is the unrejected background rate. For a Crab-like spectrum the source rate will be 0.64 counts s^{-1} per mCrab. We have estimated that we can reject background events well enough that b will be negligible except for the weakest sources.

Figure 37.5 shows a sample modulation curve of an engineering demonstration unit, which reads out the data with the APV25 ASIC selected for flight. This ASIC was designed by Imperial College and Rutherford Appleton Laboratories for use in the Compact Muon Solenoid (CMS) experiment at CERN. It provides continuous

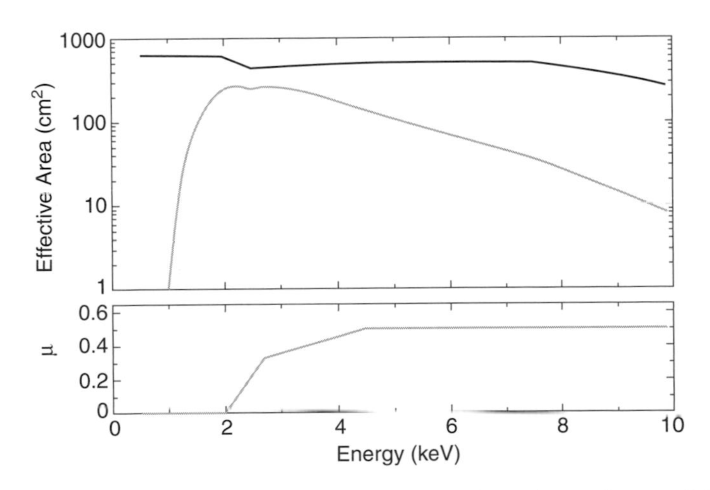

Figure 37.4 GEMS performance characteristics. Upper: the total area of the three telescopes and the effective area when the polarimeter quantum efficiencies are taken into account. Lower: the modulation for a 100% polarized incidence flux.

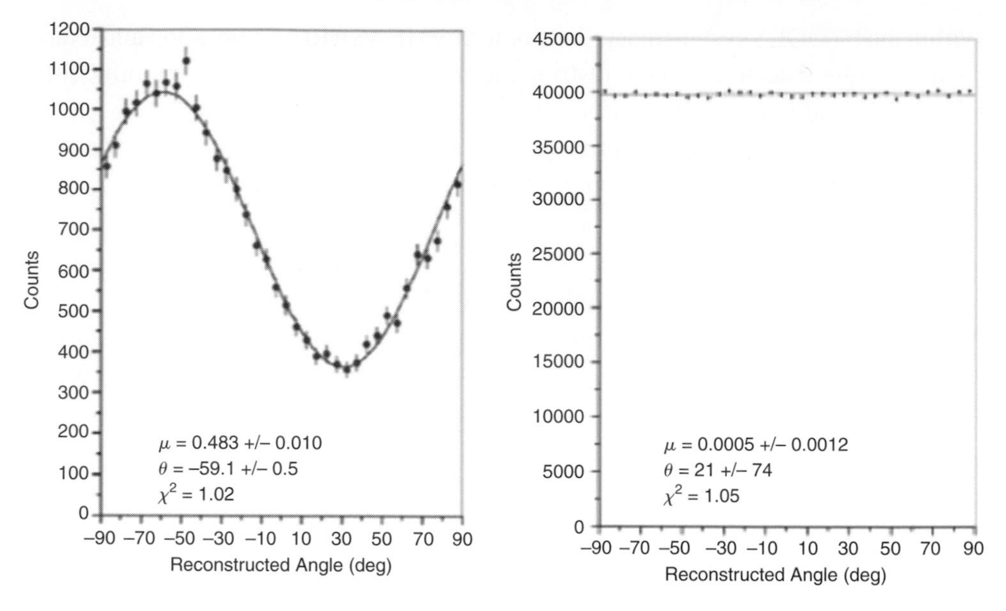

Figure 37.5 Data from polarized and unpolarized sources. Left: the response to a polarized 4.5 keV line from Ti. Right: the response to an unpolarized 5.9 keV ^{55}Fe source averaged over the simulated rotation of the detector.

sampling at 40 MHz. The unpolarized data have been analyzed as they would be from a polarimeter on a rotating spacecraft.

37.5 Conclusions

Theories of stellar and supermassive black holes, strongly magnetic neutron stars and supernovae remnants have important unresolved questions. Polarization measurements can provide guidance. GEMS is a small polarization mission, now with a launch date in 2014, that can obtain significant results and prove X-ray polarization's usefulness.

References

[1] Novick, R., Weisskopf, M. C., Berthelsdorf, R. L. & Wolff, R. S. (1972). *ApJ* **174**, L1–L8.

[2] Weisskopf, M. C., Cohen, G. G., Kestenbaum, H. L., Long, K. S., Novick, R. & Wolff, R. S. (1981). *ApJ* **208**, L125–L128.

[3] Forot, M., Laurent, P., Grenier, I. A., Gouiffes, C. & Lebrun, F. (2008). *ApJ* **688**, L29–L32.

[4] Rees, M. J. (1975). *MNRAS* **171**, 457–465.

[5] Connors, P. A., Stark, R. F. & Piran, T. (1980). *ApJ* **235**, 224–244.

[6] Li, L., Narayan, R., & McClintock, J. (2009). *ApJ* **691**, 847–865.

[7] Schnittman, J. D. & Krolik, J. H. (2009). *ApJ* (submitted).
[8] van Adelsberg, M. & Lai, D. (2006). *MNRAS* **373**, 1495–1522.
[9] Uchiyama, Y., Aharonian, F. A., Tanaka, T., Takahashi, T. & Maeda, Y. (2007). *Nature* **449**, 576–578.
[10] Black, J. K., Baker, R. G., Deines-Jones, P., Hill, J. E. & Jahoda, K. (2007) *Nucl. Instr. Meth. A* **581**, 755–760.
[11] Costa, E., Soffitta, P., Bellazzini, R., Brez, A., Lumb, N. & Spandre, G. (2001). *Nature* **411**, 662–665.
[12] Bellazzini, R., et al. (2007) *Nucl. Instr. Meth. A* **579**, 853–858.

38

Programs of X-ray polarimetry in Italy

E. Costa, A. Argan, S. Di Cosimo, S. Fabiani, M. Feroci, F. Lazzarotto, F. Muleri, A. Rubini, P. Soffitta & A. Trois

INAF/IASF-Roma

R. Bellazzini, A. Brez, M. Minuti, M. Pinchera & G. Spandre

INFN-Pisa

G. Tagliaferri, S.Basso, O. Citterio, V. Cotroneo & G. Pareschi

INAF/OAB-Milano

G. Matt

University of Roma3-Roma

Programs of X-ray polarimetry in Italy arise from the convergence of a long experience of X-ray astronomy missions with an outstanding tradition in development of radiation detectors. The gas pixel detector in the focus of X-ray optics can perform angular-resolved polarimetry with a breakthrough improvement in sensitivity, even with a moderate collecting surface. *POLARIX* makes a large use of already existing items and, in particular, of the three telescopes from the *JET-X* program. It can extend the X-ray polarimetry from one positive detection only, to tens of sources, including a few brighter extragalactics: an ambitious pathfinder on a very limited budget. Phase A study of *POLARIX*, and of four other missions, was performed in 2008 and ASI should select two missions to fly. Another pathfinder is under study: two short telescopes, designed with modern tight packing techniques, mounted as piggy-back on the Chinese mission *HXMT*.

38.1 The national context

X-ray polarimetry has been a line of research at IASF for many years. A scattering polarimeter was proposed for the *XMM* mission[1], but was not selected. An IASF team joined the collaboration headed by R. Novick for the Stellar X-ray Polarimeter[2], aboard the *Spectrum X-Gamma* mission, contributing the detectors as the sensors for both scattering and diffraction stages[3]. SXRP was completed but the

X-ray Polarimetry: A New Window in Astrophysics, eds. R. Bellazzini, E. Costa, G. Matt and G. Tagliaferri. Published by Cambridge University Press.

whole mission was not. Considering the large area of the telescope, SXRP was close to the best achievable with conventional techniques. While rejuvenating this discipline it showed the limits of the conventional approach in terms of sensitivity and of imaging. A significant step forward could only be possible by implementing the concept of photoelectric polarimetry, often proposed, but never implemented because of the difficulty of building a gas detector with the needed fine subdivision. (For a review see the paper by R. Bellazzini and G. Spandre[4] in this same volume).

In 1993 the IASF team started a collaboration with the team of INFN of Pisa that was distinguished in bringing the microstrip technique to the highest performance level and in developing new configurations such as the microgap[5] and the microwell[6]. A first demonstration of the possibility to detect the effects of the track orientation was given, in one dimension, with a microgap detector filled with a low-Z mixture[7; 8]. The further step was the passage from strips to pixels. The Pisa team made a first implementation in 2001 by using a GEM to amplify the photoelectron track and a technique of multi-layer printed circuit board to route the signals from metal pads to external electronics[9]. This demonstrated the feasibility of the imaging photoelectric polarimeter but the technical solution was still limited in number and size of pixels and cumbersome in read-out. The second dramatic step was based on an intensive application of microelectronics techniques. A dedicated ASIC chip, harbored inside the gas cell, included a complete chain of analogue and digital electronics below the projection of the metal pad and combined the functions of anode and front end electronics. This reduced the problem of routing to a few tens of lines, allowing for a chip of 2×10^4 pixels with a 80-μm pitch[10]. The last step was the implementation of the capability to auto-trigger and define a region of interest to be selectively fetched to the output. This overcame the problem of read-out dead time and allowed for a substantial increase of the dimension of the chip, to the goal of 10^5 pixels, with a 50-μm pitch[11] and high counting rate. The concentration within a single chip of so many functions made possible a very tight package with a sealed gas cell with long-term stability[12]. Prototypes of this detector have been tested with different gas mixtures down to 2 keV energy[13; 14], and have passed mechanical and radiation tests so that a design qualified for flight was eventually established. The modulation for polarized photons is consistent with simulations. The spurious modulation for unpolarized photons is below 1%: the intrinsic control of systematics is another strength of the technique.

This new device, named Gas Pixel Detector (GPD), is sensitive to the interaction point, the energy, time and polarization of photons and, in the focus of an X-ray telescope, can perform imaging polarimetry with a huge increase of sensitivity, with respect to conventional techniques. The device was proposed for the large European telescope XEUS[15; 16], subsequently evolved into the ESA/NASA/JAXA mission *IXO*[17].

But even a pathfinder with telescopes of relatively small areas would represent a break-through. The context was favorable for a partnership with the team of Osservatorio Astronomico of Brera. This team started the development of X-ray Optics in the early 1980s for the *SAX* mission[18]. The shells were produced by the technique of replica. A superpolished mandrel was coated with gold. A nickel shell was electroformed on the gold and eventually detached from the mandrel. A second cluster of telescopes, with improved angular resolution, was manufactured for the JET-X telescope of the *SPECTRUM-X-Gamma* mission[19]. After the sinking of the mission one of the optics was employed for the X-Ray Telescope of *SWIFT*[22]. Also an important role was performed by OAB and Italian industry in the design and manufacture of the telescopes of *XMM-Newton*[20]. More recently OAB has studied several improvements in the techniques for X-ray telescopes, mainly aimed at extending the band (via multi-layer coating), and to increasing the reflectivity, the surface/weight ratio and the angular resolution.

In conclusion, since 2004 there has been in Italy the complete know-how to perform a mission of X-ray polarimetry and a good menu of designs, or even of physical items, available for such a mission.

38.2 The *POLARIX* mission

The JET-X telescopes, with a focal length of 3.5 m, compatible with a small launcher, are a natural candidate for a pathfinder mission of polarimetry. Two flight units and one EQM are there. Moreover mandrels are still available so that more telescopes could be manufactured with reasonable cost and time. Since the time the first GPD was available we have tried to calculate the capabilities of a mission based on JET-X telescopes and on the existing detectors[21]. Such a mission, named *POLARIX*, has shown interesting capabilities, and the concept was included in the National Aerospace Plan of the Italian Space Agency. In 2007 ASI issued an announcement of opportunity for two small scientific missions to be flown in 2012 and 2014 respectively. *POLARIX* was one of the five missions selected to perform a phase A study. ASI is expected now to select the two missions for the flight.

The guidelines to the phase A design have been:

- To exploit at the maximum level the scientific capabilities resulting from the existing telescopes and detectors.
- To open a new window of polarimetry on a reasonable sample of astrophysical objects, including a few extragalactic ones.
- To build an observatory and an archive open to the community worldwide.
- To stay within the (ambitious) limit of 50 MEuros plus the VEGA launcher.

38.2.1 The payload

Three telescopes, built by Medialario, are available from the JET-X program: two flight units and one EQM. All of them have been calibrated and qualified. Moreover a fourth unit is performing very well as the optics of *SWIFT* X-ray telescope. From the existing mandrels two more telescopes could be manufactured and coated with a carbon layer to fill the reflectivity gap around M edges of gold. The cost would be moderate, but this *nice-to-have* option is presently out of the budget. In case it becomes viable in the future, we designed the *POLARIX* structure in such a way that it can host up to five telescopes. JET-X telescopes have been described many times[19; 22; 23; 24]. The angular resolution is $15''$ at 1.5 keV and $17''$ at 8 keV. The effective area of each telescope is 159 cm^2 at 1.5 keV and 70 cm^2 at 8 keV. The weight of each unit is 60 kg, high but acceptable for *POLARIX*. The thermal control of the bus will keep the temperature of the mirror assembly within $20\pm2^\circ$ in any part, to prevent deformations from gradients.

The bottom of the payload/bus is a plane of carbon fibre composite hosting all the focal plane instrumentation shown in Figure 38.1(b). This includes **three detectors** with their back-end electronics. Our baseline are detectors of the existing design, based on the 105 600 pixel ASIC chip and a sealed gas cell with 50 μm thick beryllium window, a 10 mm thick absorption gap and, as a baseline, a mixture of He (20%) and DME (80%). The detector will be thermally stabilized with a Peltier cooler to $15\pm2^\circ$. Each detector is mounted together with a filter wheel, including

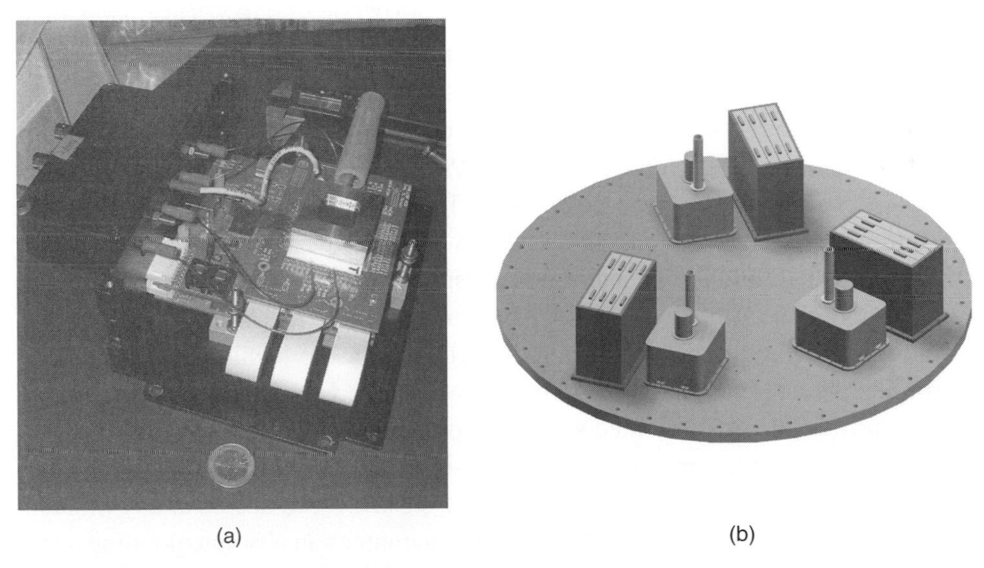

Figure 38.1 A GPD under test in laboratory (a) and the focal plane with three GPDs, each mated with the back-end electronics (b).

a closed position, a polarized X-ray source and an unpolarized one, a field reducer and a *grey* filter (in case of a very bright transient).

Beside each detector **back-end electronics** performs A/D conversion of the signals and pedestal subtraction, in the region of interest, and then creates a string associated to each event, included time tagging of the events (synchronized with GPS), and suppression of zeros. Also it performs functions of programming the chip and housekeeping monitoring, and includes high-voltage power supplies. Strings of events from each detector are delivered to common **control electronics** that organizes data for telemetry. For each photon an information of $\sim$1000 bit (depending on the track size) is transmitted including time (8 μs resolution), coordinates of region-of-interest, position and charge content of each hit pixel. In case data from a bright source exceed the capacity of the telemetry two strategies are possible:

- Record data on a large local mass memory and download the data during a subsequent observation of a faint source
- Analyze tracks on board and transmit, for each photon, the time, the energy, the reconstructed impact point and ejection angle and a few descriptors, for further off-line filtering.

Both functions are present in the control electronics, designed in collaboration with Thales Alenia Space of Milan, based on a TSC21020 DSP. The memory is dimensioned to store one day's data from a bright source.

38.2.2 The bus and the mission

The JET-X telescopes fit within the airing of the VEGA launcher. This can deliver up to two tons in an equatorial LEO, so that the cost prevails over the weight as a design driver. The bus was designed by Thales Alenia Space of Turin with the largest use of commonalities with other missions. It is a derivation of the *Prima Science* bus, that uses recurrent subsystems of Prima bus, with a mechanical structure suitable to carry astronomical telescopes. The core of the satellite is the **Service module**, that includes the mirrors, and provides all the necessary functions for the operation of the scientific instrumentation (thermal control, structure, on-board data handling, telecommunications with ground, attitude and orbit control, electrical power generation and distribution, interface with the launch system). A long cylindrical structure connects the optical bench to the **Focal Plane**. The moderate need of power allows for a fixed set-up of solar panels with a total of 6.9 m^2, oriented at 90° from the optical axis and oversized in order to accept an inclination within $\pm$20° from the sun direction. Thence *POLARIX* can access at any moment 30% of the sky. The Attitude Control System guarantees an absolute pointing accuracy of $5'$ (with a f.o.v. of $16' \times 16'$) and an absolute measurement accuracy of $10''$. An OBDH (based on an ERC32) communicates with the payload via 1553

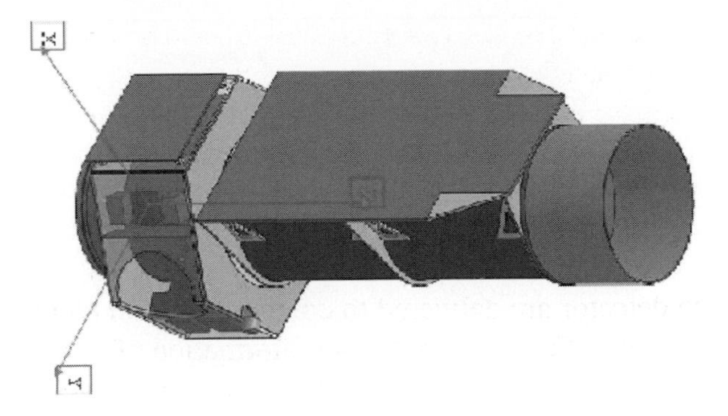

Figure 38.2 The bus of *POLARIX*: the Service Module and optics are on the left.

bus. Telecommunications with Malindi Base are in S band with two fixed antennas. Data rate is 512 kbit/s; the data volume downloadable is 3.25 Gb/day.

The total mass of *POLARIX*, including the launch adapter and a 20% margin is 925 kg. The total power (with the same margin) is 1200 W in sunlight and 540 W in eclipse.

We selected an orbit 505 $\pm$ 15 km equatorial ($< 5°$ inclination). The nominal mission duration is 1 year, plus 3 months for commissioning and 1 month for decommissioning, during which the satellite has a ground coverage variable from 8 to 11 minutes per orbit considering Malindi as the ground station in support of the satellite operations. The very low altitude is aimed to have a guaranteed mission duration of 15 months while preserving full operativity of the ACS at the re-entry time, as required by ASI.

38.2.3 The ground segment, the user segment, the data

With an equatorial LEO and an uplink/downlink from Malindi the overall management of ground-based activities architecture is very similar to BeppoSAX and to AGILE. Following the latter, in a cost saving approach, we foresee a clear separation of the **Ground Segment** (based on ASI Malindi Base and a Mission Operation Center located at Telespazio station in Fucino), and the **User Segment** (articulated in Team Data Center, Science Operation Center and Scientific Data Center) based on scientific institutes and ASI Scientific Data Center. The Observing Plan foresees a **Core Program** based on 25% of the operative phase plus data of a science verification phase. Data of the Core Program belong to the *POLARIX* team. The **Guest Observer Program** (for the remaining 75%) will be based on proposals from the worldwide community selected with a peer review process. After one year all data will become available at ASI ASDC.

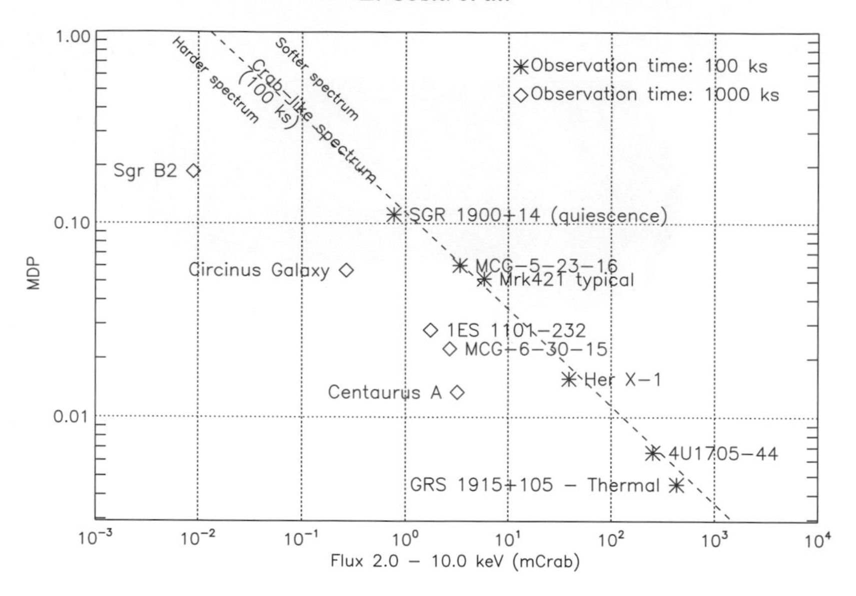

Figure 38.3 Minimum detectable polarization (at 99% level) for some representative sources for observations of one day and ten days with *POLARIX*.

38.2.4 The performance

Based on the telescope and detector figures and in the (realistic) assumption of negligible background, we can evaluate the minimum detectable polarization. In Figure 38.3 we show that with one day's observation we can completely characterize bright galactic sources. Sensitivities $\leq$1% mean that we can do energy-resolved polarimetry of BH binaries, searching for rotation of polarization with energy to determine the spin of the black hole. With 1–3 days' observation of X-ray pulsators, resolved in phase and in energy, *POLARIX* can determine the emission mechanism and fix the geometry of the system. In the extragalactic domain *POLARIX* can determine polarization of a few percent of the brightest AGN. Objects like blazars or highly obscured galaxies can be studied only to a higher degree of polarization but this is sufficient according to the general picture. Last but not least a polarization of $\geq$20% can be detected on Sgr B2, less than what we expect if the association with the GC black hole holds.

The good space resolution of the GPD and the long focal length of the telescope, which limits blurring due to inclined penetration of photons in the gas, result in an angular resolution only marginally worse than that of the telescopes. A HPW of 19.3″[25] is good enough to resolve the main features of the Crab Nebula Figure 38.4(b).

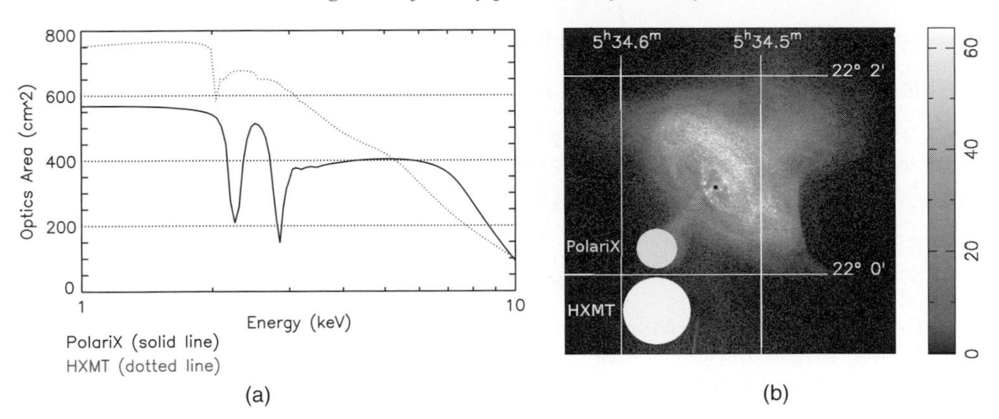

Figure 38.4 Comparison of *POLARIX* and *HXMT* in terms of effective area and angular resolution. PSF are projected on the Chandra image of the Crab.

38.3 Another pathfinder option: *HXMT*

A connection does exist between physics of hard X-ray emitters and expectations of polarization. Non-thermal processes can be singled out from the presence of hard tails but also from the existence of linear polarization, that also provides geometric information (e.g. the orientation of magnetic fields or the direction of particle acceleration). Alternatively the presence of a hard component and the absence of polarization can provide the evidence for disordered systems (e.g. fragmented magnetic fields for synchrotron, or diffuse source of seed photons in an inverse Compton). A polarimeter and the Hard X-ray Modulation Telescope of the Chinese Space Agency[26] could efficiently combine in a very high-performing mission. Inter-agency negotiations are in progress to add to *HXMT* two X-ray telescopes with a polarimeter in the focus[27; 28]. We have tight requirements on weight and on the focal length ($\leq$2.5 m). By using up-to-date technology to maximize the filling factor of the mirrors, we propose to use shells with a thickness ranging from 0.1 to 0.2 mm. Also the surfaces are coated with a thin carbon layer. The effective area in the operative range is comparable to that of *POLARIX* (Figure 38.4(a)) and so is the sensitivity, with *HXMT* better tuned to sources with soft spectrum. The shorter optics results in a larger blurring for the inclined penetration of photons. HPW, for this effect only, cannot be better than 35$''$. Moreover the thinner shells, notwithstanding improved mounting techniques, would impact on the angular resolution. HPW will be anyway within 60$''$ and a goal of 40–45$''$ is realistic. *HXMT* is still capable of some angular-resolved polarimetry on the Crab Nebula as shown in Figure 38.4(b).

38.4 Acknowledgements

Beside substantial support of INFN and INAF, the programs were supported by ASI with various scientific and technological contracts, including the Phase A contract of *POLARIX*. We acknowledge the cooperation of different teams of Thales Alenia Space, Medialario, Telespazio and ASI.

References

[1] Costa, E. et al. (1990). *Nuovo Cimento C* **13**, 431.
[2] Kaaret, P. E. et al. (1994). *Proc. of SPIE* **2010**, 22.
[3] Soffitta, P. et al. (1998). *Nucl. Instr. and Meth. A* **414**, 218.
[4] Bellazzini, R. & Spandre, G. (2009). These proceedings.
[5] Bellazzini, R. & Spandre, G. (1995). *Nucl. Instr. and Meth. A* **368**, 259.
[6] Bellazzini, R. et al. (1999). *Nucl. Instr. and Meth. A* **423**, 125.
[7] Soffitta, P. et al. (1998). *Proc. of SPIE* **2517**, 156.
[8] Soffitta, P. et al. (2001). *Nucl. Instr. and Meth. A* **469**, 164.
[9] Costa, E. et al. (2001). *Nature* **411**, 662.
[10] Bellazzini, R. et al. (2006). *Nucl. Instr. and Meth. A* **560**, 425.
[11] Bellazzini, R. et al. (2006). *Nucl. Instr. and Meth. A* **566**, 552.
[12] Bellazzini, R. et al. (2007). *Nucl. Instr. and Meth. A* **579**, 853.
[13] Bellazzini, R. et al. (2007). *Nucl. Instr. and Meth. A* **572**, 160.
[14] Muleri, F. et al. (2008). *Nucl. Instr. and Meth. A* **584**, 149.
[15] Arnaud, M. et al. (2009). *Experimental astronomy* **23**, 139.
[16] Bellazzini, R. et al. (2006). *Proc. of SPIE* **6266**, 62663Z.
[17] Bellazzini, R. et al. (2009). These proceedings.
[18] Conti, G. et al. (1994). *Proc. of SPIE* **2279**, 101.
[19] Citterio, O. et al. (1996). *Proc. of SPIE* **2805**, 56.
[20] Gondoin, P. et al. (1994). *Proc. of SPIE* **2279**, 86.
[21] Costa, E. et al. (2006). *SPIE* **6266**, 62660R.
[22] Burrows, D. N. et al. (2005). *Space Science Reviews* **120**, 165.
[23] Wells, A. A. et al. (1997). *Proc. of SPIE* **3114**, 392.
[24] Moretti, A. et al. (2004). *Proc. of SPIE* **5165**, 232.
[25] Lazzarotto, F. et al. (2009). These proceedings.
[26] Li, T. (2007). *Nuclear Physics B* **166**, 131.
[27] Costa, E. et al. (2007). *Proc. of SPIE* **6686**, 66860Z.
[28] Soffitta, P. et al. (2008). *Proc. of SPIE* **7011**, 701128.

39

A polarimeter for *IXO*

R. Bellazzini, A. Brez, M. Minuti, M. Pinchera & G. Spandre

INFN-Pisa, Italy

F. Muleri, E. Costa, S. Di Cosimo, S. Fabiani & F. Lazzarotto, A. Rubini, P. Soffitta

NAF/IASF-Rome, Italy

The X-ray POLarimeter (XPOL) is an instrument that will fly on-board the International X-ray Observatory (*IXO*). We will describe the XPOL setup in *IXO* and we will compare the *IXO* requirements with the actual prototype performance. The environmental tests performed on the XPOL prototype (thermo-vacuum, vibration and heavy ions irradiation) show that this technology is ready for a space application.

39.1 XPOL on the *IXO* focal plane

IXO is a collaboration of NASA, ESA and JAXA, and is foreseen to fly in 2020[1]. The optics area will be 2 m^2 at 2 keV with a 20-m focal length and with an angular resolution of $5'$. The focal plane of *IXO* will be a rotating platform hosting several instruments that will take data alternatively: a wide-field imager, an X-ray microcalorimeter-spectrometer, a hard-X-ray imager, a high-time-resolution spectrometer, and the polarimeter XPOL. Further, an X-ray grating spectrometer will be continuously in operation.

XPOL is a sealed gas pixel detector (GPD)[2; 3], with a 50 μm beryllium window, a photo-absorption gap of 1 cm, a gas electron multiplier (GEM) for the charge preamplification and a readout ASIC with a 15×15 mm^2 active area, covered by 105 600 hexagonal pixels with a 50-μm pitch. Each pixel has a complete electronic chain (preamplifier, shaper, sample and hold) with a very limited noise (50 el ENC). The gas used is a He20-DME80 (di-methyl ether) mixture at 1 bar. The photons that have a photoelectric interaction with the gas atoms cause the emission of a photoelectron with an angle relative to the X-ray polarization modulated as a

X-ray Polarimetry: A New Window in Astrophysics, eds. R. Bellazzini, E. Costa, G. Matt and G. Tagliaferri. Published by Cambridge University Press.

Table 39.1 *Summary of performance requirements (assuming a focal length of 20 m and 2 m^2 of area at 2 keV).*

	Performance requirements	Reqs satisfied by the actual XPOL prototype?
Field of view	$2.6' \times 2.6'$	YES
Polarization sensitivity	1% MDP (99%CL) for 1 mCrab source in 100 ks	YES
Energy range	2–10 keV	YES
Energy resolution	20% at 6 keV	YES
Angular resolution	$6''$	YES
Timing resolution	5 μs	YES
Efficiency	See Figure 39.1(a)	YES
Modulation factor	See Figure 39.1(b)	YES
Polarization angle resolution	2° in 100 ks 1 mCrab $P = 10\%$	YES
Dead time	10 μs	NO ($\tau = 200$ μs)

$\cos^2 \phi$ function. The ionization electrons left along the photoelectron track are drifted towards the GEM that multiplies them, and are collected, amplified and recorded by the pixels which store the track map. The ASIC has an auto-triggering capability: when a charge >3000 el is collected by a mini-cluster of four pixels, the ASIC holds the amplified charges and provides the trigger signal and the coordinates of the corners of a region of interest (ROI) which comprises the pixels over threshold plus a ±10 or ±20 pixels fiducial area to collect the tails of the charge distribution. The readout system reads only the ROI, not the whole ASIC, saving on average a factor of 100 in readout time. An algorithm identifies the start point of the track (impact point), calculates the direction of emission of the photoelectron from the pixels near the impact point and the charge of the whole track, which is proportional to the photon energy.

The main performance parameters of XPOL are described in Table 39.1, Figure 39.1(a) and Figure 39.1(b). The 1 keV cutoff in the efficiency curve (Figure 39.1(a)) is due to the Be window absorption. In the modulation factor plot (Figure 39.1(b)) some experimental points are compared with the Monte-Carlo prediction. The minimum detectable polarization (MDP) = 1% (99% CL) for 1 mCrab source in 100 ks, the high angular resolution ($6''$), and the spectroscopic capability in the 2–10 keV range make XPOL a real breakthrough in X-ray polarimetry.

The last column of Table 39.1 states the correspondence of the performance of the actual detector with the *IXO* requirements. The only parameter that must be improved is the dead time. The actual dead time is 200 μs and it is dominated by

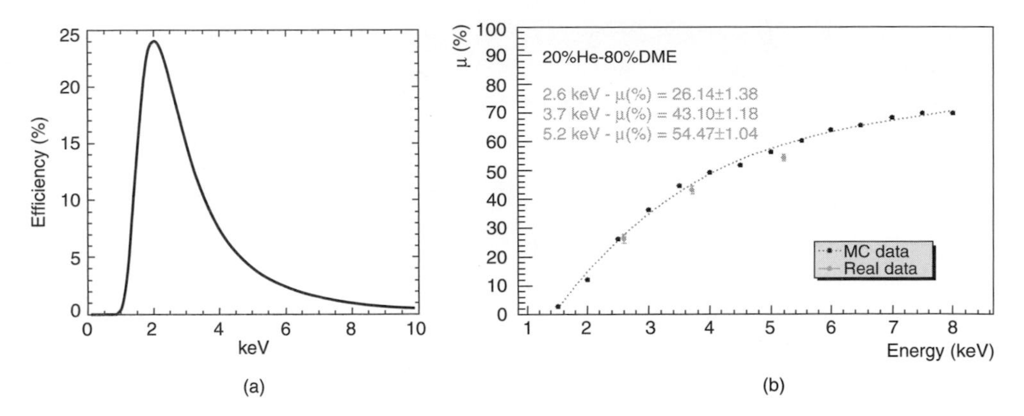

Figure 39.1 (a) Detection efficiency of XPOL with a 50 μm Be window and a He20-DME80 1 bar gas filling. (b) Modulation factor measured at 2.6 keV, 3.7 keV and 5.2 keV compared with the Monte-Carlo prediction.

the ADC readout time of the pixels of the ROI. A new ASIC is under study with a 10 μs dead time that will be obtained with a factor 2 increase of the clock rate from 10 MHz to 20 MHz and by the optimization of the ROI. Operational experience has demonstrated that the ROI can be reduced by a factor 5 without losing any relevant information. In the actual ASIC there is a second readout of the ROI to evaluate the pedestals just after the event. In the new version the increased stability of the pedestal levels will avoid this second readout, with a factor 2 gain in the frame rate. The minimum threshold level will also be lowered to a few hundred electrons to obtain a high trigger efficiency even at low GEM gain.

39.2 The XPOL configuration

The GPD is a very compact device (Figure 39.2(a)) weighing only 50 g. It is assembled by Oxford Instruments Analytical Oy (Finland), an industry with a long history of building sealed proportional counters for space. Only low-outgassing materials and adhesives are used. The GPD is left several days in vacuum at high temperature for deep outgassing, is filled with the gas mixture and sealed. The first sealed GPD is still working at nominal voltages after two years from assembly.

XPOL on *IXO* will be hosted in a protection box (Figure 39.2(b)). A filter wheel will be mounted in front of the instrument. The filter wheel will have seven positions: one open and one closed position, one diaphragm to see a faint source close to a bright one, a filter to reduce the rate of over-bright sources, two fluorescent unpolarized calibration X-ray sources and one compact polarized source.

In the *IXO* project, near the GPD, there is the back-end electronics (BEE) that provides the low and high voltages and the commands to the ASIC, and receives

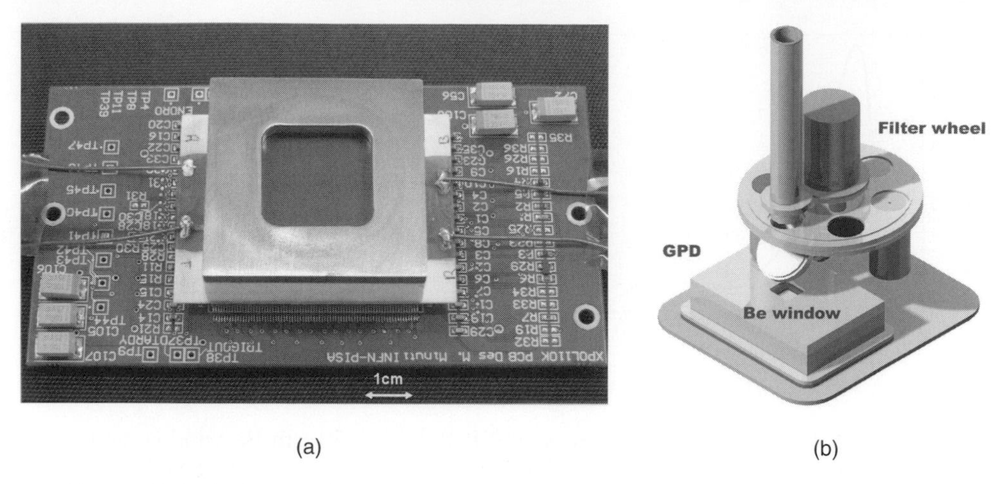

Figure 39.2 (**a**) The XPOL prototype. (**b**) Drawing of the XPOL configuration in IXO. The GPD is closed in a protection box leaving open the Be window area. The filter wheel moves over the window the slots open, diaphragmed, closed, with calibration sources (one polarized and two unpolarized)

and digitizes the ASIC signals. The BEE amplifies and digitizes the spectroscopic signal coming from the top side of the GEM and tags the event with a time stamp, and regulates the GPD temperature ($\pm 2°$C) by means of a Peltier cooling system.

The BEE communicates with a control electronic unit (CE) which stores the data in a 16 GBytes buffer memory. The CE is the interface to the spacecraft. The CE is also able to do a preprocessing of the events to reduce the data volume for the telemetry by sending to earth the track angle and the coordinates of the impact point instead of the list of coordinates and amplitudes of the pixels over threshold (factor 7 data reduction). The maximum data flow to telemetry will be 1 Mbit/s, equivalent to a 0.3 Crab of non-preprocessed data.

39.3 Space environmental tests: vibration tests

The actual XPOL prototype was tested to prove its capability to survive the launch and the space environment. XPOL was glued on an aluminum flange that ensures the necessary stiffness and the thermal path. A finite element model (ASNYS) of XPOL predicted the first resonance mode at more than 3000 Hz, well over the 20–2000 Hz test range. We mounted XPOL on the table of a TYRA-TV5220 shaker, registering the acceleration with three Piezotronics-352A24 ICP accelerometers. A sine sweep test at 0.25 g confirmed the absence of resonances in the 20–2000 Hz range. XPOL survived with no damage a random vibration test up to 11.4 g_{rms}, which is 3dB over the minimal workmanship acceptance test as defined in the NASA GEVS[4].

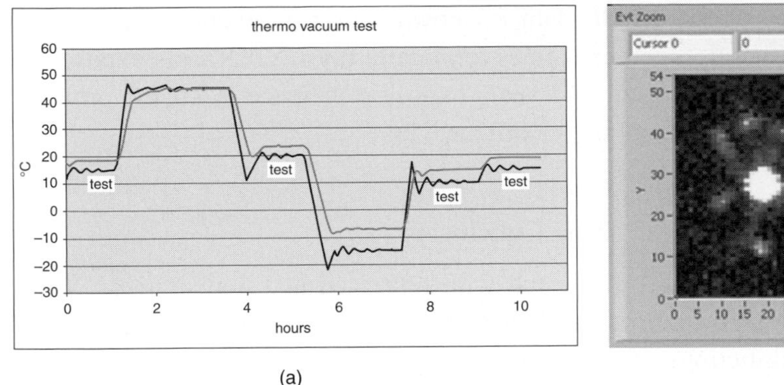

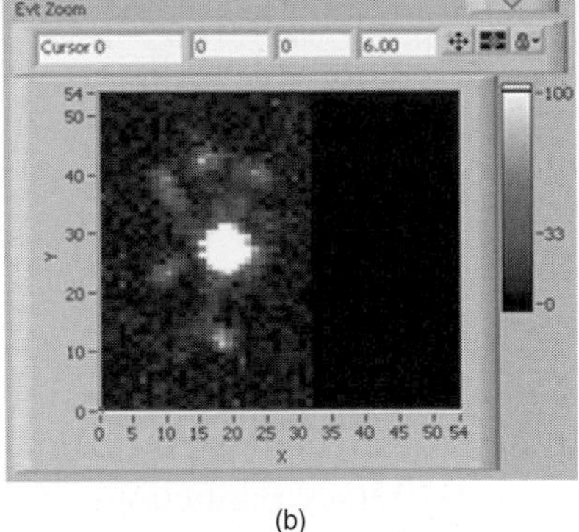

(a) (b)

Figure 39.3 (a) Thermo-vacuum cycle: the heavy line is the temperature of the aluminum interface between the cool plate of the cryostat and XPOL (driving temperature). The faint line is the temperature measured on top of the Be window. The test periods are indicated. (b) Fe ion test (500 MeV/nucleon): a single event shown by the online monitor. Scale in pixels (50 μm). The event saturates the electronics in a 5 pixels radius. Several delta rays are also visible.

39.4 Space environmental tests: thermo-vacuum test

XPOL was tested in a climatic chamber for eight cycles in the non-operative temperature range −15°C to +45°C. XPOL was tested in operation at +10°C, +15°C and +20°C with a ^{55}Fe source. XPOL survived the test showing a gain dependence on temperature of −2%/°C. After this test, XPOL was mounted over the cold head of a Cryodine cryostat, inside a vacuum vessel with the high voltage and the communication fed through. A set of resistors glued around the cold head could heat it to the desired temperature. Two PT100 thermistors measured the temperature of the interface flange below XPOL (driving temperature, heavy curve in Figure 39.3(a)) and of the Be window (faint curve). The vessel was evacuated to a pressure $\sim 10^{-5}$ mbar with a leakage check system Varian 979. XPOL reached the +45°C and the −15°C nonoperative temperatures, and was operated at +10°C, +15°C and +20°C with a ^{55}Fe source, showing the same behaviour registered in the thermo-cycle test.

39.5 Space environmental tests: heavily ionizing particles test.

A slow proton or a nucleus from cosmic rays can produce a large amount of ionization that can operate as a temporary short circuit in the GEM holes. The consequent spark can damage the GEM, if the GEM is operated at too high a gain.

We have exposed XPOL at operative voltages in the 500-MeV/nucleon Fe beam at the Heavy Ions Medical Accelerator in Chiba (HIMAC), Japan. In Figure 39.3(b) the online monitor of the XPOL data is shown. A single event saturating the electronics is registered, together with several delta rays. XPOL was exposed to 1.7×10^4 Fe ions equivalent to a 42 years exposure in space in a LEO orbit[5], without registering any damage or performance loss.

39.6 Conclusion

The XPOL prototypes have been tested with polarized and nonpolarized X-ray sources to measure all the performance. The performance figures are in agreement with theoretical predictions. XPOL was demonstrated in a relevant environment: vibration test over the minimal acceptance test as per NASA GEVS, eight thermal cycles and one thermo-vacuum cycle, exposure to Fe ions, at 500 MeV/n; this means that this technology is ready for a space application. Only the dead time parameter of the actual device does not meet the *IXO* requirements and a new ASIC version, with minor upgrades, is under study to overcome this limitation.

References

[1] Bookbinder, J. (2009).
`http://www.stsci.edu/~marel/decadal/rfi/`
`/IXO_Decadal_Document_Final-No_Budget.pdf`
[2] Bellazzini, R. et al. (2006). *Nucl. Instr. and Meth. A* **566**, 552.
[3] Bellazzini, R. et al. (2007). *Nucl. Instr. and Meth. A* **579**, 853.
[4] General Environmental Verification Specification (GEVS) (2005). NASA.
[5] Tamagawa, T. et al. (2009). These proceedings.

40

Polarimetry with ASTRO-H soft gamma-ray detector

H. Tajima
KIPAC, Stanford University

S. Takeda
Institute for Space and Astronautical Science, JAXA

on behalf of the ASTRO-H SGD team[1]

ASTRO-H is a next-generation JAXA X-ray satellite to be launched in 2014. The Soft Gamma-ray Detector (SGD) onboard ASTRO-H is a semiconductor Compton camera with a narrow field-of-view (FOV) to achieve very low background. Although the SGD is primarily a spectrometer in the 40–600 keV energy band, it is also sensitive to polarization in the 50–200 keV energy band. This paper describes instrument design, expected performance, and experimental validation of polarimetric performance of the SGD.

40.1 Introduction

ASTRO-H, the new Japanese X-ray Astronomy Satellite following Suzaku[1], is a combination of

- high energy-resolution soft X-ray spectroscopy (0.3–10 keV) provided by thin-foil X-ray optics (SXT, Soft X-ray Telescope) and a microcalorimeter array (SXS, Soft X-ray Spectrometer);
- soft X-ray imaging spectroscopy (0.5–12 keV) provided by SXT and a CCD (SXI, Soft X-ray Imager);
- hard X-ray imaging spectroscopy (3–80 keV) provided by multi-layer coating, focusing hard X-ray mirrors (HXT, Hard X-ray Telescope) and silicon (Si) and cadmium telluride (CdTe) cross-strip detectors (HXI, Hard X-ray Imager[2]);
- soft gamma-ray spectroscopy (40–600 keV) provided by semiconductor Compton camera with narrow FOV (SGD, Soft Gamma-ray Detector[3]).

The SXT-SXS and SGD systems will be developed by an international collaboration led by Japanese and US institutions.

X-ray Polarimetry: A New Window in Astrophysics, eds. R. Bellazzini, E. Costa, G. Matt and G. Tagliaferri. Published by Cambridge University Press.

The SXS will use a 6×6 format microcalorimeter array. The energy resolution is expected to be better than 7 eV. The FOV and the effective area will be, respectively, about 3 arc minutes and about 210 cm^2 combined with the ~6 m focal-length SXT. The SXT-SXS system will provide accurate measurements of the temperature and the turbulence/macroscopic motions in distant clusters up to redshift of about 1, allowing studies of the formation history of the large-scale structures of the Universe, which will eventually constrain the evolution of the dark energy.

The focal length of the HXT will be 12 m and the effective area will be larger than 200 cm^2 at 50 keV. The HXI utilizes four layers of double-sided Si strip detectors overlaid on a double-sided CdTe strip detector with a BGO ($Bi_4Ge_3O_{12}$) active shield. The extremely low background of the HXT-HXI system will improve the sensitivity in 20–80 keV range by almost two orders of magnitude compared to conventional nonimaging detectors in this energy band. Search for highly absorbed active galactic nuclei and understanding their evolution is one of the main science topics of the HXT-HXI.

The SGD also utilizes semiconductor detectors such as Si and CdTe pixel detectors with fine energy resolution (<2 keV). Narrow FOV constrained by the BGO active shield combined with good background rejection by Compton kinematics will improve the sensitivity by an order of magnitude in the 40–600 keV band compared with the currently operating instruments in Space. The SGD will allow us to study soft gamma-ray emissions from ~100 AGNs, 511 keV emissions from galactic sources and polarization from sources with a flux greater than a few hundredths of that of Crab.

The ASTRO-H mission is an official JAXA project and is in the phase-B development as of 2009. We expect to launch it in 2014.

40.2 SGD instrument design

The SGD is a Compton telescope with narrow FOV, which provides a constraint on Compton kinematics to enhance its background rejection capabilities. The SGD instrument is divided into two units that are mounted on the opposite sides of the spacecraft for better weight balance. Each unit is composed of a 4×1 array of identical Compton camera modules surrounded by BGO shield units and fine passive collimators. Figure 40.1(a) shows a conceptual drawing of the SGD unit. A BGO collimator defines 10° FOV of the telescope for high-energy photons while a fine collimator restricts thc FOV to 0.5° for low-energy photons (<150~300 keV), which is essential to minimize the CXB (cosmic X-ray background) and source confusion. Scintillation light from the BGO crystals is detected by avalanche photodiodes (APDs) allowing a compact design compared with phototubes.

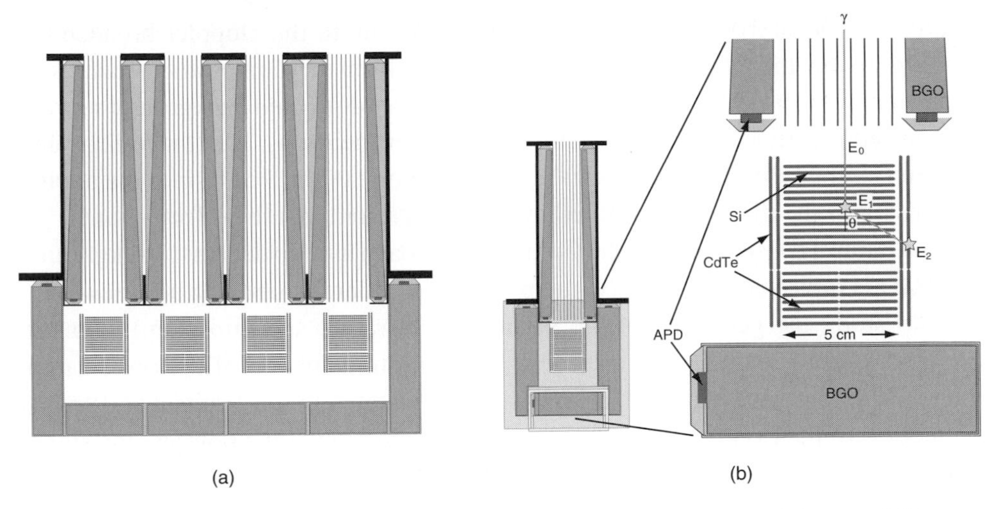

Figure 40.1 Conceptual drawing of (a) an SGD instrument and (b) an SGD Compton camera unit.

The hybrid design of the Compton camera module, illustrated in Figure 40.1(b), incorporates both pixelated Si and CdTe detectors. The Si layers enhance the efficiency below ~300 keV, and also improve the angular resolution because of smaller effect from the finite momentum of the Compton-scattering electrons (Doppler broadening) than CdTe. The Compton camera consists of 32 layers of Si sensors and eight layers of CdTe sensors surrounded by two layers of CdTe sensors. The thickness is 0.6 mm for Si and 0.75 mm for CdTe, mainly to reduce risks associated with the high bias voltages required by thicker sensors which would be preferred for higher interaction probability. The pixel size is 3.2×3.2 mm^2 for both Si and CdTe to optimize the angular resolution of the Compton kinematics and the number of channels required. The geometrical acceptance area of each Compton camera is 5.12 × 5.12 cm^2.

We require each SGD event to interact twice in a Compton camera, once by Compton scattering in a Si sensor, and then by photo-absorption in a CdTe sensor. Once the locations and energies of the two interactions are measured as shown in Figure 40.1(b), the Compton kinematics allows us to calculate the direction of the incident photon using the formula,

$$\cos\theta = 1 + \frac{m_e c^2}{E_2 + E_1} - \frac{m_e c^2}{E_2}, \tag{40.1}$$

where θ is the polar angle of the Compton scattering, and E_1 and E_2 are the energy deposited in each photon interaction. The high energy resolution of the Si and CdTe devices is essential to reduce the uncertainty of θ. The angular resolution is limited to ~8° at 100 keV and ~3° at 600 keV due to the Doppler broadening. We require that the incident photon angle inferred from the Compton kinematics be consistent with the FOV, which dramatically reduces dominant background sources such as radio-activation of the detector materials and neutrons. The low background realized by the Compton kinematics is the key feature of the SGD, since the photon sensitivity of the the SGD is limited by the background, not the effective area.

As a natural consequence of the Compton approach used to decrease background, the SGD is quite sensitive to X/γ-ray polarization, thereby opening up a new window to study the geometry of the particle acceleration and emission regions in compact objects and astrophysical jets. In addition, detection of the polarization from cosmological distances (like AGNs in flare states) can set stringent constraints on models of the Lorentz-invariance violation. Detection of the polarization signature caused by the propagation of X-rays through the Kerr space-time of stellar-mass black holes can also test Einstein's theory of General Relativity.

The Compton scattering cross section depends on the azimuth Compton scattering angle with respect to the incident polarization vector as

$$\frac{\delta\sigma}{\delta\Omega} \propto \left(\frac{E'_\gamma}{E_\gamma}\right)^2 \left(\frac{E'_\gamma}{E_\gamma} + \frac{E_\gamma}{E'_\gamma} - 2\sin^2\theta \cdot \cos\phi\right), \tag{40.2}$$

where ϕ and θ are the azimuth and polar Compton scattering angles, and E_γ and E'_γ are incident and scattered photon energies. It shows that the ϕ modulation is largest at $\theta = 90°$, i.e. perpendicular to the incident direction.

40.3 Expected performance

Effective area, polarization sensitivity and non-X-ray backgrounds are evaluated by GEANT4-based Monte-Carlo simulations. The solid line in Figure 40.2(a) shows the effective area as a function of the incident energy. Maximum effective area of more than 30 cm^2 is realized at around 80–100 keV, which corresponds to ~15% reconstruction efficiency since the geometrical area of the SGD is 210 cm^2. The effective area at low energies is suppressed due to the photo-absorption in Si while the loss at high energies is due to multiple-Compton events, which can be recovered by an improved reconstruction algorithm. The dotted line in Figure 40.2(a) shows the inverse of minimum detectable polarization (MDP) in arbitrary units assuming no background. The polarization sensitivity falls off slower at low energies and faster at high energies due to lower modulation factor resulting from more forward scatterings at higher energies. This result indicates that the SGD is sensitive to the polarization in the 50–200 keV energy band.

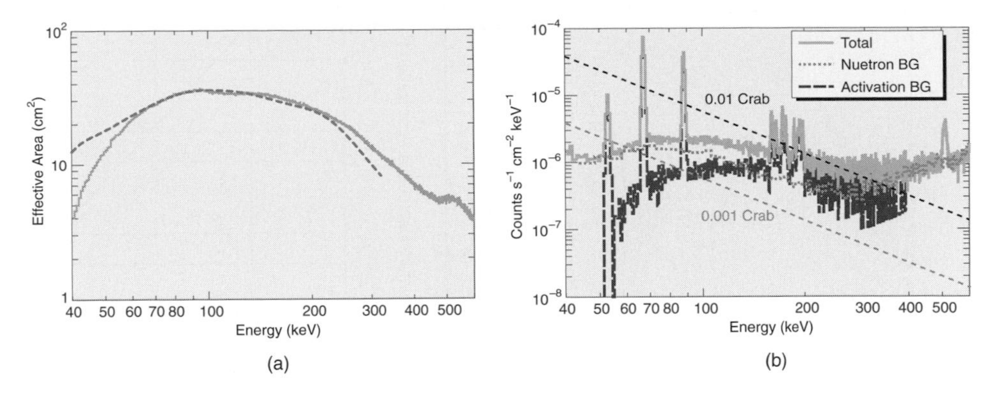

Figure 40.2 (a) Effective area (solid) and inverse of MDP in arbitrary unit (dashed) as a function of incident energy. (b) Background flux as a function of reconstructed energy.

Main in-orbit background components of the SGD are expected to be activations induced during the SAA and elastic scatterings of albedo neutrons, at the expected orbit of ASTRO-H (altitude of 550 km with an inclination angle of 31°). These background events can be heavily suppressed by a combination of multi-layer configuration, active shield, and the background rejection based on the Compton kinematics. The remaining background level is estimated to be much lower than any past instrument as shown in Figure 40.2(b). The neutron background (dotted curve) is estimated by simulations assuming the neutron spectrum described in Reference [4]. The flux of the neutron background is scaled by a factor of two based on the background studies of the Suzaku hard X-ray detector[5]. The spectrum of the activation background (dashed curve) is estimated from experimental results on the radioactivities induced by monoenergetic protons[6]. The flux is scaled by a rejection factor expected from Compton kinematics constraints. The signal fluxes corresponding to 1/100 and 1/1000 of the Crab brightness are overlaid in black and dotted straight lines, respectively. This clearly illustrates that the expected background in the SGD varies from 1/1000 to 1/100 of the Crab brightness in the 50–200 keV band.

The polarization signature of the incident gamma-rays is detected by the modulation of the azimuth angle distribution of Compton scattering in the SGD as shown in Figure 40.3(a) for a 100%-polarized source. A fit with $AVG[1+Q\cos 2(\phi-\chi_0)]$ yields $Q = 56.7 \pm 1.0\%$, where Q is the modulation factor which is proportional to the polarization degree and χ_0 is the angle of the polarization vector. Using the modulation factor obtained here and the background level described above, we can calculate the MDP analytically assuming no systematic effects from uneven background and detector response. Figure 40.3(b) shows the 3σ MDP as a function of

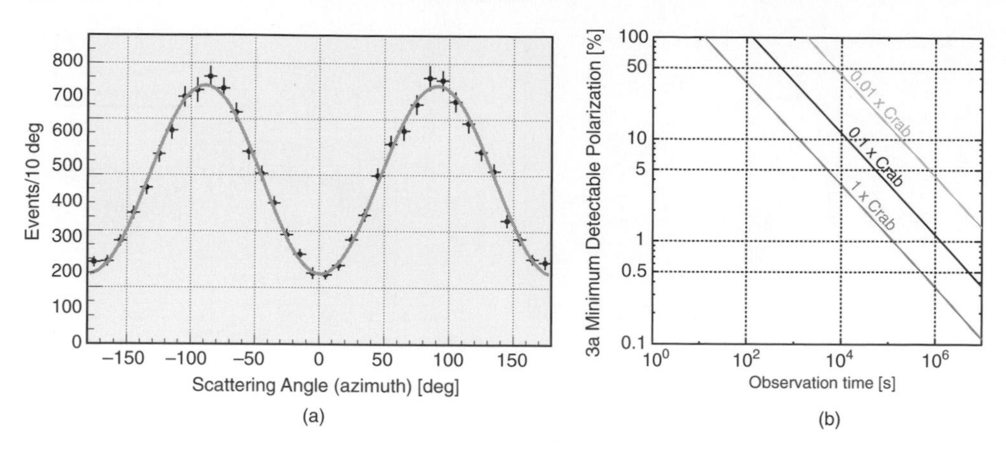

Figure 40.3 (a) Efficiency-corrected azimuth angle distribution of Compton scatterings from a source with a brightness of Crab and 100% linear polarization in a 10 ks observation. (b) 3σ MDP as a function of observation time for sources with 1, 1/10 and 1/100 of the Crab brightness.

the observation time for sources with 1, 0.1 and 0.01 of the Crab flux, which can be parametrized as $3.5\%\sqrt{10^4/t_{\mathrm{obs}}}$, $3.6\%\sqrt{10^5/t_{\mathrm{obs}}}$ and $4.3\%\sqrt{10^6/t_{\mathrm{obs}}}$, respectively, where t_{obs} is the observation time in seconds. We can conclude that the SGD can detect polarization from sources down to a few$\times 1/100$ of the Crab flux with a polarization degree of several percent in a few$\times$100 ks of observation time.

40.4 Experimental validation

Most polarimeters rotate instruments to minimize systematic effects which may arise from uneven detector responses. Since the SGD is primarily a spectrometer, it does not rotate the instrument to avoid complex mechanical design. Because of this, it is crucial to obtain correct azimuthal response of the instrument from the MC simulation. We plan to validate and calibrate the MC simulation using nonpolarized astronomical sources and also using beam tests. As a demonstration of this validation and calibration process, we have performed a beam test of a prototype Compton camera using nearly 100%-polarized 250-keV beam at the SPring-8 synchrotron photon facility.

The prototype Compton camera consists of one layer of Si sensor followed by four layers of CdTe sensors surrounded by one layer of CdTe sensors. The 250-keV synchrotron beam with 99.99% polarization is scattered in an aluminum block by 90° with a ~3° acceptance window in order to illuminate the entire Si sensor surface. The resulting photon beam has an energy of ~170 keV and a polarization degree of ~92.5%. Detailed description of the experimental setup and results

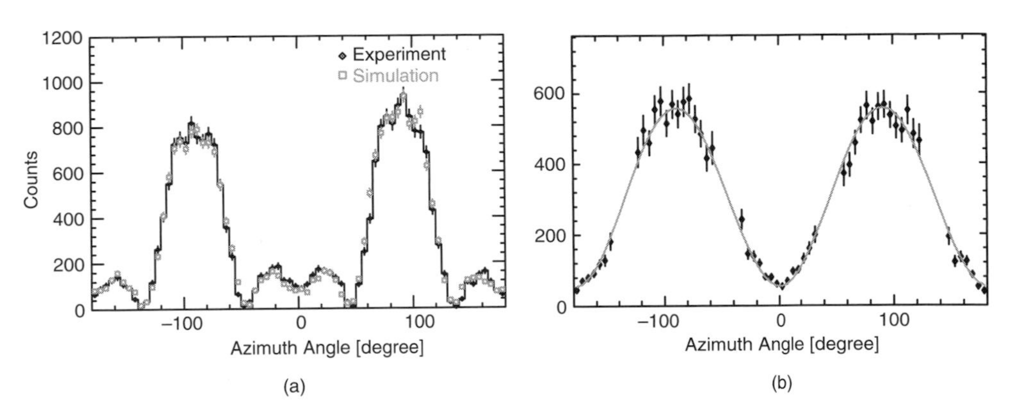

Figure 40.4 (a) Comparison of experimental (solid diamond) and simulation (open square) results of the azimuth angle distribution without any instrument response corrections. (b) Experimental azimuth angle distribution with instrument response corrections obtained from the MC simulation with a model fit.

is given elsewhere[7]. Figure 40.4(a) compares the experimental and simulation results of the azimuth angle distribution without any instrument response corrections. Note that dips at $\pm 45°$ and $\pm 135°$ are due to gaps at the four corners of the experimental setup. Figure 40.4(b) shows the experimental azimuth angle distribution with instrument response corrections obtained from the MC simulation. A fit to the above formula yields $Q = 82.9 \pm 0.8\%$ while a similar procedure performed on the simulation yields $Q = 85.6 \pm 2.7\%$. This result validates our modeling of instrument geometry and response at $\sim 3\%$ level.

Since the SGD Compton camera is square shaped, measurements of the polarization property may systematically depend on the polarization angle. Figure 40.5 shows (a) the modulation factor as a function of the polarization angle and (b) the difference between the polarization angle derived from the experimental data and that of the incident photons as a function of the incident polarization angle. Both plots show no visible systematic effect from the incident polarization angle.

Fake polarization signature in the azimuth angle distribution for zero polarization sources degrades the polarization sensitivity of the instrument and also produces systematic effects on measurements of polarization properties. Such fake polarization may be caused by uneven instrument response, wrong modeling of the instrument response and/or uneven backgrounds. The effect of uneven backgrounds needs to be carefully studied using several faint sources with no expected polarization on orbit. Here we study the effect of uneven instrument response and its modeling by using unpolarized data synthesized from two samples of experimental data with orthogonal polarization. A fit to the synthesized data yields $Q = 3.0 \pm 1.5\%$, corresponding to an apparent polarization of $3.3 \pm 1.7\%$, which

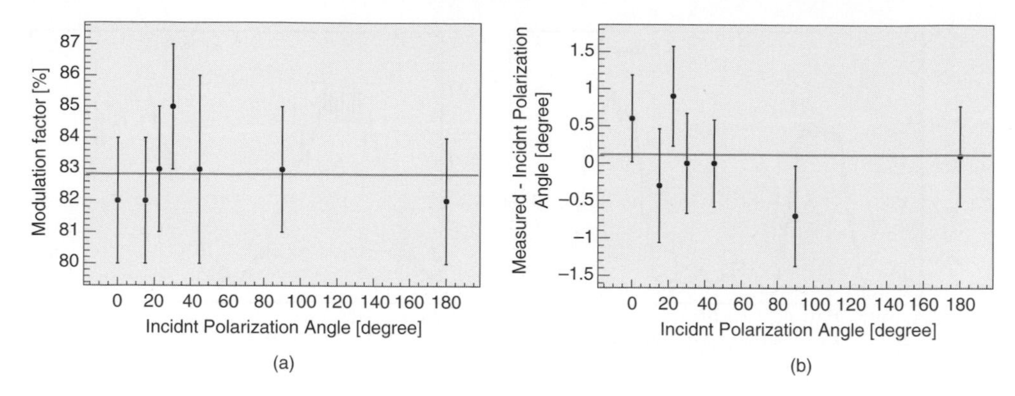

Figure 40.5 (a) Modulation factor as a function of the polarization angle. (b) The difference between the polarization angle derived from the experimental data and the polarization angle of the incident photons as a function of the incident polarization angle.

is comparable to uncertainties of the modulation factor (i.e. polarization degree) described above.

40.5 Summary

To summarize, the SGD onboard ASTRO-H will be a very sensitive polarimeter in the 50–200 keV band as well as the most sensitive spectrometer in the 80–600 keV band after the expected launch in 2014. The SGD will be sensitive to $\lesssim 10\%$ polarization for sources with several $\times 0.01$ of the Crab flux, which includes many Galactic sources and some AGNs in flare states. Since ASTRO-H is a general X-ray observatory, it will observe thousands of astronomical sources in its lifetime regardless of prospect for the polarization. Observations of many unpolarized sources with different degree of brightness will allow us to calibrate the instrument response and to study carefully systematic effects from backgrounds. Observations of many sources also present opportunities for a great discovery of the polarization from unexpected sources.

Notes

1. ASTRO-H SGD team includes S. Nakahira, K. Yamaoka (Aoyama-gakuin University), K. Doutsu, K. Fukami, Y. Fukazawa, Y. Hanabata, K. Hayashi, K. Hiragi, M. Matsuoka, T. Mizuno, S. Nishino, Y. Umeki (Hiroshima University), T. Fukuyama, S. Ishikawa, J. Katsuta, M. Kokubun, M. Koseki, H. Odaka, S. Saito, C. Sasaki, R. Sato, S. Sugimoto, T. Takahashi, S. Watanabe (Institute for Space and Astronautical Science), T. Endo, J. Harayama, T. Kozu, M. Tashiro, Y. Terada, Y. Yaji (Saitama University) G. Madejski, T. Tanaka, Y. Uchiyama (Stanford University), T. Enoto, T. Kitaguchi, K. Makishima,

K. Nakajima, K. Nakazawa, H. Nishioka, H. Noda, S. Torii, S. Yamada, T. Yuasa (University of Tokyo), J. Kataoka, T. Miura, M. Yoshino (Waseda University)

References

[1] Takahashi, T. et al. (2008). *Proc. SPIE* **7011**, 14T.
[2] Kokubun, M. et al. (2008). *Proc. SPIE* **7011**, 21K.
[3] Tajima, H. et al. (2005). *IEEE Trans. Nucl. Sci.* **52**, 2749.
[4] Armstrong, T. W. et al. (1973). *J. Geophys. Res.* **78**, 2715.
[5] Fukazawa, Y et al. (2009). *Pub. of Astro. Soc. of Japan* **61**, S17.
[6] Murakami, M. et al. (2003). *IEEE Trans. Nucl. Sci.* **50**, 1013.
[7] Takeda, S. et al. To be submitted to *Nucl. Inst. and Meth. A.*

41

The Energetic X-ray Imaging Survey Telescope and its polarization sensitivity

A. B. Garson III & H. Krawczynski

Washington University in St. Louis and McDonnell Center for the Space Sciences

for the EXIST Team (Grindley et al.)

EXIST (Energetic X-ray Imaging Survey Telescope) is a proposed space-borne observatory that combines a wide-field-of-view X-ray telescope (5–600 keV) with a pointed optical/infrared telescope, and possibly with a soft-X-ray telescope contributed by Italian collaborators. The primary science drivers of EXIST are the study of the high-redshift Universe and the epoch of re-ionization through the detection and follow-up observations of high-redshift gamma-ray bursts (GRBs) at $z \sim 10$, the study of supermassive black holes in galaxies (including heavily obscured and dormant black holes), and the study of stellar-mass and intermediate-mass black hole populations in the Milky Way galaxy and in the Local Group. In this contribution, we discuss the polarimetric capabilities of the EXIST hard X-ray telescope. Based on a pointed five-day observation (or based on four-months all-sky survey observations), EXIST can detect the hard X-ray polarization of 100 mCrab sources for polarization degrees down to 6%. The wide field of view of EXIST will make it possible to measure the polarization of transient events like GRBs and flaring galactic and extragalactic sources. We discuss the scientific potential of the hard-X-ray polarimetric measurements. The EXIST observations would allow us to (i) obtain qualitatively new constraints on the locale of particle acceleration in the vicinity of compact objects, (ii) gain key insights into the structure of jets from GRBs and active galactic nuclei, (iii) test high-order QED predictions in the extreme magnetic fields of neutron stars, and (iv) search for quantum gravity signatures (the helicity-dependence of the speed of light) with unprecedented sensitivity.

X-ray Polarimetry: A New Window in Astrophysics, eds. R. Bellazzini, E. Costa, G. Matt and G. Tagliaferri. Published by Cambridge University Press.

41.1 Introduction

EXIST is a proposed medium class mission that combines a high-energy telescope (HET) with a 1.1-m-class optical/infrared telescope (IRT), and an optional pointed soft X-ray imager (SXI) contributed by the Italian collaborators[1; 2; 3]. The design of the observatory is shown in Figure 41.1. The EXIST mission would conduct the most sensitive full-sky survey for black holes on all scales, stellar to supermassive:

- EXIST observations and autonomous identification of high-redshift ($z \sim 5$–15) Gamma Ray Bursts (GRBs) would enable us to study the birth of stellar mass black holes in the very early Universe.
- EXIST would make a census of supermassive black holes (SMBHs) in galaxies to constrain their properties and their role in galaxy evolution.
- EXIST would acquire detailed time resolved information on the X-ray-emission properties of the stellar- and intermediate-mass black hole populations in the Galaxy and Local Group.

In this paper, we will discuss the capabilities of EXIST's HET to measure the polarization of hard X-rays. In Section 41.2 we give a brief technical description of the design of the HET. In Section 41.3 we discuss polarization measurements with the HET. In Section 41.4, science topics that can be addressed with the

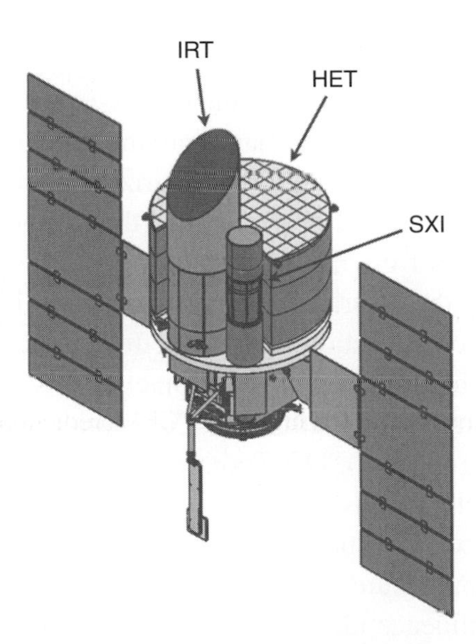

Figure 41.1 Design of the EXIST observatory. The HET with a detector area of 4.5 m^2 operates over the energy range from 5 keV to 600 keV and can scan the entire sky in two orbits. The EXIST mission includes a 1.1-m diameter infrared telescope and an optional soft X-ray imager. A detailed description of the EXIST design is given in [3].

EXIST polarization measurements are reviewed, and in Section 41.5 the results are summarized.

41.2 The EXIST high-energy telescope: technical description

The HET is a wide field of view hard X-ray telescope made with 4.5 m^2 of 0.5 cm thick Cadmium Zinc Telluride (CZT) detectors viewing the sky through a 7.5 m^2 coded mask that can survey the entire sky every 190 minutes. The detector assembly is made of 11 264 CZT detectors, each of which will be contacted with 1024 pixels at a pitch of 600 μm. The detectors will be read out with a 1024-channel ASIC similar to the ASIC used for the NuStar X-ray telescope[5]. The tungsten coded mask has a hybrid design with thin/fine mask elements ($1.25\times1.25\times0.3$ mm^3) and coarse/thick mask elements ($15\times15\times3$ mm^3). The hybrid mask design makes it possible to combine a wide bandpass (5–600 keV) with excellent angular resolution at low energies ($E_\gamma < 150$ keV) while avoiding the auto-collimation effects of a thick monolithic mask. The low-energy and high-energy angular resolutions are 2.4 arcmin and 20 arcsec, respectively. The HET is equipped with a graded-Z collimator and an active anticoincidence rear shield that doubles as GRB spectrometer in the 200 keV–5 MeV energy range[6].

41.3 Hard X-ray polarimetry with EXIST

The HET can measure the polarization of incident X-rays by making use of the fact that photons Compton scatter preferentially into the direction perpendicular to the orientation of the electric field vector. The polarization analysis uses events with two or more triggered CZT detector pixels, and the azimuthal scattering angle is determined for each event from the two pixels with the largest amplitude signals. The energy threshold of the polarization measurements is determined by Compton kinematics, and by the cross sections for photoeffect and Compton interactions in the CZT. In addition, the detection of multi-pixel events with the CZT detectors requires that the mean free pathlength of the Compton-scattered photons be on the order of the pixel pitch (600 μm). These contraints result in a low energy threshold of ~80 keV for incident gammas.

Analysis of multi-pixel Compton scattering events requires the knowledge of which pixel's interaction occured first. At primary photon energies below $m_e c^2/2$ $\approx$ 255 keV, a high-confidence identification of the first interaction is possible, as such photons lose less than half of their energy in the first interaction. For higher-energy events one can use sophisticated sequencing algorithms to determine the pixel most likely associated with the first interaction[7], and/or restrict the analysis to certain subsets (e.g. to events with all energy deposits in source pixels with source exposure, or, with all energy deposits in background pixels with no source exposure due to mask blockage).

41.3.1 Estimate of the polarization sensitivity based on detector simulations

We estimated the polarization sensitivity of EXIST based on a simulation study with the GEANT4 code[8], the low-energy electromagnetic processes package GLEPS[9], and an in-house-developed detector-simulation package. Polarized and unpolarized events with photon energies between 30 keV and 500 keV and with a Crab spectrum[10] were simulated. We assume a $>$100 keV background level of 0.03 $\text{cm}^{-2}\ \text{s}^{-1}$. The photons were assumed to hit the CZT detectors under normal incidence. The detector simulations accounted for the electronic readout noise and for thermal electron diffusion (holes were not simulated).

Charge sharing events (events with a single photoeffect interaction that triggered $\geq$2 pixels due to the effect of charge sharing) contain little information about the polarization of the incident radiation and should be excluded from the polarization analysis. An extremely effective cut is to exclude events from the analysis with the two largest energy deposits in two horizontally or vertically adjacent pixels. The cut suppresses $>$99% of the charge sharing events and retains 64% of the multi-pixel events caused by Compton interactions.

The left panel of Figure 41.2 shows the distribution of azimuthal scattering angles (computed from the two pixels with the highest energy deposits) from the results of a simulated 5-day observation of a source with a flux of 100 mCrab and with a polarization fraction of 6% – after subtracting the distribution obtained for an unpolarized signal. The figure demonstrates that the EXIST CZT detectors can indeed be used to measure the polarization of X-rays. The modulation factor for a 100% polarized signal was found to be 38%.

The right panel of Figure 41.2 shows the distribution of the energies of the multi-pixel events surviving the charge-sharing suppression cut. We fit the resulting distribution of azimuthal scattering angles by a template with the phase and the modulation amplitude as free parameters. The sensitivity limit is determined by varying the amplitude of the polarization until the amplitude of the modulation deviates by five standard deviations from zero.

Based on a pointed five-day observation, EXIST can detect the hard X-ray polarization of 100 mCrab sources for polarization degrees down to 6%. Obtaining the same sensitivity in all-sky survey mode would require an observation of four months. The wide field of view of EXIST will make it possible to measure the polarization of transient events like GRBs and flaring galactic and extragalactic sources. Assuming a GRB with a fluence of 5×10^{-6} erg cm^{-2}, we calculate a minimum detectable polarization fraction of 20%.

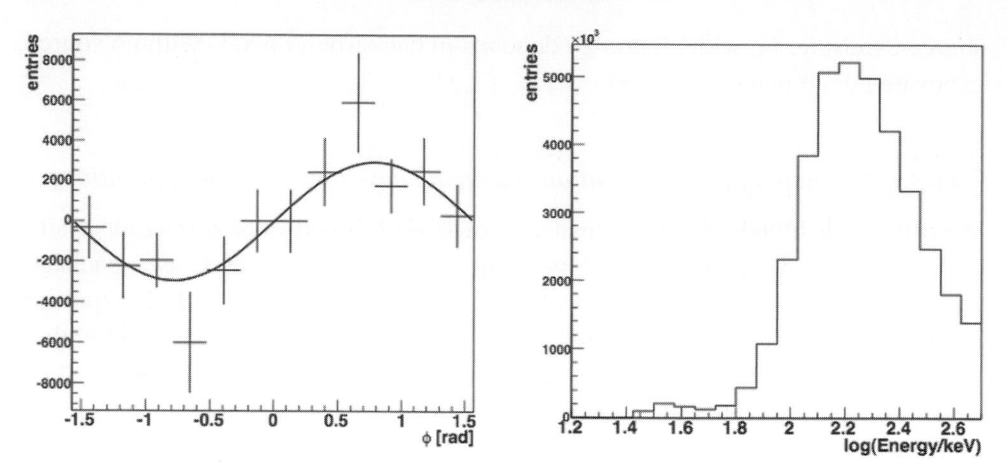

Figure 41.2 The left panel shows the net modulation of the distribution of the azimuthal scattering angle for a simulated five-day observation of a 100 mCrab source exhibiting a polarization fraction of 6% (after the cut for events with the largest energy deposits in horizontally or vertically adjacent pixels). The modulation resembles a sinus-curve with the exception of the bins at ∼0.65 which correspond to events with the two highest pixel signals in diagonally adjacent bins. The sinus curve is shown for guiding the eye only. The right-hand side shows the energy distribution of the events. Most events that can be used for the polarization analysis have energies between 80 keV and 250 keV.

The EXIST measurements will be subject to much smaller systematic uncertainties than those obtained with current and previous missions, because the number of independent detector elements is very large (11 million pixels), and the distance between pixels is very small (600 microns) compared to the size of the instrument. If there are large-scale gradients across the detector plane, they will have little impact on the polarization measurements as the latter use only pixels that are rather close to each other. Gain variations of individual pixels of one 4 cm^2 CZT detector can be determined and corrected for with high precision, making use of the background data acquired during the mission. In summary, EXIST should suffer from systematic uncertainties much less than previously flown instruments due to the scale of the detector array.

41.4 Hard X-ray polarimetry with EXIST: science topics

For the Crab Nebula, Weisskopf *et al.* reported a 2.6 keV and 5.2 keV polarization fraction of ∼20% and a polarization direction 30° oblique to the orientation of the X-ray jet[11]. Dean *et al.* measured a high polarization fraction of (46±10)% of the 100 keV–1 MeV energy range, and a polarization direction parallel to the orientation of the X-ray jet[12]. EXIST may find that an increase of the polarization fraction with photon energy is a common phenomenon.

Prime targets for polarization studies with EXIST are bright galactic sources. Galactic binary black holes exhibit different emission states. EXIST's large energy bandpass from 5 keV to 600 keV and excellent timing capabilities are ideally suited to constrain the emission state and to track the evolution of the emission components. EXIST will be able to scrutinize the polarization of the hard X-rays from mass accreting neutron stars (e.g. Hercules X-1, Cen X-3), young high-field pulsars (e.g. B1509-58), and magnetars. Photon-splitting (a higher order QED effect) is expected to produce strong polarization leading up to a high-energy cut-off[13] in high-field pulsars[14] and magnetars[15]. This effect cannot be observed in terrestrial laboratories and would provide a test of QED in extreme conditions.

X-ray polarization may reveal the inner structure of the jets from GRBs[16; 17; 18; 19; 20; 21] and can be used to constrain fundamental physics beyond the standard model (see [22] for a review), e.g. quantum gravity theories, and other theories (see e.g. [23]). The detection of an energy dependent change of the polarization direction for a large number of different GRBs could give positive evidence for a birefringence of the vacuum. Currently the best limits on the helicity dependent variation of the speed of light come from observations of polarized UV emission from a GRB afterglow[24]. The sensitivity of searches for first order variations of the speed of light improve with the observation frequency squared. The EXIST observation could thus improve by a factor of 109 over the limits from UV observations, and by a factor of 104 over limits derived from X-ray polarimetry observations at $\sim$1 keV with soft-X-ray polarimeters.

41.5 Summary and conclusions

The EXIST hard X-ray survey telescope will have excellent hard X-ray polarization capabilities. The polarization measurements use Compton scattered photons that trigger more than one pixel of EXIST's finely pixellated CZT detectors and would ideally complement polarization results obtained in the soft X-ray regime from gas pixel detectors[25] and/or time projection chambers[26], and in the γ-ray regime with Compton telescopes like the Advanced Compton Telescope. With its large field of view, EXIST will be able to obtain polarization measurements of transient phenomena like GRBs. The polarization measurements would make unique contributions to the astrophysics of compact objects and to astroparticle physics topics.

A dedicated calibration program should be performed to minimize systematic errors. Detailed studies of cross-talk properties (including electronic and weighting potential cross talk), as well as demonstrations that the polarization of signals with zero and nonzero polarization fractions can be measured accurately, are of particular importance.

Acknowledgements

The Washington University group acknowledges support by NASA (grant NNX07AH37G).

References

[1] Grindlay, J. E. (2007). *AIP Conference Proceedings* **921**, 211–216.
[2] The Official EXIST website `http://exist.gsfc.nasa.gov/`
[3] Grindlay, J. E. et al. (2009). In preparation.
[4] The Beyond Einstein Program `http://universe.nasa.gov`
[5] Harrison, F. A., Christensen, F., & Craig, W. (2005). *Experimental Astronomy* **20**, 131.
[6] Garson, A. B., Krawczynski, H., Grindlay, J. E., et al. (2006). *A&A* **456**, 379.
[7] Xu, D., He, Z., Lehrer, C. E., et al. (2004). *Proc. SPIE* **5540**, 144.
[8] Agostinelli, S., Allison, J., Amako, K., et al. (2003). *NIMA* **506**, 250.
[9] McConnel, M. L. & Kippen, R. M. (2004). *Bulletin of the AAS* **36**, 1206.
[10] Ling, J. C. & Wheaton, W. M. A (2003). *ApJ* **398**, 334.
[11] Weisskopf, M. C., Silver, E. H., Kestenbaum, H. L., et al. (1978). *ApJL* **220**, L117.
[12] Dean, A. J., Clark, D. J., Stephen, J. B., et al. (2008). *Science* **321**, 1183.
[13] Baring, M. G. & Harding, A. K. (2001). *ApJ* **547**, 929.
[14] Harding, A. K., Baring, M. G. & Gonthier, P. L. (1997). *ApJ* **476**, 246.
[15] Wang, C., Lai, D. (2009). *MNRAS* accepted [arXiv:0903.2094].
[16] Coburn, W. & Boggs, S. E. (2002). *Nature* **423**, 415.
[17] Eichler, D. & Levinson, A. (2003). *ApJL* **596**, 147.
[18] Granot, J. & Konigl, A. (2003). *ApJL* **593**, 83.
[19] Lyutikov, M., Pariev, V. I., & Blandford, R. D. (2003). *ApJ* **597**, 998.
[20] McGlynn, S., Clark, D. J., Dean, A. J., et al. (2007). *A&A* **466**, 433.
[21] Gotz, D., Laurent, P., Lebrun, F., et al. (2009). *ApJL* **695**, L208.
[22] Mattingly, D. (2005) *Living Reviews in Relativity*, `http://relativity.livingreviews.org/Articles/lrr-2005-5/`
[23] Rubbia, A., Sakharov, A. (2008). *Astroparticle Physics* **29**, 20.
[24] Fan, Y., Wei, D., & Xu, D. (2007). *MNRAS* **376**, 1857.
[25] Costa, E., Bellazzini, R., Bregeo, J., et al. (2008). *Proc. SPIE* **7011**, 15.
[26] Hill, J. E., Barthelmy, S., Black, J. K., et al. (2007). *Proc. SPIE* **6686**, 29.

42

PoGOLite: a balloon-borne soft gamma-ray polarimeter

M. Pearce

KTH, Dept. of Physics,
The Oskar Klein Centre for Cosmoparticle Physics, Stockholm

on behalf of the PoGOLite Collaboration[1]

The physical processes postulated to explain the high-energy emission mechanisms of compact astrophysical sources are in many cases predicted to result in polarized soft gamma-rays. The polarisation arises naturally for synchrotron radiation in large-scale ordered magnetic fields and for photons propagating through a strong magnetic field. Polarization can also result from anisotropic Compton scattering. In all cases, the orientation of the polarization plane is a powerful probe of the physical environment around compact astrophysical sources. Observations with PoGOLite will help resolve the source geometry for many classes of astrophysical objects. PoGOLite applies well-type phoswich technology to polarization measurements in the 25–80 keV energy range. The instrument uses Compton scattering and photoelectric absorption in an array of detector cells made of plastic and BGO scintillators, surrounded by a BGO side anticoincidence shield. A pathfinder balloon flight is scheduled for 2010 from the Esrange facility in the north of Sweden with the Crab and Cygnus X-1 as the main observational targets.

42.1 Introduction

Despite the wealth of sources accessible to polarization measurements and the importance of these measurements, there has been a paucity of missions with dedicated instrumentation. The most recent was a measurement of the Crab at 2.6 keV and 5.2 keV by an experiment on the OSO-8 satellite in 1976[1; 2]. Measurements using instruments on-board the INTEGRAL satellite have reinvigorated the field of late[3; 4]. At soft gamma-ray energies, nonthermal processes are likely to produce high degrees of polarization. PoGOLite[5] is a balloon-borne soft

X-ray Polarimetry: A New Window in Astrophysics, eds. R. Bellazzini, E. Costa, G. Matt and G. Tagliaferri.
Published by Cambridge University Press.

gamma-ray (25–80 keV) polarimeter which is optimized for the study of point-like astrophysical objects such as pulsars, and accreting black holes. The instrument is designed to allow a minimum detectable polarization (MDP) of better than 10% for a 200 mCrab source during a six-hour-long balloon flight. The effective area at 50 keV is $\sim$200 cm^2, with a corresponding modulation factor of $\sim$30%. The science goals are thoroughly reviewed elsewhere[5]. Initial targets with the reduced volume PoGOLite pathfinder to be flown in 2010 will be the Crab (the nebula component) and Cygnus X-1 in its hard state. With the complete instrument, phase-resolved polarization will be measured for the Crab Pulsar. Later targets are Cygnus X-1, Her X-1 and variable sources such as Mkn501, V0332+53, 4U0115+63, and GRS1915.

42.2 Instrument design

Compton scattering and photoabsorption events are identified in an array of phoswich detector cells (PDC) made of plastic and BGO scintillators, surrounded by a segmented BGO side anticoincidence shield (SAS), as shown in Figure 42.1.

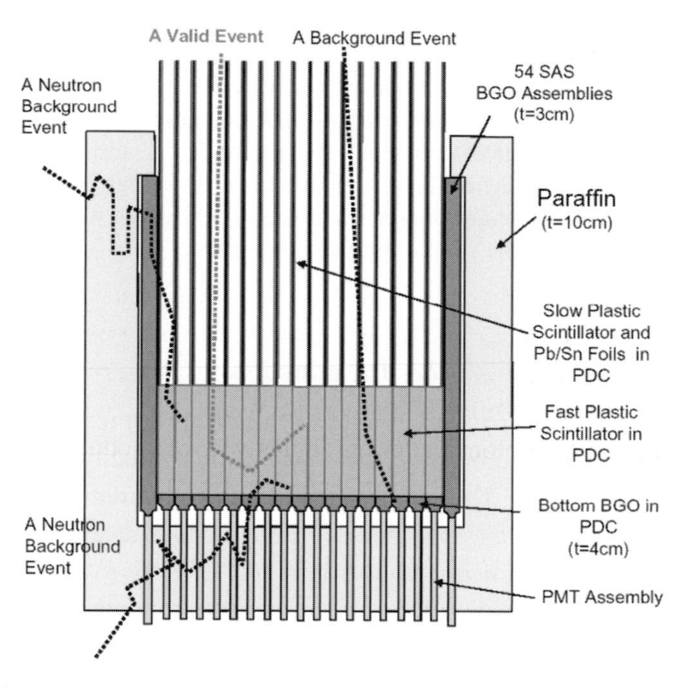

Figure 42.1 Schematic cross section of the PoGOLite instrument (not to scale) showing valid and background photon interactions. Possible background atmospheric neutron interactions are also shown. Each PDC is $\sim$1 m long. The side anticoincidence shield (SAS) is segmented to allow background asymmetries to be studied in flight. (M. Pears *et al.*)

The full-size PoGOLite instrument consists of 217 PDC units. For the pathfinder flight, a reduced volume instrument will be flown consisting of 61 PDC units.

Each PDC is composed of a thin-walled tube (well) of slow plastic scintillator at the top (fluorescence decay time ~280 ns, length 60 cm), a solid rod of fast plastic scintillator (decay time ~2 ns, length 20 cm), and a short bismuth germanate oxide (BGO) crystal at the bottom (decay time ~300 ns, length 4 cm), all viewed by one photomultiplier tube (PMT). The wells serve as a charged particle anticoincidence, the fast scintillator rods as active photon detectors, and the bottom BGOs act as a lower anticoincidence. Each well is sheathed in thin layers of tin and lead foils to provide passive collimation.

Figure 42.1 shows a simplified cross-section of the instrument with possible photon interactions indicated. Gamma-rays entering within the field-of-view of the instrument (1.2 msr, $2^\circ \times 2^\circ$ FWHM, defined by the slow plastic collimators) will hit one of the fast plastic scintillators and may be Compton scattered, with a probability that depends on the photon energy. The scattered photon may escape, be photoabsorbed in another detector or undergo a second scattering. Electrons resulting from a photoabsorption will deposit their energy in the plastic scintillator and produce a signal at the PMT. A 25-keV Compton scattering event will result in a 1–3 keV energy deposit in the fast plastic scintillator, requiring single photoelectron detection. The PMT is designed to have ~0.05 photoelectron ripple for a gain of 10^6. The detection of an energy deposit compatible with photoabsorption will initiate high-speed waveform sampling of PMT outputs from all PDCs with signals above threshold. Valid Compton scattering events will be selected from these waveforms after the completion of a flight. The locations of the PDCs in which the Compton scatter and photoabsorption are detected determine the azimuthal Compton scattering angle. The geometry of the PDC arrangement limits the polar scattering angle to approximately $90^\circ \pm 30^\circ$, roughly orthogonal to the incident direction. Little of the energy of an incident gamma-ray photon is lost at the Compton scattering site(s). Most of the energy is deposited at the photoabsorption site. This makes it straightforward to differentiate Compton scattering sites from photoabsorption sites. The azimuthal Compton scattering angles will be modulated by the polarization of the photon. The polarization plane can be derived from the azimuthal distribution of scattering angles. The degree of polarization (%) is determined by the ratio of the measured counting rate modulation around the azimuth to that predicted for a 100% polarized beam (from simulations calibrated with experiments using polarized photon beams).

42.3 Expected performance

Figure 42.2 shows signal and background event rates for the 217 PDC set-up. The background event rates due to the albedo neutrons and gamma-rays are compared

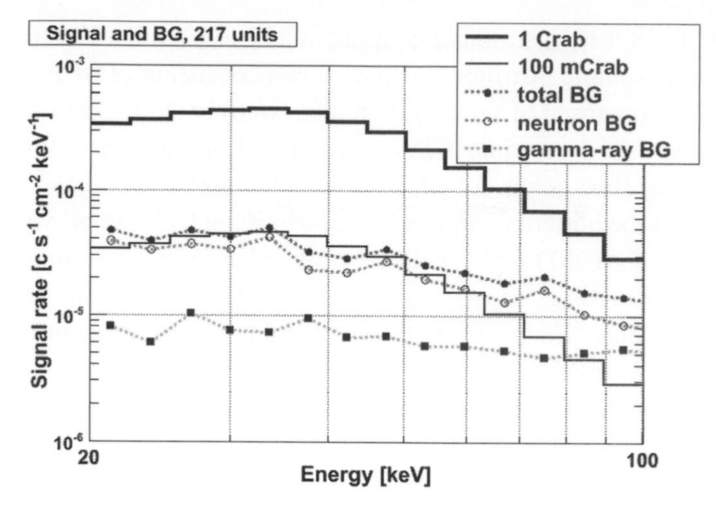

Figure 42.2 Valid and background event rates for the 217 PDC set-up with an overburden of 4 g/cm^2.

with the valid event rates expected for 1 Crab and 100 mCrab sources. Background from charged cosmic-ray particles and locally produced neutrons is negligible. The background rate is mostly due to albedo neutrons[6] and gamma-rays. Their fluxes and angular distributions have been modeled using available observational data[5]. The instrument performance characteristics have been determined from a combination of the test results obtained in polarized synchrotron beams[7; 8] and results from Geant4 simulations.

Potential backgrounds to the polarization measurement come from extraneous gamma-ray sources within the field of view, as well as neutrons and gamma-rays that penetrate the collimators, anticoincidence systems and polyethylene shield. PoGOLite employs four background suppression schemes: all nonzero PDC waveforms are stored between -0.6 and $+0.9$ μs of the trigger; an active collimator limits the field-of-view to 1.25 msr; signals are recorded in all BGO crystal assemblies of the SAS; and a neutron shield is provided by the polyethylene blanket and the slow plastic scintillator. Assuming 19% polarization as measured in the X-ray band, the PoGOLite pathfinder can detect polarization (7σ significance) and determine the position angle with a precision of approximately 5°. The full-size PoGOLite configuration can measure the energy dependence of the position angle with an accuracy of about 3°.

42.4 The PoGOLite pathfinder mission in 2010

The PoGOLite pathfinder is currently in a construction and test phase. The detector system is shown in Figure 42.3. The testing scheme for the different

Figure 42.3 The PoGOLite pathfinder instrument with 61 PDCs surrounded by a segmented BGO side anticoincidence system ($\sim$150 kg).

scintillator detectors is described elsewhere in these proceedings[9], as are the science goals[10].

42.4.1 Attitude control system

The PoGOLite field-of-view, 2°× 2° (FWHM), needs to be accurately aligned to observation targets during the balloon flight. Simulations have shown that to maximize the effective area for observations, and thereby reach the target MDP, alignment is needed to within approximately 5% of the field-of-view. The orientation of the gondola is redundantly surveyed in three dimensions by five types of attitude sensors: custom-built star trackers, a differential GPS system, three-axis MEMS and fibre optic gyroscopes, three-axis accelerometers, and magnetometers. Two star trackers are foreseen. One will have a field-of-view similar to that of the polarimeter and perform full star-pattern matching ('slow'). The other will have a larger field-of-view and be optimized for single-star tracking ('fast'). The attitude control loop is driven primarily by data from the fast star tracker, gyroscopes and accelerometers. Absolute attitude information provided by the differential GPS and the slow star tracker is used periodically to calibrate the control loop against bias and drift. Sensor outputs are processed by a dedicated computer built around a custom real time processor[11]. Control signals are sent to the polarimeter elevation

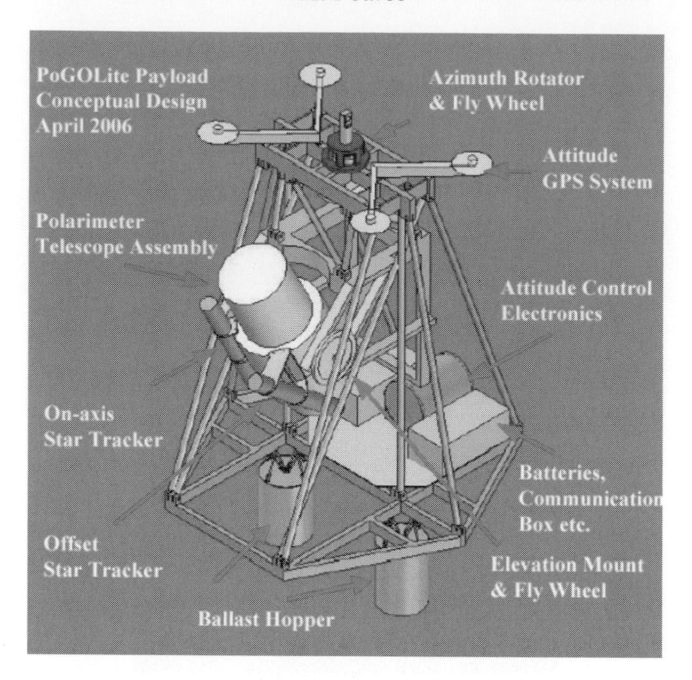

Figure 42.4 Stylized overview of the PoGOLite gondola design. The footprint is approximately 3 m $\times$ 3 m, and the height $\sim$4 m.

motor, and an azimuthal flywheel. The angular momentum of the flywheel can be dumped to the balloon when the flywheel velocity saturates. The gondola is based on a modular frame with replaceable components, as shown in Figure 42.4. The polarimeter can rotate around the viewing axis to minimize systematic bias in polarization measurements.

42.4.2 *Maiden balloon flight, and future plans*

The PoGOLite pathfinder will be launched from the Esrange space centre in the north of Sweden. A nominal launch date of August 15th 2010 is chosen due to optimum wind conditions ('turn-around') to enhance the flight duration, as well as to maximize the Crab-Sun separation which will reduce solar background and simplify star tracking. The total scientific payload weight is estimated to be about 1 tonne, giving a total suspended weight of 1.7 tonnes, and an overall flight weight of 3.5 tonnes. Flying on a million cubic metre balloon, PoGOLite will reach an altitude of at least 40 km, with a resulting vertical overburden of <5 g/cm^2. The primary aim of this maiden flight will be observations of the Crab (six hours needed on target) and background observations with a focus on neutrons. A longer flight will also allow observations of Cygnus X-1, as shown in Figure 42.5. In addition, potential backgrounds from auroral X-rays will be monitored[12].

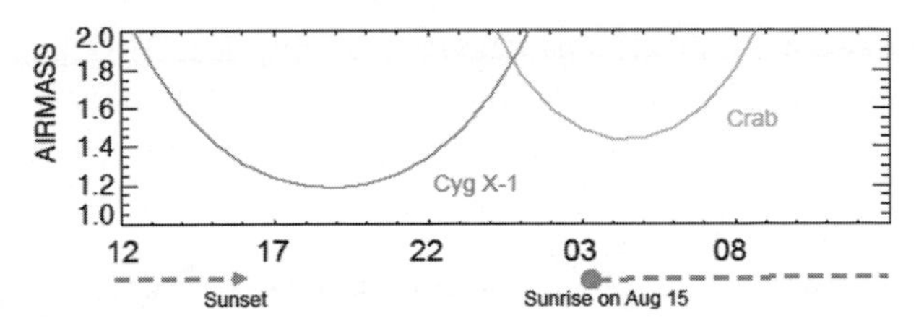

Figure 42.5 The visibility of the Crab and Cygnus X-1 during the pathfinder flight. Local Esrange time is shown on the x-axis.

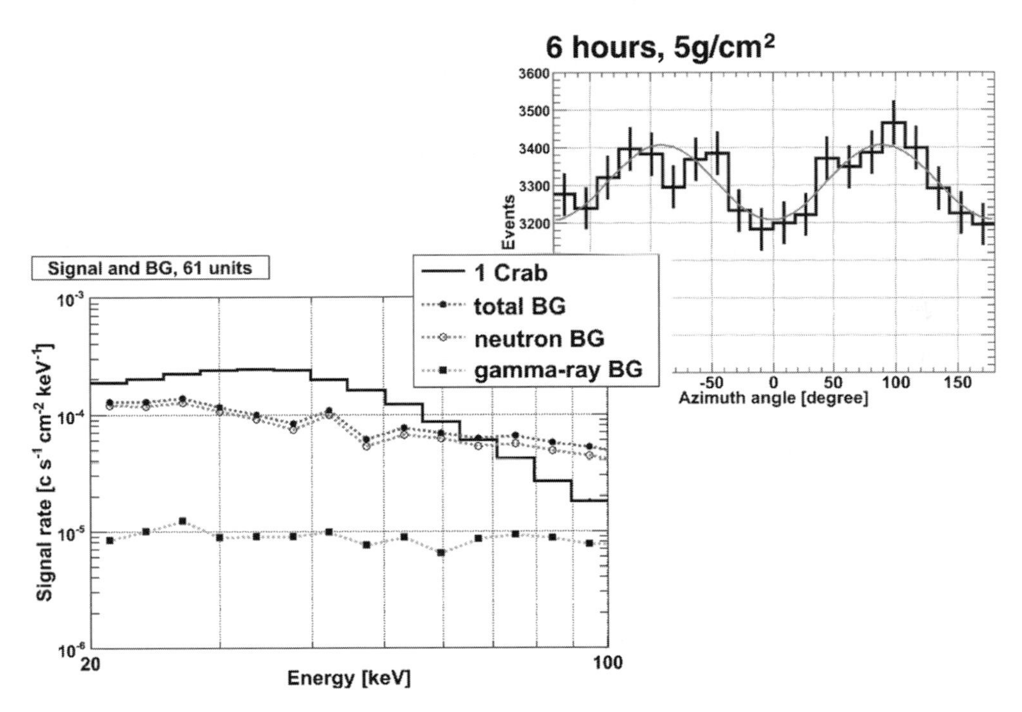

Figure 42.6 Upper: Expected modulation curve for the Crab nebula (the P1 and P2 time windows of the Crab pulsar are excluded) for a six-hour-long observation during the pathfinder flight. An overburden of 5 g/cm^2 is assumed. Lower: Expected signal and background levels for the pathfinder flight.

Long duration flights from Esrange to Western Canada (or complete circum-navigations, hopefully) are foreseen once the observation technique is proven. Figure 42.6 illustrates the expected performance of the pathfinder instrument.

Notes

1. The PoGOLite Collaboration: M. Pearce, M. Jackson, M. Kiss, W. Klamra, S. Larsson, C. Marini Bettolo, F. Ryde, S. Rydström (KTH and The Oskar Klein Centre for Cosmoparticle Physics, Stockholm, Sweden); M. Axelsson (Stockholm University and

the Oskar Klein Centre for Cosmoparticle Physics, Sweden); H-G. Florén, G. Olofsson (Stockholm University, Stockholm, Sweden); Y. Fukazawa, T. Mizuno, H. Takahashi, T. Tanaka, H. Yoshida (Hiroshima University, Higashi-Hiroshima, Japan); S. Gunji (Yamagata University, Yamagata, Japan); T. Kamae, G. Madejski, H. Tajima (SLAC and KIPAC, Menlo Park, USA); Y. Kanai, N. Kawai, M. Ueno (Tokyo Institute of Technology, Meguro-ku, Tokyo, Japan); J. Kataoka, K. Maeda, Y. Miyamoto (Waseda University, Japan); J-E. Strömberg (DST Control AB, Linköping, Sweden); T. Takahashi (JAXA, ISAS, Sagamihara, Japan); T. Thurston (The Thurston Co., Seattle, USA); G. Varner (University of Hawaii, Honolulu, Hawaii, USA)

References

[1] Weisskopf, M. C., et al. (1976). *Ap. J* **208**, L125.
[2] Weisskopf, M. C., et al. (1978). *Ap. J* **220**, L117.
[3] Dean, A. J., et al. (2008). *Science* **321**, 1183.
[4] Forot, M., et al. (2008). *Ap. J* **688**, L29.
[5] Kamae, T., et al. (2008). *Astropart. Phys.* **30**, 72.
[6] Kiss, M. (2009). These proceedings.
[7] Kanai, Y., et al. (2007). *Nucl. Instr. Meth.* **A570**, 61.
[8] Mizuno, T., et al. (2009). *Nucl. Instr. and Meth.* **A600**, 609.
[9] Jackson, M. S., et al. (2009). These proceedings.
[10] Axelsson, M. (2009). These proceedings.
[11] `http://www.dst.se/products_OEM.html`
[12] Larsson, S., et al. (2007). *Proc. 18th ESA-PAC Symposium*, Visby, Sweden.

43

Studies of neutron background rejection in the PoGOLite polarimeter

M. Kiss

Royal Institute of Technology (KTH), Dept. of Physics,
and the Oskar Klein Centre for Cosmoparticle Physics,
AlbaNova University Centre, Sweden

on behalf of the PoGOLite collaboration

The Polarized Gamma-ray Observer (PoGOLite) is a balloon-borne polarimeter based on measuring anisotropy in the azimuthal scattering angle distribution of photons in the energy range 25–80 keV. This is achieved through coincident detection of Compton scattering and photoelectric absorption within a close-packed array of phoswich detector cells (PDCs). Each PDC contains a plastic scintillator rod (main detector component), a plastic scintillator tube (active collimator) and a BGO crystal (anticoincidence shield).

A significant in-flight background is expected from atmospheric neutrons as well as from neutrons produced by interactions of cosmic rays with mechanical structures surrounding the instrument. Although this background can be reduced by introducing suitable shielding materials such as polyethylene, the shield geometry must be optimized through simulations in order to yield sufficient shielding with an acceptable increase in weight.

Geant4-based Monte-Carlo simulations have shown that a 10 cm thick polyethylene shield surrounding the PoGOLite instrument is required to sufficiently reduce the background, i.e. fake polarization events from atmospheric neutrons. In order to validate these simulations, a beam test was carried out, at which 14 MeV neutrons were used to irradiate a simple detector array with four plastic scintillators and three BGO crystals. The array was configured to mimic the PoGOLite detector geometry and also featured a polyethylene neutron shield. Here, we present details of the neutron beam test and our simulation thereof, which demonstrate that the treatment of neutron interactions within the Geant4 framework is reliable. Such simulations can therefore be used to assess in-flight neutron background in balloon-borne instruments, such as the PoGOLite polarimeter.

X-ray Polarimetry: A New Window in Astrophysics, eds. R. Bellazzini, E. Costa, G. Matt and G. Tagliaferri. Published by Cambridge University Press.

43.1 Introduction

The PoGOLite balloon-borne polarimeter is designed to measure as low as 10% polarization from a 200 mCrab source in a six-hour flight by detecting Compton scattering and photoelectric absorption in an array of 217 well-type phoswich detector cells. Each cell contains a plastic scintillator tube (60 cm long), a plastic scintillator rod (20 cm) and a BGO crystal (4 cm), providing active collimation, photon detection and bottom anticoincidence, respectively. The measurement concept is shown in Figure 43.1. A detailed description of the instrument and its scientific goals can be found in [1].

A 61-unit version of the instrument is currently under construction and is scheduled for its first flight from Esrange in northern Sweden in August 2010[2]. This 'pathfinder' instrument will be used to assess the background at float altitude ($\sim$40 km, atmospheric overburden $\sim$5 g/cm^2) as well as to measure the polarization of the Crab nebula with a 7σ significance and possibly Cygnus X-1 in the hard spectral state.

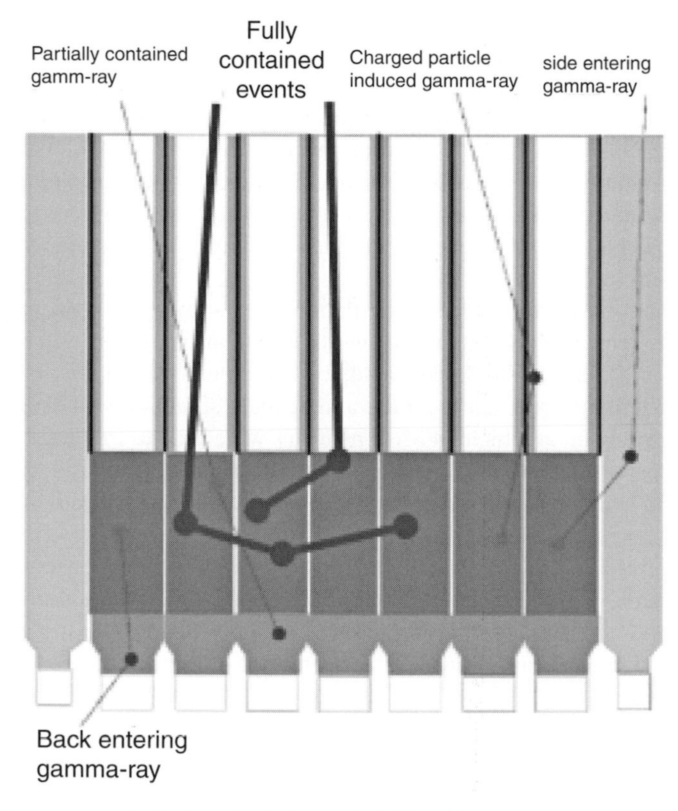

Figure 43.1 Simplified sketch of the PoGOLite detector array (not to scale, all units not shown) and different event types. The array is about 0.4 m wide and 1 m tall.

43.2 Background to neutron studies

For Compton-based polarimeters, background can be induced by neutrons scattering into multiple detector cells, thus causing fake polarization events. To better understand the performance of the instrument, it is therefore crucial to study the background caused by atmospheric neutrons in simulations. Such studies require detailed information about the atmospheric conditions at float altitude, including neutron spectra. Although there are numerous measurements of neutron spectra near ground level (e.g. the Mt. Washington neutron monitor) and with satellites (for example CGRO[3] and OGO 6[4]), balloon measurements remain scarce, often dating back to the sixties, seventies and eighties, and covering a few locations only. The situation is further complicated by the anisotropy in the neutron albedo and modulation caused by the solar cycles. A compilation of some of the available balloon-altitude neutron studies can be found in [5].

A detailed Geant4 Monte-Carlo simulation of the in-flight background of PoGOLite has been completed. This simulation included not only cosmic-ray and gamma-ray background, but also albedo neutrons with spectra based on [5], as well as structure-induced particles. (Albedo neutrons arise from cosmic rays interacting with the atmosphere, whereas structure-induced neutrons are produced when cosmic rays interact with passive materials surrounding the instrument.) Although the structure-induced background was found to be negligible, in accordance with simulations[6] and numerical calculations based on empirical formulas[7], the neutron albedo background turned out to be severe, exceeding the gamma-ray background by more than an order of magnitude in the PoGOLite energy range. In response to this discovery, the payload design was modified to include a polyethylene neutron shield surrounding the instrument. Our simulations demonstrated that the neutron background could be reduced to less than a 100 mCrab level in the PoGOLite energy range with a 10 cm shield thickness[6].

For simulating the neutron interactions, the 'QGSP_BERT_HP' physics list had been used. In order to validate these simulations, a neutron beam test was carried out[8] with a simple detector array mimicking the PoGOLite design, and the measured results were compared with simulation data obtained using this physics list. An agreement would validate the treatment of neutron interactions in Geant4 with this physics list, which strengthens our claim that the in-flight background level of PoGOLite becomes manageable with the introduction of the 10 cm thick polyethylene shield.

43.3 Neutron beam test

The beam test utilized a neutron generator based on the deuterium-tritium reaction (^{2}H + ^{3}H $\rightarrow$ ^{4}He + n), which produces mono-energetic 14 MeV neutrons.

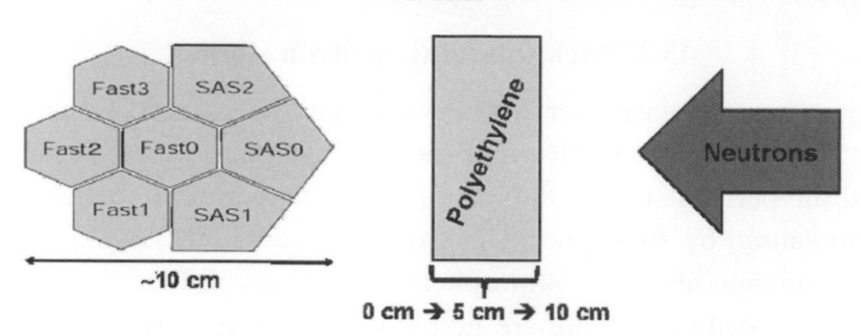

Figure 43.2 The detector array used in the neutron beam test.

Simulations[9] have shown that such neutrons are most likely to cause fake polarization events: at lower energies, the neutrons will not scatter into multiple cells and therefore will not give a polarization signal, and at higher energies, the cross-section for detection in the BGO anticoincidence system is high, so the signal can easily be rejected.

The detector array used for this test comprised four plastic scintillators ('Fast') with bottom BGO crystals and three segments of the BGO side anticoincidence shield ('SAS'); this is sketched in Figure 43.2.

The array was irradiated with neutrons as indicated, and the count rate in the central unit ('Fast0') was monitored for different thicknesses of the polyethylene shield (0 cm, 5 cm, 10 cm) as well as with three different event selection criteria:

No vetoing (passive BGOs): all hits in the central unit are counted
Active vetoing (BGOs only): hits in the central unit are only counted if there is no coincident hit in the anticoincidence ('SAS') units
Full active vetoing: hits in the central unit are only counted if there is no coincident hit in any other unit

The detector was calibrated with 59.5 keV photons from ^{241}Am and events from the fast scintillator of the central unit ('Fast0') were selected using pulse shape discrimination as described in [8]. By comparing the neutron count rate (number of counts per second of live time) for different event selection criteria, the performance of the anticoincidence system can be evaluated. Results can then be compared with simulations in order to test if the treatment of neutron interactions within the Geant4 framework is reliable.

43.4 Neutron beam test simulation

The detector geometry (plastic scintillators, BGO crystals, polyethylene shields) was recreated as accurately as possible in Geant4. Mono-energetic 14 MeV neutrons were generated and individually tracked through the detector geometry where different interactions can take place, with probabilities dictated by cross-section tables

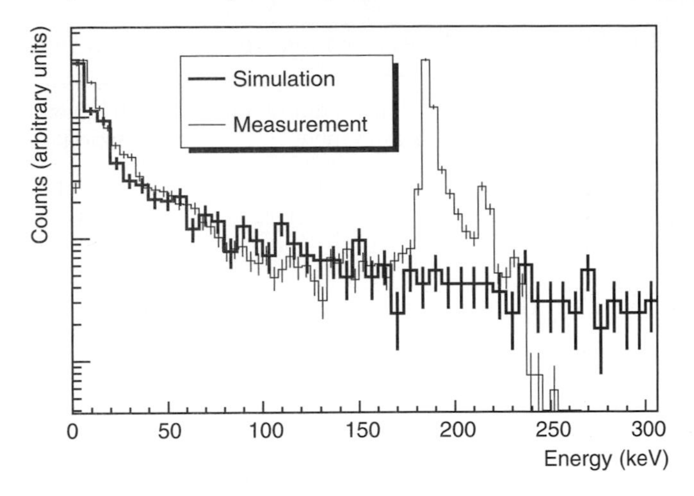

Figure 43.3 Simulated (thick solid line) and measured (thin solid line) neutron spectra for the measurement with 10 cm polyethylene and full active vetoing. The remaining configurations show similar likeness when compared. Distributions have been normalized to the value at 10 keV. The peaks in the measured spectrum are not caused by neutron interactions but by saturation in the electronics, and are therefore not seen in the simulation.

in Geant4.[1] In interactions where secondary particles are produced, these are tracked until they either lose all their energy or leave the simulation region. For interactions in the detector array, the energy deposition in each of the detector cells is recorded, along with the particle type depositing the energy. This allows quenching[10; 11] to be individually applied for the different particle types.[2] The total energy spectra are then obtained by adding the contribution from each of the interacting particle types. An example of such a spectrum and the corresponding one from the measurement, is shown in Figure 43.3, and a more quantitative comparison between the relative count rates in the PoGOLite energy range obtained in the measurement and the simulation is given in Table 43.1. Additional details on the simulation and data analysis can be found in [8].

43.5 Conclusions

Both the spectra and the relative neutron counts rates show good agreement, demonstrating that the treatment of neutron interactions in Geant4 with 'QGSP_BERT_HP' is reliable. This validates our in-flight background simulations and demonstrates that the background is manageable.

The PoGOLite pathfinder will be launched in August 2010 and will measure the polarization of the Crab nebula and assess in-flight background levels. Results can be used to optimize the shield of the full-size instrument.

Table 43.1 *Comparison of measured and simulated neutron rates.*

Configuration	Relative neutron rate: measurement	Relative neutron rate: simulation
0 cm – no vetoing	1	1
0 cm – active vetoing	0.59 ± 0.08	0.63 ± 0.03
0 cm – full active vetoing	0.41 ± 0.06	0.38 ± 0.02
5 cm – no vetoing	1	1
5 cm – active vetoing	0.57 ± 0.09	0.61 ± 0.03
5 cm – full active vetoing	0.39 ± 0.07	0.38 ± 0.02
10 cm – no vetoing	1	1
10 cm – active vetoing	0.56 ± 0.10	0.60 ± 0.03
10 cm – full active vetoing	0.38 ± 0.08	0.38 ± 0.03

Notes

1. In this simulation, G4NDL3.12 (Geant4 Neutron Data Library) was used, which includes both thermal and non-thermal cross-sections for neutrons.
2. Quenching is a process causing protons and heavier nuclei to produce less scintillation light than a photon or an electron depositing the same amount of energy. In these cases, some of the deposited energy is converted to heat instead.

References

[1] Kamae, T., et al. (2008). *Astropart. Phys.* **30**, 72–84.
[2] Pearce, M., et al. (2009). These proceedings.
[3] Morris, D. J., et al. (1995). *J. Geophys. Res.* **100 (A7)**, 12243–12249.
[4] Lockwood, J. A., et al. (1973). *J. Geophys. Res.* **78 (34)**, 7978–7985.
[5] Armstrong, T. W., et al. (1973). *J. Geophys. Res.* **78 (16)**, 2715–2726.
[6] Kazejev, J. (2007). *Studies of Neutron Backgrounds for PoGOLite – a Balloon-borne Gamma-ray Polarimeter* (Royal Institute of Technology (KTH) Master's Thesis, available at `http://www.particle.kth.se/pogolite`).
[7] Cugnon, J., et al. (1997). *Nucl. Phys.* **A625**, 729–757.
[8] Kiss, M. (2008). *Studies of PoGOLite Performance and Background Rejection Capabilities* (Royal Institute of Technology (KTH) Licentiate Thesis, ISBN: 978-91-7415-029-2, also available at `http://www.particle.kth.se`).
[9] Myrsten, K. (2008). *Polarized Gamma-ray Observer (PoGOLite) Pathfinder Experiment: Neutron Shield Design and Limits on Polarization Measurements* (Royal Institute of Technology (KTH) Master's Thesis, available at `http://www.particle.kth.se/pogolite`).
[10] Zimmerman, E. J. (1955). *Phys. Rev.* **99**, 1199–1203.
[11] Verbinski, V. V., et al. (1968). *Nucl. Instr. and Meth.* **65**, 8–25.

44

Observing polarized X-rays with PoGOLite

M. Axelsson

Stockholm University and the Oskar Klein Centre for CosmoParticle Physics, AlbaNova University Center, Stockholm

on behalf of the PoGOLite Collaboration

X-ray polarimetry is a powerful probe of astrophysical emission mechanisms, yet remains essentially unexplored. PoGOLite, a balloon-borne X-ray polarimeter currently under construction, will observe both pulsar and X-ray binary targets. We here discuss the expected scientific gains of these measurements.

44.1 Introduction

Linear polarization in X- and gamma-rays is an important diagnostic of many astrophysical sources, foremost giving information about their geometry, magnetic fields, and radiation mechanisms. However, very few X-ray polarization measurements have been made. PoGOLite is a dedicated balloon-borne X-ray polarimeter using Compton scattering, sensitive in the 25–80 keV energy range. Here we present the two main scientific targets for the PoGOLite pathfinder, scheduled for launch in August 2010 (further information on PoGOLite is provided in [9]).

44.2 X-ray emission from pulsars

Many pulsars are strong sources of X-rays. Yet it is not clear how this emission is produced. Several models exist, each of which tie the high-energy emission to different regions around the neutron star. In the polar-cap model the emission arises close to the surface of the neutron star, whereas the slot-gap and outer-gap models place the emission region in the outer magnetosphere. This is schematically illustrated in Figure 44.1.

X-ray Polarimetry: A New Window in Astrophysics, eds. R. Bellazzini, E. Costa, G. Matt and G. Tagliaferri. Published by Cambridge University Press.

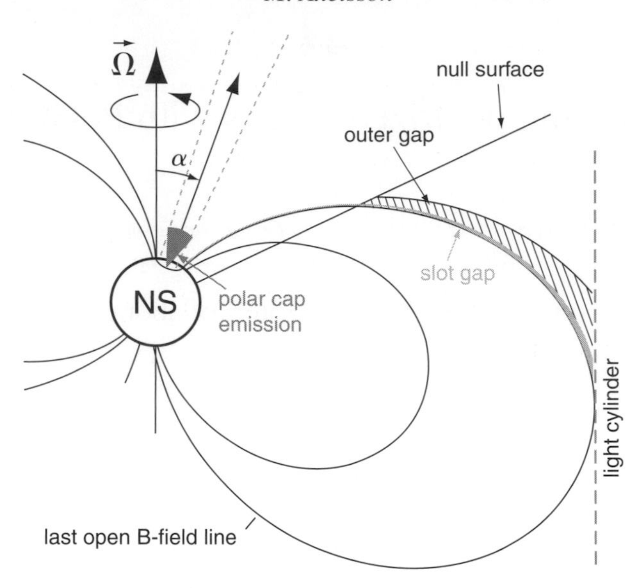

Figure 44.1 Cartoon picture of a pulsar showing the different emission regions of the polar cap, slot gap and outer gap models. The polar cap emission arises close to the neutron star surface, and is radiated in a narrow beam. The outer gap extends from the null surface to the light cylinder, and the slot gap emission arises along the entire length of the last open magnetic field lines.

The *Fermi* Gamma-ray Space Telescope was launched in June 2008, and has started to probe the gamma-ray emission from pulsars. The results point to an origin in the outer magnetosphere, and thus do not favour the polar-cap model[1; 7].

The first target of PoGOLite will be the Crab pulsar and nebula. The angular resolution of the instrument does not allow us to spatially resolve these two components, although the temporal accuracy of PoGOLite is sufficient to separate on- and off-peak emission.

X-ray polarization from the Crab nebula was first detected in 1972[8], at 2.6 and 5.2 keV. The results showed a polarization angle of $\sim 155°$, and a polarization degree of $(15 \pm 6)\%$. Recent analysis of INTEGRAL observations in the 0.1–1 MeV range showed $(46 \pm 10)\%$ polarization with a polarization angle of 123 ± 11 degrees[3].

The X-ray emission from rotation-powered pulsars is often assumed to arise in the same region as the gamma-ray emission. Yet the recent *Fermi* results from the Vela pulsar show that there are significant differences between the X-ray and gamma-ray behavior[1; 7]. It is not therefore clear whether the emission regions are linked. Observations of additional pulsar properties (e.g. polarization) may improve our understanding of these regions. The different theoretical models predict strongly different polarization properties, indicating that polarization provides a powerful tool to probe the emission region of the X-ray radiation. A six hour flight by PoGOLite will allow us to distinguish between the polarization signatures predicted by the models[5]. The energy range of PoGOLite, 25–80 keV, will also bridge the gap between the existing observations, allowing any possible evolution of the polarization with energy to be studied.

44.3 Accreting black holes

Polarized X-ray emission may also arise in accreting black holes. In the so-called hard state, the accretion is generally believed to occur through a two-component flow. At large radii, matter forms a geometrically thin, optically thick accretion disc. Close to the black hole, the thin disc is instead replaced by a quasi-spherical hot inner flow.

Soft X-ray photons are Comptonized in the hot flow, producing a characteristic power law in the radiation spectrum. The hard photons may subsequently reflect off the thin disc, giving rise to net polarization. The degree of polarization depends both on the relative strength of the reflected component, as well as its intrinsic degree of polarization. Theoretical models predict a polarization degree of up to 30% in the reflected component[6], and observations show that the relative strength of this component is ~30% (Figure 44.2). The total net polarization may therefore be ~10%. Our simulations show that PoGOLite will detect this level of polarization in a strong source[2].

The prime target in this case is Cygnus X-1, perhaps the most well-known of all Galactic X-ray binaries. It is generally accepted that the system harbors a black hole of ~10 solar masses, and the source is both persistent and bright. A detection

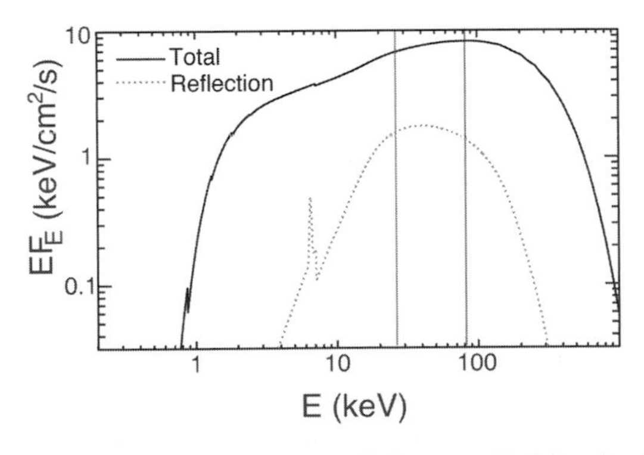

Figure 44.2 Typical radiation spectrum of Cygnus X-1 in the hard state. The contribution from the reflected component is shown (dotted line) as well as the total emission (solid line). The vertical lines show the energy range of PoGOLite. In this range, the reflected component contributes up to 30% of the total flux.

of polarization in the hard state of Cygnus X-1 would lend strong support to the two-component accretion-flow scenario. The observed degree of polarization is also dependent on the inclination of the system, giving polarimetry the potential to determine this parameter, which is often difficult by other means.

Although fainter, Cygnus X-1 is also a strong source of hard X-rays in its soft state. In this case the radiation spectrum rules out the emission arising through Comptonization in a hot inner flow. Rather, the hard X-rays are believed to be produced through single inverse-Compton scattering in a nonthermal electron distribution. Such a scenario may also give rise to polarization in the observed emission. Detection of polarization in the soft state of Cygnus X-1 may both test the emission scenario and give possible constraints on the electron distribution.

44.4 Summary

Polarimetry provides an essentially unexplored tool to probe the X-ray emission mechanisms in a variety of astrophysical objects. Here we have focused on two such sources: rotation-powered pulsars and accreting black holes. Theoretical considerations show that both sources may display significant degrees of X-ray polarization, with the polarimetric signature varying strongly between models. Therefore, polarization measurements will lead to an understanding of the physical processes behind the X-ray production.

PoGOLite is a balloon-borne hard X-ray (25–80 keV) polarimeter designed to study these sources. Our simulations show that it will be able to discriminate between the theoretical models for X-ray emission in the Crab pulsar, and detect polarization as low as $\sim$5% in the accreting black hole Cygnus X-1. A first pathfinder mission will be flown from northern Sweden in August 2010, observing these targets.

References

[1] Abdo, A. et al. (2009). *Astrophys. J.* **696**, 1084.
[2] Axelsson, M. et al. (2007). *Astropart. Phys.* **28**, 327.
[3] Dean, A. J. et al. (2007). *Science* **321**, 1183.
[4] Dyks, J., Harding, A. & Rudak, B. (2004). *Astrophys. J.* **606**, 1125.
[5] Kamae, T. et al. (2008). *Astropart. Phys.* **30**, 72.
[6] Matt, G. (1993). *Month. Not. Roy. Astron. Soc.* **260**, 663.
[7] McGlynn, S. et al. (2009). These proceedings.
[8] Novick, R. et al. (1972). *Astrophys. J.* **174**, L1.
[9] Pearce, M. et al. (2009). These proceedings.

45

Pre-flight qualification tests of the PoGOLite detector system

M. S. Jackson & M. Kiss

KTH, Dept. of Physics, and Oskar Klein Centre, AlbaNova University Centre, Stockholm

on behalf of the PoGOLite collaboration

Tests were performed on each section of each PoGOLite detector in order to characterize its behaviour, as well as to choose which detectors will be used in flight and in what configuration. We present the method and results of the tests of these detectors, as well as the strategy used for placing them in the instrument.

45.1 Introduction

The Polarized Gamma-ray Observer (PoGOLite)[1; 2] is a balloon-borne, Compton-based polarimeter, with an energy range of 25–80 keV. In the pathfinder instrument to be flown in August 2010, the detector system will employ 61 phoswich detector cells (PDC) and 30 side anticoincidence shield (SAS) detectors situated in an unbroken ring around the PDCs. The full size PoGOLite instrument will contain 217 PDCs. The previous tests and simulations of the detector system are explored in more detail in [3].

The 61 PDC and 30 SAS detectors must be arranged in a way which optimizes the detection efficiency of valid events, while also allowing for the virtually complete rejection of background. For this purpose, the light yield of each component of each PDC and SAS unit was measured, using radioactive sources with particles and energies to which the detector materials are most sensitive. The light yield of a detector indicates its efficiency and is given by the peak channel number in the spectrum.

45.2 PDC testing

The PDCs each comprise a fast scintillator, a long, hollow, slow scintillator tube above the fast scintillator at the entrance of the instrument, and a BGO crystal below,

X-ray Polarimetry: A New Window in Astrophysics, eds. R. Bellazzini, E. Costa, G. Matt and G. Tagliaferri. Published by Cambridge University Press.

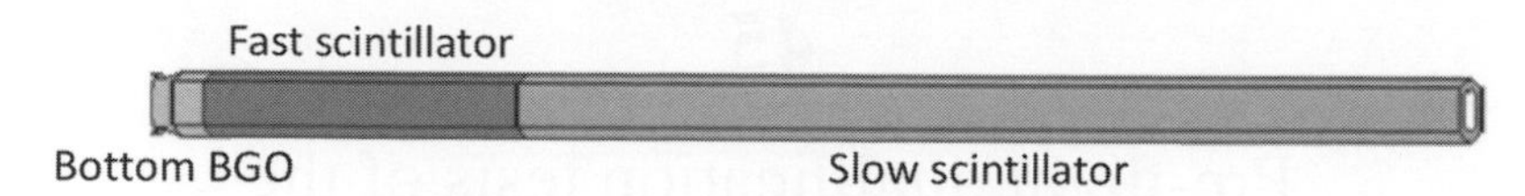

Figure 45.1 Diagram of a PDC unit, showing the slow scintillator, the fast scintillator, and the bottom BGO. The total length of each PDC is approximately 1 m.

as shown in Figure 45.1. The fast scintillators detect Compton scattering events and subsequent photoelectric absorptions in adjacent detectors, for the purpose of determining the scattering direction and thus the polarization of the incoming photons. The slow scintillator tube limits the field of view by rejecting all off-axis events, and the narrow field of view gives the instrument excellent background rejection capabilities. The bottom BGO in the PDCs, along with the BGO crystals in the SAS units, exclude cosmic rays and other off-axis high energy particles, further enhancing the background rejection.

45.2.1 Slow scintillator tests

The slow scintillator of each PDC was irradiated with electrons from ^{90}Sr at five equidistant points along its length. The tests of the slow scintillator resulted in spectra as shown at the top of Figure 45.2, which yielded peak channel numbers as shown in the graph at the bottom of Figure 45.2. These results were checked for consistency by comparing with previous work[4].

45.2.2 Fast scintillator tests

The fast scintillator of each PDC was irradiated at a point 4 cm from the top of the scintillator with photons from ^{241}Am. The tests of the fast scintillators yielded spectra showing the photoabsorption peak from ^{241}Am at 60 keV, as well as a Compton edge and low-energy source photons, and these features are consistent with the results of previous tests.

45.2.3 BGO crystal (PDC) tests

The BGO crystal in each PDC was irradiated with photons from ^{137}Cs at approximately its central point. An example of the results of the present BGO tests is shown in the left panel of Figure 45.3, and a comparison with previous work[5] using the BGO crystal which had not yet been added to the PDC is shown in the right panel of Figure 45.3. The hump to the right of the photoabsorption peak in the left graph arises from the hexagonal geometry of the BGO crystal and the fact

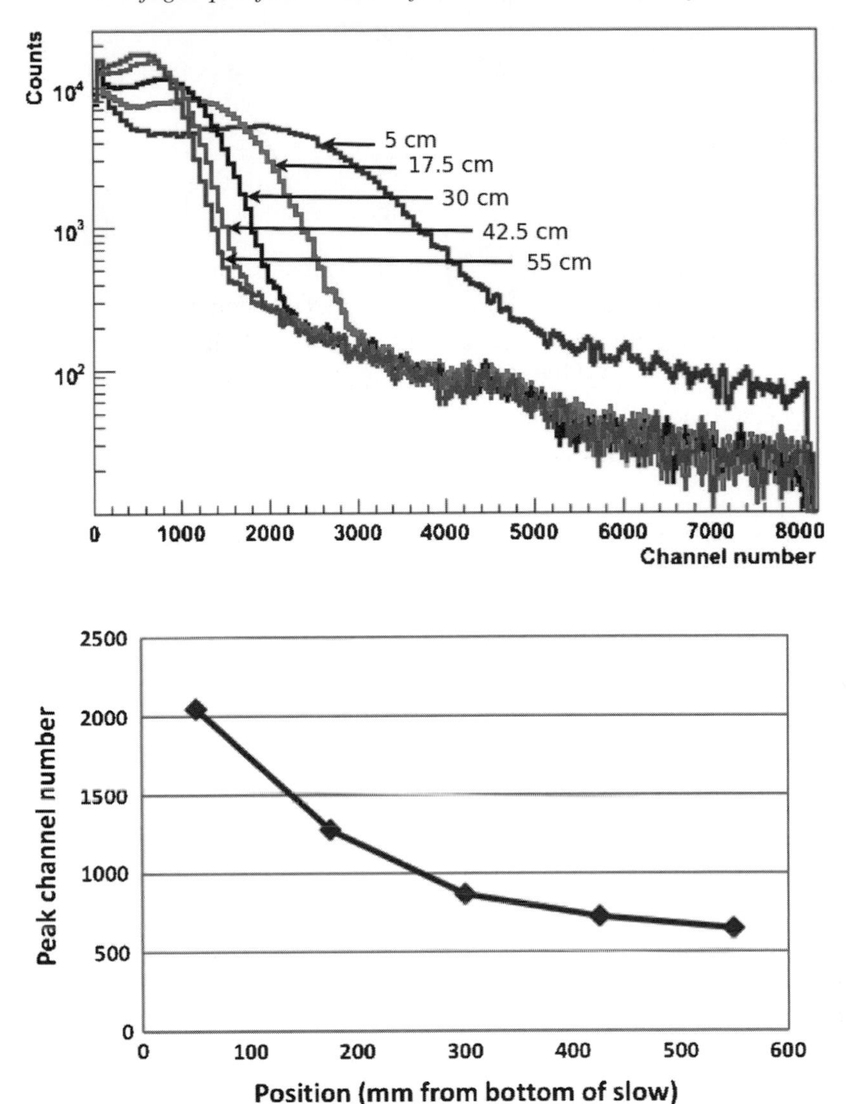

Figure 45.2 Results from the present tests of the slow scintillator: *Top:* Spectra resulting from the irradiation of the slow scintillator at five points along its length. The distance from the bottom of the slow scintillator is indicated for each curve. *Bottom:* Graph of peak channel number vs. position along the slow scintillator.

that the crystal was irradiated from the side, and this feature does not appear in the graph on the right side because the BGO crystal was irradiated from above in the previous test. The channel numbers along the bottom are arbitrary and correspond to particle energy.

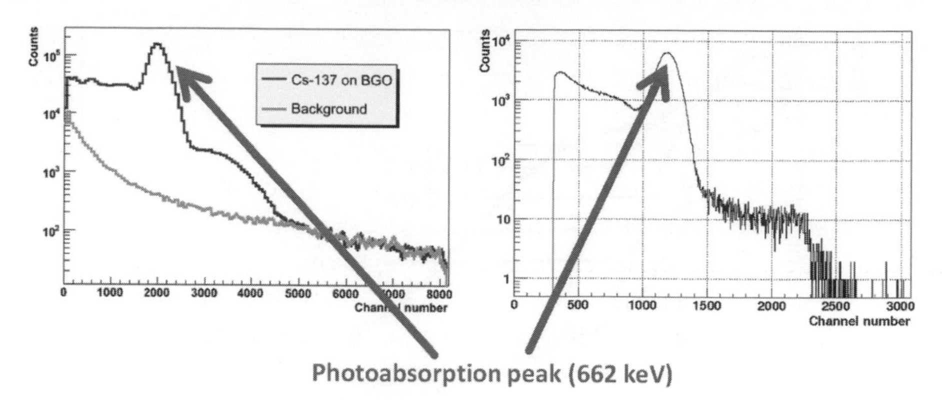

Figure 45.3 Example spectra from the present tests (*left panel*) and from previous tests[5] of the PDC BGO (*right panel*).

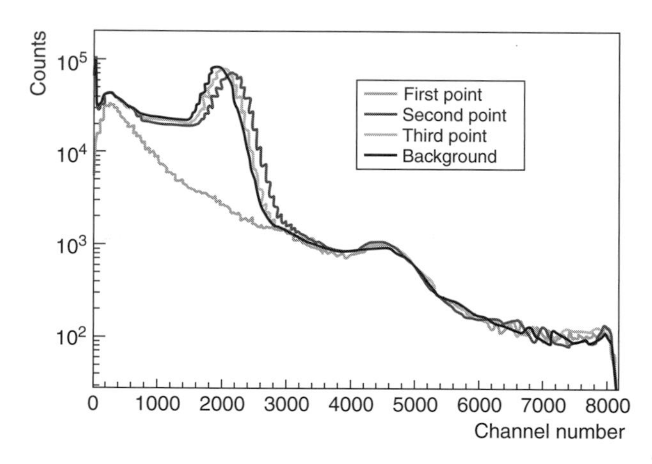

Figure 45.4 Spectra resulting from the irradiation of the SAS at the centre point of each crystal.

45.3 SAS testing

Each of the three BGO crystals comprising each SAS unit was irradiated at its approximate centre point with photons from ^{137}Cs. The purpose of these tests was to check the glue joints between the crystals as well as the reflective $BaSO_4$ coating. The tests resulted in spectra and peak channel numbers similar to those shown in Figure 45.4. The similarity of the graphs for the three crystals indicates both that the glue joints are sound and that very little light is lost through the coating.

45.4 Selection procedures and results

PDCs were ranked according to the light yields in the individual components. The PDCs with better slow scintillator rankings were placed in the outside list (24

PDCs to be placed in the outer ring) and those with better fast scintillator rankings were placed in the inside list (the remaining 37 PDCs). This was done because the highest rate of background is expected in the outer ring and thus the ability to reject background is crucial, and also so that if the background is higher than expected, the outside ring of PDC units can function as an extra layer of anticoincidence.

The SAS units were sorted according to the peak channel numbers for the three crystals. The best were chosen for flight and arranged in a way which places the best next to the worst, so that the average light yield among any two adjacent crystals is approximately constant. This will ensure an approximately uniform level of background rejection in all directions.

References

[1] Kamae, T., et al. (2008). PoGOLite – A high sensitivity balloon-borne soft gamma-ray polarimeter, *Astroparticle Physics* **30**, 72.

[2] Pearce, M., these proceedings.

[3] Kiss, M. (2008). *Studies of PoGOLite Performance and Background Rejection Capabilities*, KTH Licentiate thesis (2008), available from www.particle.kth.se/pogolite.

[4] Andersson, V. (2005). *Tests of Scintillators and PMTs for the Polarized Gamma-ray Observer (PoGO)*, KTH Masters thesis (2005), available from www.particle.kth.se/pogolite.

[5] Marini Bettolo, C. (2008). *PoGOLite – The Polarised Gamma-ray Observer*, KTH Licentiate thesis (2008), available from www.particle.kth.se/pogolite.

46
The Gamma-RAy Polarimeter Experiment (GRAPE) balloon payload

P. F. Bloser, M. L. McConnell, J. S. Legere, T. P. Connor & J. M. Ryan
University of New Hampshire, USA

The Gamma-RAy Polarimeter Experiment (GRAPE) is a concept for an astronomical, hard X-ray, Compton polarimeter operating in the 50–500 keV energy band. The instrument has been optimized for wide-field polarization measurements of transient outbursts from energetic astrophysical objects such as gamma-ray bursts and solar flares. The GRAPE instrument is composed of identical modules, each of which consists of an array of scintillator elements read out by a multi-anode photomultiplier tube (MAPMT). Incident photons Compton-scatter in plastic scintillator elements and are subsequently absorbed in inorganic scintillator elements; a net polarization signal is revealed by a characteristic asymmetry in the azimuthal scattering angles. We have constructed a prototype GRAPE module, containing a single CsI(Na) calorimeter element, which has been calibrated using a polarized hard X-ray beam and flown on an engineering balloon test flight. A full-scale scientific balloon payload, consisting of up to 36 modules, is currently under development. The first flight, a one-day flight scheduled for 2011, will verify the expected scientific performance with a pointed observation of the Crab Nebula. We will then propose long-duration balloon flights to observe gamma-ray bursts and solar flares.

46.1 Introduction

The Gamma-RAy Polarimeter Experiment (GRAPE) is a scintillator-based Compton polarimeter designed to observe polarized astrophysical phenomena in the hard X-ray energy band (50–500 keV). Although intended primarily for observations of bright, transient events such as gamma-ray bursts (GRBs) and solar flares, GRAPE may also be operated in a collimated, pointed mode. The basic instrument concept

X-ray Polarimetry: A New Window in Astrophysics, eds. R. Bellazzini, E. Costa, G. Matt and G. Tagliaferri. Published by Cambridge University Press.

has been validated in laboratory experiments and during a calibration campaign using a polarized X-ray beam, and the hardware has been flight-tested on an engineering balloon flight. We are currently developing a full-scale scientific payload for a one-day balloon flight, and plan to expand this instrument for long-duration balloon flights in the coming years while pursuing opportunities for an orbital platform.

The primary science goal of GRAPE is to measure polarization from energetic transient sources, such as GRBs and solar flares, in order to study the physics of particle acceleration. The prompt gamma-ray emission in GRBs is believed to result from particles accelerated by shocks in a highly relativistic jet aimed at the observer. Polarization measurements will help to distinguish between two basic classes of models[1; 2]. In "physical" models, a globally ordered magnetic field produces high levels of polarization via synchrotron radiation, with only a weak dependence on viewing geometry. In "geometric" models a randomly tangled magnetic field can yield high polarization only if the event is viewed very close to the edge of the jet, due to relativistic beaming. These models can be distinguished by observing a large sample of GRBs and measuring the fraction displaying significant polarization[1]. Solar flares represent a process of explosive energy release in the magnetized plasma of the solar corona, with electrons being accelerated to hundreds of MeV. The details of how the Sun releases this energy and efficiently transfers it into accelerated particles are unknown. Polarization measurements of the hard X-ray emission from flares, due to bremsstrahlung from these energetic electrons, are expected to be useful in determining the beaming, or anisotropy, of these electrons[3; 4], which will provide clues about electron acceleration and transport mechanisms.

46.2 The GRAPE instrument

The design of the GRAPE polarimeter is based on the fact that photons are preferentially Compton-scattered at right angles to the incident electric field vector, or polarization vector[5]. The measurement principle is shown in Figure 46.1. Photons scatter in an array of low-Z, organic scintillator elements and are absorbed in high-Z, inorganic scintillator elements. The relative positions of the triggered scintillators determine the azimuthal scatter angle ϕ; a polarization signal is found by creating a histogram of the measured ϕ values and, after correcting for geometric effects, fitting for the expected azimuthal modulation pattern[5]. The polarimeter is characterized by the modulation factor, μ_{100}, measured for 100% polarized radiation.

The detailed GRAPE design has evolved over the course of a ten-year development project at the University of New Hampshire[6; 7; 8; 9; 10; 11]. The most recent prototype, the third Science Model (SM3), employs an array of 60 plastic

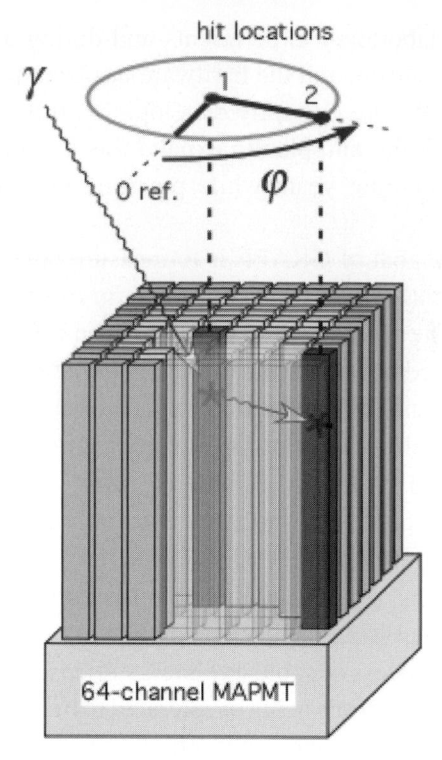

Figure 46.1 Principle of operation of the GRAPE polarimeter.

scintillator elements, each 5 mm × 5 mm × 50 mm, surrounding a single CsI(Na) calorimeter element, 10 mm × 10 mm × 50 mm, all read out by a Hamamatsu H8500 multi-anode photomultiplier tube (MAPMT), see Figure 46.2. The SM3 module has been tested in the laboratory using a ~60% polarized hard X-ray beam (~288 keV) created by scattering 662 keV photons at 90° in a block of plastic. The polarized data were corrected for geometric effects using unpolarized 356 keV photons from a ^{133}Ba source. For this 60%-polarized beam we measured a modulation factor $\mu \sim 0.35$, in agreement with Monte-Carlo simulations[9].

In order to develop GRAPE beyond a laboratory prototype and prepare for balloon-flight measurements, we developed an engineering model (EM1) consisting of the SM3 module combined with custom electronics and data acquisition software running on a PC/104 computer (Figure 46.3;[11]). The electronics boards fit within the footprint of the MAPMT, allowing for multiple modules to be tiled into a compact array. An important feature of the EM1 electronics is the implementation of pulse-shape discrimination to eliminate events triggered by optical cross talk between the CsI(Na) calorimeter and the plastic elements. This is possible due to the long decay time of the CsI(Na) scintillation light compared to plastic[11].

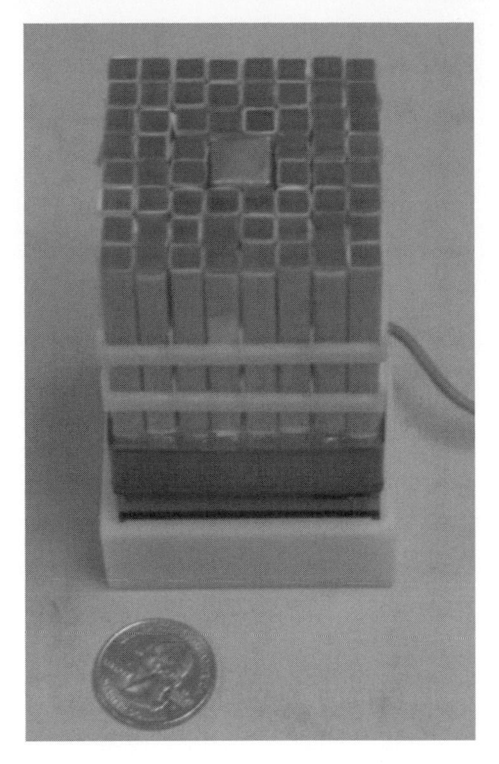

Figure 46.2 GRAPE Science Model 3.

46.3 Beam calibration

The GRAPE EM1 was calibrated in December 2006 using a hard X-ray beam at the Advanced Photon Source (APS) at Argonne National Laboratory in order to measure the modulation factor for nearly 100% polarized radiation and to validate our Monte-Carlo simulations[11]. The instrument was exposed to two energies, 69.5 keV and 129.5 keV, with a polarization fraction calculated to be $97 \pm 2\%$. As the beam was less than 1 mm in diameter it was necessary to expose one plastic element at a time using an X-Y translation table. Alternate elements were exposed in a checkerboard pattern due to time constraints. After completing a scan, EM1 was rotated 90° and the scan was repeated. In this way we were able to create an "unpolarized" exposure by combining data from both orientations relative to the polarization vector; these unpolarized data were used for the removal of geometric effects. The data for each plastic exposure were corrected for exposure time and beam flux and combined into azimuthal scatter angle histograms for the entire instrument.

The fitted azimuthal modulation histogram for 129.5 keV is shown in Figure 46.4. The fitted modulation factor, assuming 97% polarized radiation, is $\mu_{97} = 0.48 \pm$

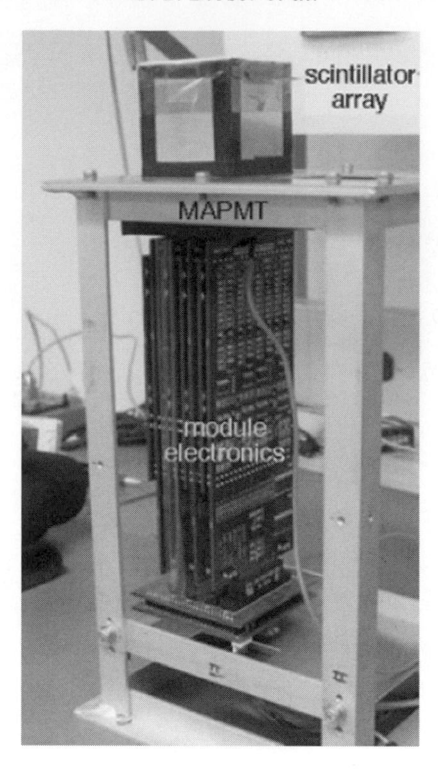

Figure 46.3 GRAPE Engineering Model 1.

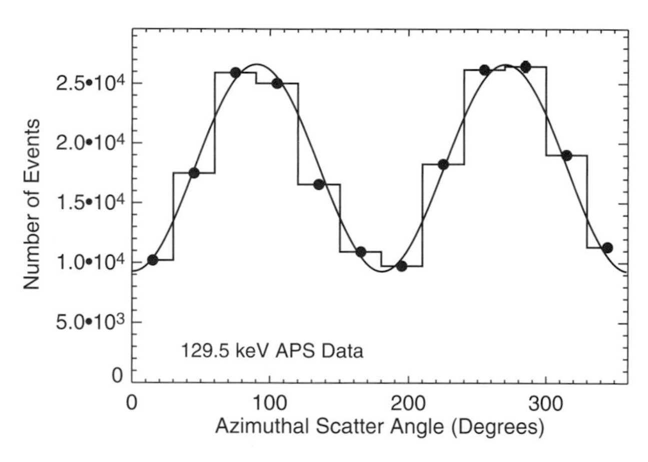

Figure 46.4 APS beam data at 129.5 keV. We find $\mu_{97} = 0.48 \pm 0.03$.

0.03. The beam exposure was then simulated using MGEANT enhanced by the GLEPS package for including the effects of polarization.[1] The simulated modulation histogram is shown in Figure 46.5. For the simulation we find $\mu_{97} = 0.49 \pm 0.02$,

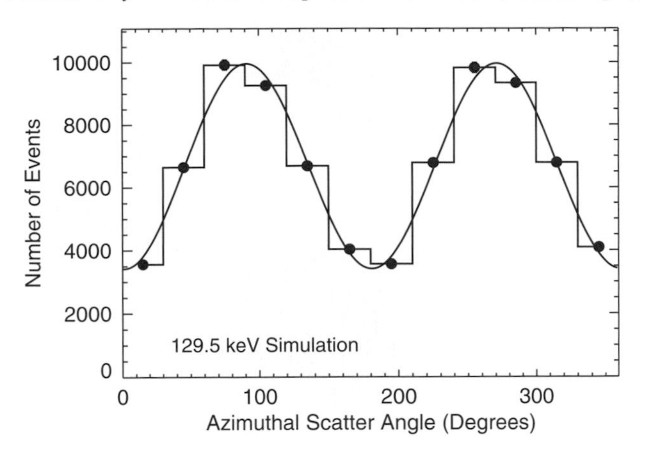

Figure 46.5 Simulated APS histogram at 129.5 keV. We find $\mu_{97} = 0.49 \pm 0.02$.

in excellent agreement with the data. These results show that GRAPE is a sensitive polarimeter, and that our Monte-Carlo tools accurately reproduce its performance.

46.4 Engineering balloon flight

The EM1 instrument was fitted with passive and active plastic shielding, sealed within a pressure vessel, and flown on an engineering balloon flight from Palestine, TX in June 2007 in order to flight-qualify the hardware and measure background. The payload spent 5.5 hours at a float altitude of ~39 km. The "good" count rate, consisting of events with one triggered plastic in coincidence with the calorimeter and no trigger in the active shields, was ~1.1 cts s^{-1} (50–300 keV) and was stable throughout the flight. The background spectrum was roughly consistent with a simple Monte-Carlo simulation of the expected counts due to atmospheric and cosmic gamma rays. More detailed background simulations are in progress.

46.5 GRAPE science payload

A full GRAPE scientific balloon payload is currently under development. For this payload, each polarimeter module will have 28 CsI(Tl) calorimeter elements, each 5 mm × 5 mm × 50 mm, placed around the edge of the H8500 MAPMT and surrounding the 6 × 6 array of plastic elements (as in Figure 46.1). This will roughly double the efficiency of each module compared to SM3. The first flight is scheduled for the Fall of 2011 from Ft. Sumner, NM, and will carry at least a 4 × 4 array of GRAPE modules. As this will be a "conventional," one-day flight, the instrument will be collimated and pointed at the Crab Nebula. Subsequent long-duration balloon (LDB) flights will expand the array to 5 × 5 and then 6 × 6

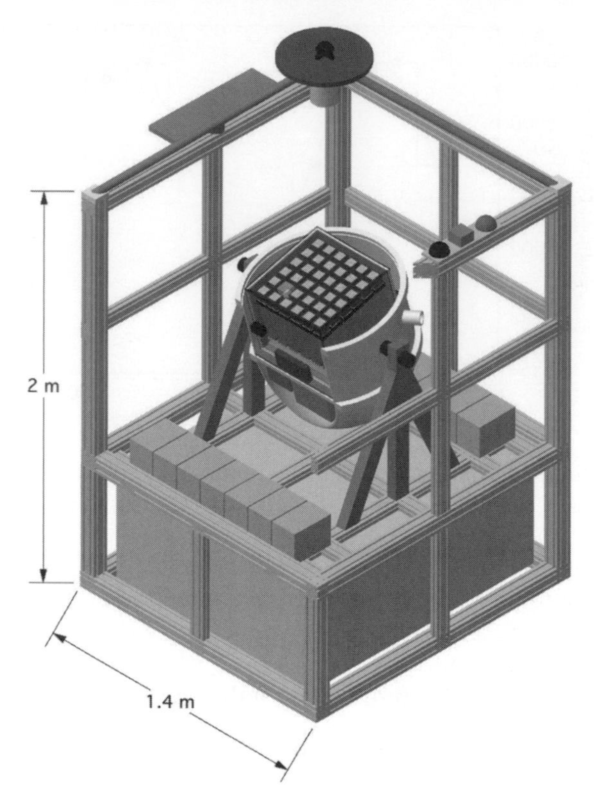

Figure 46.6 GRAPE science balloon payload, with a 6×6 array of polarimeter modules.

modules (Figure 46.6) and will fly for up to 30 days at a time from Antarctica in order to observe multiple GRBs. Based on the recorded background during the engineering balloon flight of EM1, we have used Monte-Carlo simulations to estimate the minimum detectable polarization (MDP;[5]) for GRBs, as a function of their 20–100 keV fluence, achievable by the LDB payloads (Figure 46.7). We have assumed an atmospheric depth of 3.5 g cm^{-2} and a zenith angle of 30°. For reference we indicate the approximate fluence of GRB 041219A, for which time-varying polarization was reported at a level greater than 20% by the IBIS instrument on INTEGRAL[12]. The GRAPE LDB payload would have a 3σ MDP of $\sim$9% for a similar GRB.

Over the course of a series of LDB flights we would expect to achieve significant polarization measurements for several GRBs, and thus begin to distinguish the competing GRB models described in Section 46.1. To compile a large sample of measurements, however, the GRAPE instrument would need to fly on a space-based platform. One such platform, the POlarimeters for Energetic Transients (POET)

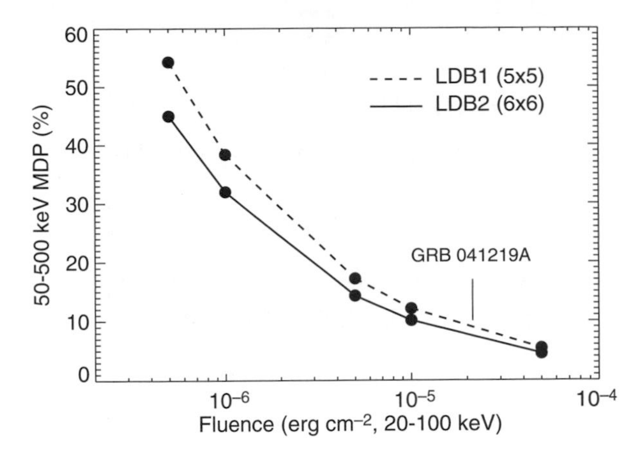

Figure 46.7 Expected sensitivity (3σ MDP) of the GRAPE LDB science payloads to GRBs. The approximate fluence of GRB 041219A is indicated.

mission, was proposed as a NASA Small Explorer mission in 2008[13]. Such a mission would provide measurements with MDP of 20% or better for $\sim$80 GRBs in two years and would definitively test GRB and solar-flare emission models.

Notes

1. `http://public.lanl.gov/mkippen/actsim/glecs/`

References

[1] Toma, K., Sakamoto, T., Zhang, B., Hill, J. E., McConnell, M. L., Bloser, P. F., Yamazaki, R., Ioka, K., and Nakamura, T. (2009). *ApJ* **698**, 1042–1053.

[2] Waxman, E. (2003). *Nature* **423**, 388–389.

[3] Bai, T. and Ramaty, R. (1978). *ApJ* **219**, 705–726.

[4] Leach, J. and Petrosian, V. (1983). *ApJ* **269**, 715–727.

[5] Lei, F., Dean, A. J., and Hills, G. L. (1997). *Space Sci. Rev.* **82**, 309–388.

[6] McConnell, M. L., et al. (1998). *AIP Conf. Ser.* **428**, 889–893.

[7] McConnell, M. L., et al. (1999). *Proc. SPIE* **3764**, 70–78.

[8] McConnell, M. L., et al. (2004). *Proc. SPIE* **5165**, 334–345.

[9] Legere, J. S., et al. (2005). *Proc. SPIE* **5898**, 58981G.

[10] Bloser, P. F., et al. (2006). *Ch. Jour. Astron. Astrophys.* **6**, 393–397.

[11] Bloser, P. F., Legere, J. S., McConnell, M. L., Macri, J. R., Bancroft, C. M., Connor, T. P., and Ruan, J. M. (2009). *NIM A* **600**, 424–433.

[12] Götz, D., et al. (2009). *ApJ* **695**, L208–L212.

[13] McConnell, M. L., et al. (2009). These proceedings.

47

POLAR: an instrument dedicated to GRB polarization measurement

N. Produit

ISDC and Geneva Observatory

on behalf of the POLAR collaboration[1]

POLAR is an instrument designed to measure polarization of gamma-ray bursts (GRB). The concept is based on building the simplest and the smallest possible instrument that can perform this measurement properly. Compton scattering is used. Energy measurement and localization of the GRB is performed only at a level sufficient to maintain the systematic errors at an acceptable level. If the information of the GRB spectrum and of the GRB localization is known by other means, it could be used to reduce further the systematic errors. The instrument is very light and very compact but has enough acceptance to enable measurement of more than 10 GRB per year with a systematic error less then 10%.

47.1 Introduction

Polarization of the prompt signal is a key ingredient to understand the gamma-ray-burst phenomenon. In fact very different scenarios (for example Poynting-flux-driven or baryon-driven models) differ widely on the predicted polarization level[1] and, at the same time, agree on most of the other measurable parameters.

The bulk of the energy in the prompt signal of GRB is emitted around 100 keV so this is the part of the spectrum that contains the most valuable information. At this energy, the Compton effect dominates. Fortunately the Compton cross section is dependent on the incoming photon polarization, being maximal for scattering angles perpendicular to the polarization direction.

The desirable features for a GRB polarimeter are:

- It should be a space instrument because gamma-rays do not penetrate the atmosphere.
- It should feature a large angular dependence to collect as many GRB as possible.

X-ray Polarimetry: A New Window in Astrophysics, eds. R. Bellazzini, E. Costa, G. Matt and G. Tagliaferri.
Published by Cambridge University Press.

- It should favour Compton scattering.
- It should be capable of reconstructing Compton-event geometry.
- It should be able to reject noise coming from cosmic rays and low-energy trapped particles.
- It should minimize systematics in the spatial frequency of 180° where the signal resides.

Other features are desirable but not essential: energy reconstruction is useful to constrain the Compton geometry or reject noise, imaging capabilities are useful to decrease systematics due to nonuniform angular response of the instrument and to reject diffuse noise.

Other scaling factors are evident and constrained by external parameters: effective area and field of view are constrained by total weight. The total available power also limits the effective area by limiting the number of channels and also has an impact on the possible level of sophistication of the signal processing and noise rejection.

47.2 The POLAR concept

The POLAR concept is based on some drastic a priori decisions: the detector is meant to measure polarization and polarization only, it has to be as small as possible to demand minimal requirements on the platform. We consider that knowledge of GRB position and spectrum are nowadays a commodity provided by existing satellites.

Light material favours Compton interaction so it was chosen. It would be useful to capture in heavy material the photon after the initial Compton scattering but, in a wide-acceptance instrument, it is not possible to arrange the geometry so that the useful photons will see heavy material only after the initial Compton scattering. Introducing heavy material in the detector will inevitably remove photons that incur photo-electric absorption as the initial interaction. For this reason, we decided that the target will exclusively consist of light material. We understand that most of the events will not be fully contained and therefore the energy reconstruction will be very imprecise. Systematic error introduced by a poor energy reconstruction can be modelled very precisely by Monte-Carlo if the GRB spectrum is known from other instruments. Monte-Carlo shows also that systematic errors stay manageable if a typical band spectrum is used to reconstruct polarization in the case that the spectrum is unknown but reasonable. The plastic scintillator has all the desirable features of being sensitive to electromagnetic energy deposition, of being very fast (ns resolution), of producing a large signal (one optical photon for 100 eV energy deposition) and of being a stable, light and solid material that has been successfully used for extended periods of time in space without degradation.

47.3 Noise rejection

There are different types of noise and different strategies to remove them. First, the diffuse gamma rays constitute an unavoidable background. This background is unpolarized unless it arrives at the detector after a scattering in the earth atmosphere or on the satellite platform. It has to be subtracted by modelling its contribution to the GRB signal. Extensive Monte-Carlo and using data outside of the GRB will reduce this background to its statistical variation. Fluctuation of this background is ultimately what limits the instrument in case of very long or faint GRB.

Background events induced by cosmic rays are very easily removed by the event topology (tracks) or by the total energy deposition (>2 MeV/g/cm^2). This rejection can be accomplished already at the initial trigger level by our multiple-threshold electronics without inducing appreciable dead time.

Low-energy trapped particles interact in the shielding structure (2 mm of carbon) and then create single energy deposition events most of the time. Those events do not pass the trigger criteria that ask for two energy depositions in a 50 ns time coincidence. In the extreme cases of the South Atlantic anomaly, random coincidences can introduce spurious events. We consider that the detector cannot work in this region or in the polar caps. This decreases the efficiency of the detector by some percent depending on the orbit. Simulation has shown us that the geostationary orbit is also not a possible orbit for this detector due to the excessive electron population.

47.4 The POLAR detector

The detector consists of 25 identical modules (see Figure 47.1). A module consists of a multichannel photomultiplier (MAPMT) (H8500 from Hamamatsu). The active target is made of 64 plastic scintillator bars $6\times6\times200$ mm, each one wrapped in a highly reflective material (Vikuiti ESR^{TM} from 3M). The front-end electronics has the same lateral size as the MAPMT and is soldered directly to it. The module is enclosed in carbon fiber for structural support and absorption of low-energy electrons.

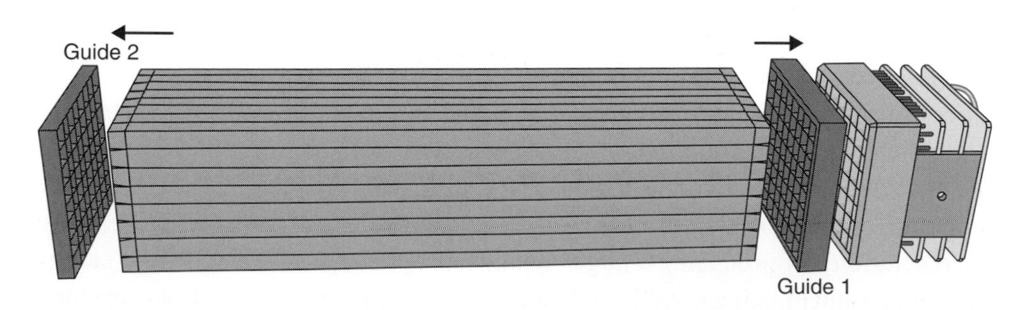

Figure 47.1 One POLAR module.

The quality of the polishing of the bars has been found to be crucial to ensure uniform light collection along the bar by internal reflection. Both ends of the bars are cut into a truncated pyramid form to enter a support (marked 'guide' on Figure 47.1). It ensures precise centering of the scintillator on the sensitive area of the MAPMT. This support was realized by plastic injection moulding of a very precise master, the final mechanical accuracy is 50 microns. This piece turns out to be very important to limit the cross talk between bars induced by light rays expanding at the output of the scintillator through the glass of the MAPMT before reaching the segmented photo-cathode. Furthermore, this piece ensures mechanical stability of the full target.

The electronics consists of three PCBs, the first PCB provides the voltage-divider chain as well as 64-signal adaptation and decoupling. The second PCB contains the front-end electronic chip (IDEAS 64 PMT) that carries out thresholding, sample-and-hold and analog serialization. This PCB contains also a FPGA (IGLOO) that performs triggering by requiring two hits (5 keV threshold), rejecting tracks and collaborating with the central trigger in the case only one hit is recorded. The FPGA also drives a fast ADC, a FIFO memory and I^2C readout. The third PCB contains connectors and some ancillary logic.

The 25 modules are put together in a carbon fiber and aluminum mechanical frame. The mechanical frame also provides room for the central computer, the central trigger, the low-voltage power supply, the high-voltage power supply and the slow control.

Monte-Carlo predictions of the performance of the POLAR detector[2] can be summarized as: effective area around 200 cm^2 and modulation factor around 30% depending on angle of arrival and spectrum of the GRB. Recently we have demonstrated with Monte-Carlo that POLAR will be able to trigger itself on GRB and to measure the GRB coordinate with an accuracy of few degrees. This coordinate measurement is not of a sufficient quality to be useful for alerting purposes but is sufficient to insure that the position determination is not the dominant systematic effect of the measurement[3].

47.5 Status of POLAR module tests

We have used preliminary modules in our polarized test bench facility. This facility uses a Na-22 positron source that creates correlated photons of 511 keV. One of the photons is analyzed using two scintillators ('S1' and 'tag' on Figure 47.2). This simple apparatus provides us with an unpolarized beam (requiring coincidence with just S1) or a polarized beam (requiring coincidence with S1 and tag). The results are compared with a Monte-Carlo code made specially for this configuration and found to qualitatively agree, but more work is needed.

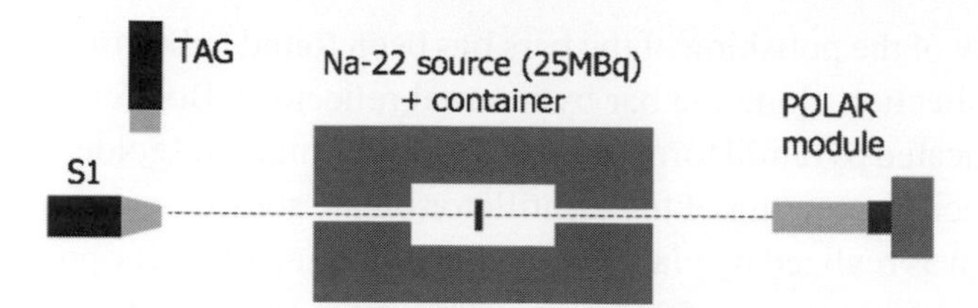

Figure 47.2 Polarized test bench.

After studying in detail all the subcomponents of a POLAR module, we are now building the first POLAR module putting together all our expertise. We use 64 scintillator bars prepared according to our polishing, cutting and acceptance procedures, two final plastic supports, a H8500 MAPMT and a stack of three PCBs of the new version of the electronics.

In our electronics, an IGLOO FPGA plays a central role. This chip is not space qualified. Therefore we irradiated it with low-energy protons in the PIF[4] facility in the Paul Scherrer Institute in Switzerland to a level of 20 krad. We experienced some bit-flip in the internal memories but the chip power consumption was stable during all the irradiation and the chip was still re-programmable after the irradiation.

The first POLAR module will be subjected to further spatialization tests and its performance will be evaluated using our polarized test bench facility before the end of the year. We are also planning to perform a test using polarized photons from a synchrotron source.

47.6 Conclusions

POLAR is a detector with the unique purpose of performing the best polarization measurements with the least constraints on a satellite platform. We are working towards flying the POLAR detector on the Chinese space station around year 2012.

Notes

1. POLAR collaboration: N. Produit, D. Haas, W. Hajdas, H. Hofer, G. Lamanna, C. Leluc, R. Marcinkowski, A. Mtchedlishvili, M. Pohl, S. Orsi, D. Rapin, D. Rybka, E. Suarez, J.P. Vialle

References

[1] Toma, K., et al. (2009). *ApJ*, **698**, 1042.
[2] Produit, N., et al. (2005). *Nucl. Instr. and Meth.* **A550**, 616.
[3] Suarez-Garcia, E., et al. (2009). Submitted to *Astroparticle Physics*.
[4] Hajdas, W., et al. (1996). *Nucl. Instr. and Meth.* **B113**, 54.

48

Polarization detection capability of GRIPS

J. Greiner & G. Kanbach
MPE Garching, Germany
A. Zoglauer
Space Sciences Laboratory, UC Berkeley, USA

We present the polarization capabilities of GRIPS (see www.grips-mission.eu), a proposed next-generation Compton-scattering and pair-creation telescope.

48.1 Introduction

GRIPS, Gamma-Ray Burst Investigation via Polarimetry and Spectroscopy, had been proposed in 2007 in response to the ESA Cosmic Vision call as a new-generation Compton-and-pair telescope[8]. Though it was not selected for further study, a variety of investigations are being performed to improve the concept and to verify the performance.

With the Compton scattering being dependent on the polarization of the incoming photons, any Compton telescope is, per se, a decent polarimeter. Beyond this, such detectors can be tailored to have a particularly high polarization sensitivity by obeying some simple design principles.

Polarization is the last property of high-energy electromagnetic radiation which has not been utilized to its full extent, and promises to uniquely determine the emission processes of a variety of astrophysical sources, among them pulsars, anomalous X-ray pulsars (AXP) and soft-gamma repeaters (SGR), or gamma-ray bursts (GRB).

GRIPS would carry two major telescopes: the gamma-ray monitor (GRM) and the X-ray monitor. The GRM is a combined Compton-scattering and pair-creation telescope for the energy range 0.2–50 MeV. It will thus follow the successful concepts of imaging high-energy photons used in COMPTEL (0.7–30 MeV) as well as EGRET ($>$30 MeV) and *Fermi* ($>$100 MeV) but combines them into

X-ray Polarimetry: A New Window in Astrophysics, eds. R. Bellazzini, E. Costa, G. Matt and G. Tagliaferri.
Published by Cambridge University Press.

one instrument. The following deals exclusively with the GRM concept, and its capability to measure polarization at unprecedented sensitivity.

48.2 The gamma-ray monitor instrument concept

The GRM, like previous Compton and pair telescopes, will employ two separate detectors to accomplish this task: a tracker (D1), in which the initial Compton scatter or pair conversion takes place, and a calorimeter (D2), which absorbs and measures the energy of the secondaries. In the case of Compton interactions, the incident photon scatters off an electron in the tracker. The interaction position and the energy transferred to the electron are measured. The scattered-photon interaction point and energy are recorded in the calorimeter. From the positions and energies of the two interactions the incident-photon angle is computed from the Compton equation. The primary-photon incident direction is then constrained to an event circle on the sky. For incident energies above about 2 MeV the recoil electron usually receives enough energy to penetrate several layers, allowing it to be tracked. This further constrains the incident direction of the photon to a short arc on the event circle. GRM will be able to determine source locations to better than 1°(radius, 3σ).

The differential Klein-Nishina cross-section for Compton scattering contains a strong dependence on the polarization of the incident γ-ray photon. Scattered photons are emitted preferentially perpendicular to the direction of the electric field vector of the incoming photon. The strongest azimuthal modulation in the distribution of scattered photons will be achieved for low gamma-ray energies and Compton scatter angles around 90°. This will make a Compton telescope with a calorimeter covering a large solid angle a unique polarimeter. The sensitivity of an

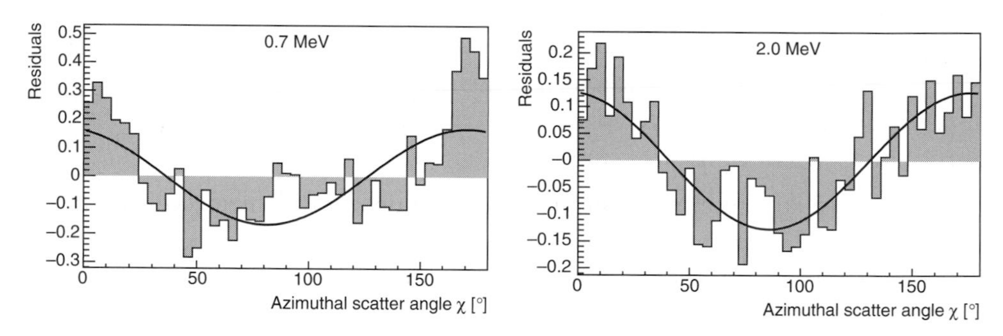

Figure 48.1 Measured polarization response of the MEGA prototype for two different energies. Within measurement errors and statistics, all values are in agreement with GEANT4 simulations: the polarization angle of 90° is reproduced to 82%±24% (0.7 MeV) and 86%±11% (2.0 MeV), respectively. The measured modulation is 0.17±0.04 and 0.13±0.03 as compared to the simulated values 0.19 (at 0.7 MeV) and 0.14 (2.0 MeV). For more details see [16] and[4].

instrument to polarization is given by the ratio of the amplitude of the azimuthal scatter angle distribution and its average, which is called the modulation factor μ. The modulation is a function of incident-photon energy E, and the Compton scatter angle θ, between the incident and scattered-photon directions.

A prototype of such a telescope concept, called MEGA, had been built and calibrated in 2001–2004. Though MEGA had not been optimized for polarization, its properties as measured in a calibration campaign at the High-Intensity Gamma-Source, Durham, NC were very promising (Figure 48.1).

48.3 The simulated performance

The design of a new high-energy gamma-ray telescope must be based on numerical simulations as well as experimental detector developments. The baseline design and input to the Monte-Carlo simulations of the GRM consists of the detector head with the central stack of double-sided Si-strip detectors (tracker D1) surrounded by the pixellated calorimeter (D2) and an anticoincidence system (ACS) made of plastic scintillator. Below the gamma-ray detector is the GRM electronics and a generic spacecraft bus. In our simulations, the bus of the Advanced Compton Telescope study was used[5].

The D1 detector consists of 64 layers each containing a mosaic of 8×8 double-sided Si strip detectors of area $10 \times 10\,\mathrm{cm}^2$. The layers are spaced with 5 mm distance. The D2 calorimeter is made of $LaBr_3$ prisms ($5 \times 5\,\mathrm{mm}^2$ cross-section) which are read out with Si Drift Diode (SDD) photodetectors. The upper half of the D2 side walls feature scintillators of 2 cm length and the lower half has 4 cm thick walls. The side wall crystals are read out by one SDD each. The bottom calorimeter is 8 cm thick and is read out on both ends of the crystals to achieve a depth resolution of the energy deposits.

The whole detector is surrounded by a plastic scintillator counter that acts as ACS against charged particles. Read out of the scintillation light, which is collected with embedded wavelength-shifting fibres to ensure the ACS uniformity, will be done with Si APD detectors.

Most of the structural material that holds the detector elements will be fabricated with carbon-fibre compounds in order to reduce the background of activation radioactivity produced by cosmic rays in aluminum.

The simulations were carried out with the MGGPOD suite[15] which allows for a detailed simulation of the primary and secondary (activation) background in the chosen low-earth orbit. The modelling of the instrument functions, the reconstruction of the Compton and pair data, and the final performance calculations were carried out with the MEGAlib software suite[17] which had been developed in the course of the MEGA prototype development[3].

MEGAlib contains a geometry and detector description tool that was used to set up the detailed modelling of the GRM with its detector types and characteristics. The geometry file is then used by the MGGPOD simulation tool to generate artificial events. The event reconstruction algorithms for the various interactions are implemented in different approaches (Chi-square and Bayesian). The high-level data analysis tools allow response-matrix calculation, image reconstruction (list-mode likelihood algorithm), detector resolution and sensitivity determination, spectra retrieval, polarization modulation determination, etc. Based on many billions of simulated events we have now derived a good understanding of the properties of GRIPS.

Various source and spectral scenarios have been simulated: a series of GRB spectra and intensities which span the range of burst properties known from BATSE as well as time-constant line and continuum sources were simulated and analyzed. In all cases the environmental orbital radiation background and the instrumental activation were taken into account. The small intrinsic radioactivity of the proposed scintillator material $LaBr_3$ (0.8-2.0 cts/sec/cm^3 depending on the compound) is negligible compared to other background components and has been ignored in the simulations.

GRIPS is a nearly perfect polarimeter (Figure 48.2, right panel): the well-type geometry allows the detection of Compton events with large scatter angles which carry most of the polarization information. GRIPS' polarization sensitivity will be best at 200–400 keV where the polarization information of the photons is best preserved through the Compton scatter process.

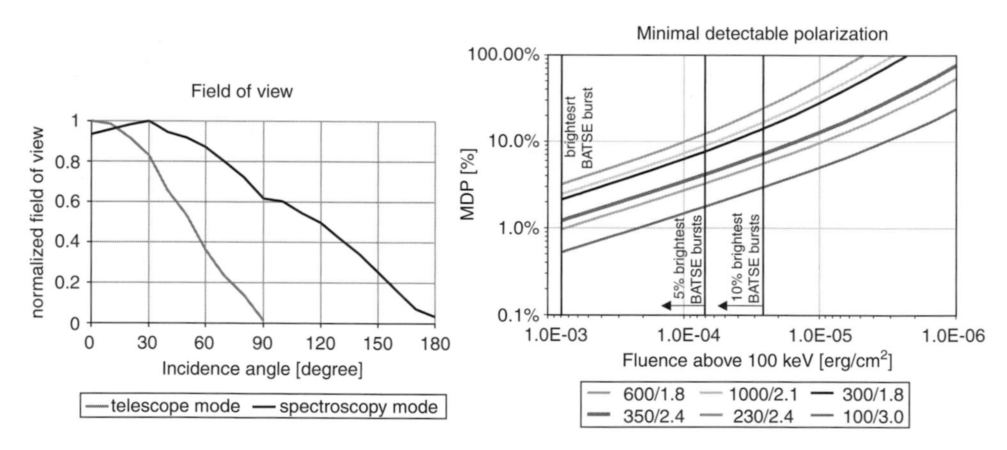

Figure 48.2 Left: field of view of the gamma-ray monitor. Right: polarization sensitivity of GRM. Various models of GRB spectra are shown which differ in their break energy and their high-energy power-law slope; see legend for the parameter pair for each model. Note that at the bright end the minimum detectable polarization changes much more slowly than the fluence.

48.4 Polarization as a diagnostic of the GRB emission mechanism

The link between the gamma-ray production mechanism in GRBs and the degree of linear polarization is expected to severely constrain models. A significant level of polarization can be produced in GRBs by either synchrotron emission or by inverse Compton scattering. The fractional polarization produced by synchrotron emission in a perfectly aligned magnetic field can be as high as 70–75%. An ordered magnetic field of this type would not be produced in shocks but could be advected from the central engine[7; 6; 12]. It should be possible to distinguish between synchrotron radiation from an ordered magnetic field advected from the central engine and Compton drag. Only a small fraction of GRBs should be highly polarized from Compton drag because they have narrower jets, whereas the synchrotron radiation from an ordered magnetic field should be a general feature of all GRBs.

GRIPS will allow us to measure the polarization of the prompt gamma-ray burst emission to a few percent accuracy for about 10% of all detected GRBs. Moreover, the superior polarization sensitivity will even securely measure whether or not the percentage polarization varies with energy, angle and/or time over the burst duration of a dozen brightest GRBs.

48.5 Pulsars, AXPs, and SGRs

Pulsars are an excellent laboratory for the study of particle acceleration, radiation processes and fundamental physics in environments characterized by strong gravity, strong magnetic fields, high densities and relativity. Above ~50 MeV, EGRET has detected 6–10 pulsars and candidates[14], and [13] reported for the COMPTEL survey 4–5 pulsars corresponding to EGRET detections and one high-magnetic-field pulsar (PSR B1509-58) with emission up to 30 MeV.

Estimates for the number of possible pulsar detections by GRIPS were previously based on our empirical knowledge of pulsar efficiencies and spectra, on new radio surveys that now contain about 2000 pulsar detections and on the extrapolations afforded by theories of high-energy emission from rotating, magnetized neutron stars. These theories are still quite disparate. For energies above several 10 MeV and for the sensitivity of the Large Area Telescope on *Fermi*, predictions range from ~60 detections[9] to about 800[10]. Since June 2008 the *Fermi* Gamma Ray Space Telescope (FGST, Atwood et al., 2009) opened this new chapter in the observation of energetic (above ~100 MeV) galactic and extragalactic sources. The sensitivity improvement of a factor of ~30 compared to EGRET, led to the detection of many point sources (more than 200 highly significant $> 10\sigma$ detections[2]), and the detection of pulsars[1] followed quickly after the commissioning of *FGST* in orbit. Presently, the pulsar catalogue contains more than 40 pulsars. Among them roughly 60% have also been seen at radio wavelengths (with ∼40% young pulsars, ∼20% old, millisecond pulsars and ∼40% pulsars which are radio quiet).

GRIPS' sensitivity for wide-band spectra above ∼10 MeV (normal pulsars) and below 1 MeV (high B-field pulsars and AXPs) exceeds the COMPTEL sensitivity by a similar factor than that going from EGRET to *Fermi*. From the EGRET/COMPTEL relation we therefore estimate the number of GRIPS pulsars to be about 50–70% of the *Fermi* pulsars and expect to detect 60–70 pulsars in the GRIPS energy range >10 MeV. GRIPS-determined light curves and phase-resolved spectra in the 1–50 MeV range will provide decisive insights into the pulsar magnetosphere and the acceleration processes located there. Polarization is a unique characteristic of particles radiating in strong magnetic fields. Pulsars and their surrounding pulsar wind nebulae are highly polarized. GRIPS will be sensitive enough to measure polarization below a few MeV from several pulsars.

'Magnetars' appear to observers in the forms of AXPs or, possibly related, SGRs. Pulsed radiation with a thermal spectrum at soft X-rays (1–10 keV) and an extremely hard power-law up to nearly 1 MeV has been observed from 6–8 AXPs[11]. The continuous all-sky survey of GRIPS promises to capture unique data for high-energy neutron-star astrophysics.

References

[1] Abdo, A.A., et al. (2008). *Science* **322**, 1218.
[2] Abdo, A.A., et al. (2009). *ApJ Suppl.* **183**, 46.
[3] Andritschke, R. (2005). *Exp. Astron.* **20**, 395.
[4] Andritschke, R. (2006). *PhD thesis*, Munich.
[5] Boggs, S.E., Coburn, W., Kalemci, E. (2006). *ApJ* **638**, 1129.
[6] Granot, J. (2003). *ApJ* **596**, L17.
[7] Granot, J., Königl, A. (2003). *ApJ* **594**, L83.
[8] Greiner, J., Iyudin, A., Kanbach, G., et al. (2009). *Exp. Astron.* **23**, 91–120.
[9] Harding, A.K., Grenier, I.A., Gonthier, P.L. (2007). *ApSS* **309**, 221.
[10] Jiang, Z.J., Zhang, L. (2006). *ApJ* **643**, 1130.
[11] Kuiper, L., Hermsen, W., den Hartog, P.R., Collmar, W. (2006). *ApJ* **645**, 556.
[12] Lyutikov, M., Pariev, V.I., Blandford, R.D. (2003). *ApJ* **597**, 998.
[13] Schönfelder, V. Bennett, K., Blom, J.J. *et al.* (2000). *AAS* **143**, 145.
[14] Thompson, D.J., Bertsch, D.L., Hartman, R.C., Sreekumar, P., Kanbach, G. (1999). *Bull of the AAS*. **31**, 732.
[15] Weidenspointner, G., Harris, M.J., Sturner, S. et al. (2005). *ApJS* **156**, 69.
[16] Zoglauer, A. (2005). *PhD thesis*, TU Munich.
[17] Zoglauer, A., Andritschke, R., Schopper, F. (2006). *New Astron. Rev.* **50**, Issues 7/8, 629.

49

X-ray and γ-ray polarimetry small-satellite mission PolariS

K. Hayashida

Osaka University

for the PolariS Working Group

PolariS (Polarimetry Satellite) is a Japanese small-satellite mission dedicated to polarimetry of X-ray and γ-ray sources. We aim to perform wide-band X-ray (2–80 keV) polarimetry of sources brighter than 10 mCrab, employing three hard X-ray mirrors and two types of polarimeters. X-ray and γ-ray polarimetry of transient sources with wide-field polarimeters is the second purpose. Most of the components have their prototype used or planned to be used in balloon and other satellite missions. Conceptual design is in progress for target launch in the mid 2010s.

49.1 PolariS concept

PolariS (Polarimetry Satellite) is a Japanese small satellite mission dedicated for polarimetry of X-ray and γ-ray sources. Design of the PolariS is now being discussed by the PolariS working group, which consists of 34 members from 11 institutes mainly in Japan, though informal international collaboration is under way.

The main purpose of the PolariS mission is wide-band X-ray (2–80 keV) polarimetry of sources brighter than 10 mCrab. We expect X-ray polarimetry with small satellites to be realized within several years. When measuring X-ray polarization of various type of sources, we consider the energy dependence of the polarization (degree and direction) to be essential to study the geometry and emission mechanisms of different objects, as has been pointed out by theoretical calculation. Furthermore, the hard X-ray band above 10 keV is of particular importance, since the physics we are exploring with polarimetry is in most cases nonthermal. We thus aim to get wide-band X-ray polarimetry. In this respect, PolariS will be

X-ray Polarimetry: A New Window in Astrophysics, eds. R. Bellazzini, E. Costa, G. Matt and G. Tagliaferri. Published by Cambridge University Press.

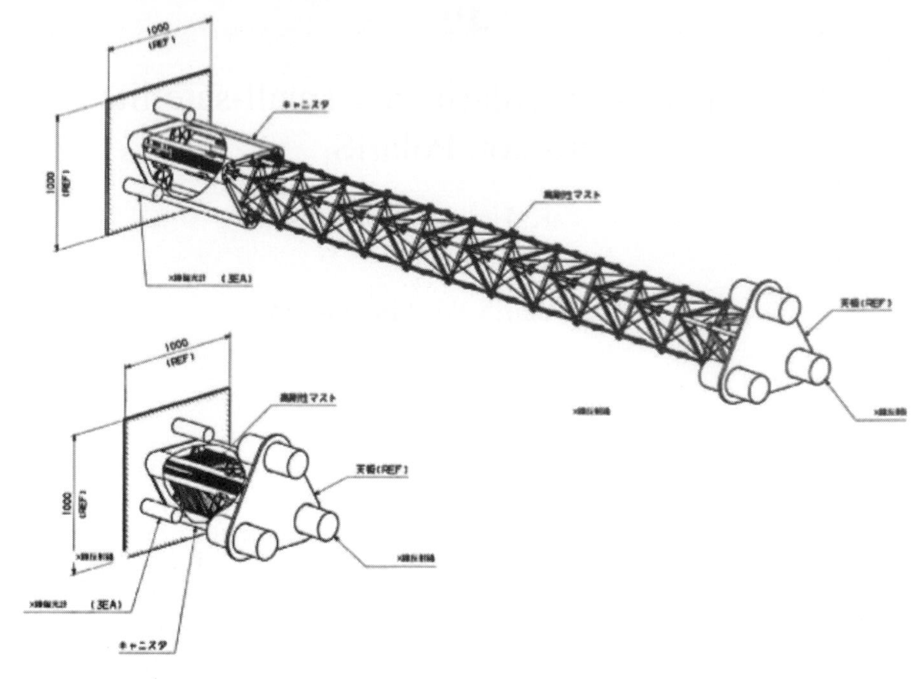

Figure 49.1 Extensible optical bench for PolariS 6 m mirrors. Two mirrors are for scattering-imaging polarimeters and one for gas-imaging polarimeter. The optical bench will be placed on the small satellite common bus, which is 1 m^3 cubic, not included in this figure.

complementary to other small satellite missions currently planned, for example, GEMS and POLARIX, both of which focus on the energy band below 10 keV.

Wide-band X-ray polarimetry will be enabled by employing hard X-ray (multi-layer-coating mirror) telescopes, and (at least) two kinds of focal-plane polarimeters. In the current design, we employ three telescopes, two for scattering-imaging polarimeters for the higher-energy X-ray band, and one for a gas-imaging polarimeter for the lower-energy band. The focal length of the mirrors is 6 m, for which an extensible optical bench will be employed (see Figure 49.1).

The second purpose of the PolariS mission is to measure the X-ray and γ-ray polarization of transient sources, such as γ-ray bursts. For this purpose, we will put wide-field scattering polarimeters without optics on board PolariS.

In this paper, we outline current design of PolariS.

49.2 PolariS design and components

PolariS is designed to be launched as one of the JAXA small-satellite series, which is characterized by a combination of a common-bus system and various mission systems on it. The common-bus system is not exactly the same for different missions, but it will be 1 m^3 cubic in size and 200 kg in weight. On this bus system, we will build the mission part of PolariS. Currently we estimate the mission part weight to be 166 kg, and the power consumption to be 106 W.

The satellite orbit will be a typical low earth orbit, at an altitude of about 550 km and an inclination of about 30°. We direct the telescope to the target source by a three axis attitude control system with an accuracy of 1 arcmin. However, we plan to rotate the satellite around the telescope axis to minimize possible systematics for polarization direction and degree. The rotation speed will be 0.1 rpm.

49.2.1 Multi-layer mirrors

In order to realize polarimetry above 10 keV, there are two design options, one with optics and the other without. We evaluate that it is difficult to reduce background to measure 10-mCrab source-polarization without optics (as collecting devices). Hard X-ray telescopes developed by Nagoya-U team for years are a prime candidate for the collecting and imaging device. The telescopes have experienced various balloon flights and will be employed for the ASTRO-H mission, too. The PolariS mirror will basically be the half-size model of the ASTRO-H mirror. The half size is not only in the focal length (12 m for ASTRO-H, 6 m for PolariS), but also for inner diameter of the mirrors. In the preliminary design, the effective area of one telescope (we will use three) is 120 cm^2 below 10 keV, 40 cm^2 at 30 keV, and sharp cut-off below 80 keV. Target weight for one mirror is 15 kg. We will save the production resources by using the ASTRO-H heritage as much as possible.

Since Bragg reflection is the working principle of the multi-layer mirrors, there has been a little worry that polarization-dependent reflectivity may introduce artificial polarization. On this point, recent experiments[1] show that the upper limit of this effect is as small as 0.8%.

49.2.2 Extensible optical bench

A light-weight and rigid, extensible optical bench (EOB) is an important component of PolariS. We have started the design based on the triangle mast which was used for the HARUKA (radio telescope in space) mission and which will be employed in the ASTRO-H mission. Preliminary evaluation gives a weight of 68 kg with a characteristic frequency lower than 1 Hz, and an extension accuracy of better than 1 mm. These values satisfy the requirements of PolariS.

49.2.3 Gas imaging polarimeter

One gas micro-electrode polarimeter will be placed on the focal plane of a telescope to cover the lower-energy band above 2 keV. Three groups in Japan have succeeded in polarimetry on the ground with different types of gas micro-electrode polarimeter: capillary plates, μ-PIC, and GEM. Recently, international collaboration to combine the gas polarimeter system developed in Italy and the GEM developed by RIKEN group has started[2; 3]. This collaboration may be a prime candidate for the PolariS gas-imaging polarimeter.

49.2.4 Scattering imaging polarimeters

Scattering polarimeters with imaging capability placed on the focal plane of two telescopes take charge of polarimetry in the hard X-ray band. The prototype of this scattering-imaging polarimeter is the polarimeter for the PHENEX balloon experiment[4; 5], which has been independently developed but is similar to the GRAPE polarimeter[6]. A multi-anode photomultiplier tube (MAPMT) and two kinds of scintillator are used to track scattering position and direction of incoming photons.

However, we need to modify the polarimeter from balloon experiments to satellite use in several points. The polarimeter must be tolerant to vibration and shock during the launch. We need to cover the energy range below 20 keV. Imaging capability of 1–2 arcmin ($\sim$2–4 mm) is desired. Incident area is 10 arcmin $\times$ 10 arcmin (18 mm $\times$ 18 mm). Considering these points, the current design of the PolariS scattering polarimeter uses the MAPMT of Hamamatsu H8711 series model instead of H8500.

We will employ five MAPMTs for one unit of polarimeter; plastic scintillator pillars are placed on one MAPMT at the center, while absorber (such as CsI(Tl), GSO, or BGO) scintillator pillars are placed on the surrounding four MAPMTs. This configuration will reduce the cross-talk problem between target (plastic) scintillators and absorber scintillators.

In order to cover the lower-energy band below 20 keV, we are considering a possibility that a light element (such as Be) passive target is placed above the plastic scintillators. This option is effective for bright point sources. Nevertheless, scattering taking place inside the passive target leads to a single-hit event. Polarization sensitivity and background rejection efficiency are different from those for double-hit events, i.e. where scattering takes place inside the plastic scintillator and recoiled electron signals are detected.

Simulation

We have started GEANT4 simulation to optimize the design of the PolariS scattering-imaging polarimeter. As a first step, we assumed a configuration with a Be target of 20 mm height. For absorber, 16 CsI(Tl) scintillators each of 4 mm $\times$ 4 mm $\times$ 60 mm and 16 plastic scintillators each of 4 mm $\times$ 4 mm $\times$ 40 mm are assumed.

Figure 49.2 *Left*. Plastic scintillator block used in preliminary experiments. This block consists of 4.3 mm × 4.3 mm × 40 mm plastic scintillator pillars, assembled with ESR sheets. *Right*. Hamamatsu H8711-200-100 MAPMT.

With this configuration, for 40 keV incidence, we get $\eta = 3.6\%$, $M = 50\%$ for double-hit events, and $\eta = 20.2\%$, $M = 57\%$ for single-hit events. The modulation factor is large enough, but the fraction of double-hit events is not. We need to improve this point.

We also evaluated the background level with GEANT4 simulation. If we do not employ anti-coincidence, the background level is as high as the Crab nebula. Introducing scintillators for anti-coincidence, as an example putting BGO scintillator pillars on surrounding positions of the surrounding four MAPMTs, reduces the background level of 30 mCrab. This value is for single-hit plus double-hit events. This value should not be regarded as an optimized one, and further reduction is being studied. Nevertheless, with this configuration, we expect 10–30 keV MDP of 10 mCrab sources is 9% with 1.5 Ms exposure. Significant improvement of MDP of 5% will be achieved if cross-talk between adjacent channels is reduced or compensated.

Preliminary experiments

We have assembled MAPMTs of Hamamatsu H8711-200-100 with plastic and absorber scintillator pillars. This type of MAPMT has a so-called ultra-bialkali photoelectric surface which has a peak QE of 43% (See [7] for X-ray test of a similar MAPMT model). For absorber scintillators, we tested CsI(Tl), GSO, and BGO. The spectral pulse height is maximum for CsI(Tl), while GSO gives about 40% of CsI(Tl). The spectral peak was less clear with BGO. Since GSO is employed in SUZAKU HXD and is known to be harder to radio activation than CsI(Tl), GSO is a prime candidate.

We also irradiated collimated X-rays from ^{109}Cd or ^{241}Am to a plastic channel to see scattering events to surrounding MAPMTs with GSO. It was found that cross-talk between adjacent channels of a MAPMT was stronger than in the case of the H8500, probably reflecting the smaller pitch size. We thus developed a data reduction algorithm to correct this effect and succeeded in detecting scattering events (recoiled electron signal in plastic and scattered photon signal in GSO) even for a 22-keV X-ray incidence. Detailed results will be reported elsewhere.

49.2.5 Wide-field X-ray γ-ray polarimeters

We will employ wide-field scattering polarimeters for transient objects. The design has not yet been determined, but the GAP polarimeter for the IKAROS mission[8] will be a prototype for it.

49.3 Summary

We are designing PolariS as one of the JAXA small satellite series. Our target launch will be in the mid 2010s, after ASTRO-H.

References

[1] Katsuta, J. et al. (2009). *Nucl. Instr. & Meth.* **A603**, 393–400.
[2] Hayato, A. et al. (2009). These proceedings.
[3] Tamagawa, T. et al. (2009). These proceedings.
[4] Kishimoto, Y. et al. (2007). *IEEE Trans. on Nucl. Science* **54**, 561–566.
[5] Gunji, S. et al. (2009). These proceedings.
[6] Bloser, P. et al. (2009). These proceedings.
[7] Toizumi, T. et al. (2009). *Nucl. Instr. & Meth.* **A604**, 168–173.
[8] Yonetoku, D. et al. (2009). These proceedings.

50

GAP aboard the solar-powered sail mission

D. Yonetoku, T. Murakami, H. Fujimoto & T. Sakashita

Kanazawa University

S. Gunji, N. Toukairin & Y. Tanaka

Yamagata University

T. Mihara

RIKEN

S. Kubo

Clear Pulse

and GAP team

We are now developing a gamma-ray burst polarimeter, named "GAP", to detect the polarization from the prompt emission of gamma-ray bursts. The GAP instrument is scheduled to launch in 2010 aboard the solar-powered sail satellite, IKAROS. GAP is a small instrument with a net weight of 4.0 kg, but we will perform reliable polarimetry with systematic uncertainty of a few percent level due to the highly symmetrical structure and the capability of coincidence event sampling. At present, we have already completed the integration of the flight model, and we will make several environmental tests and calibrations.

50.1 Introduction

Gamma-ray bursts (GRBs) are the most energetic explosions in the Universe. In the case of the brightest GRBs, the isotropic luminosity reaches 10^{54} erg s^{-1} during a short time duration, so the efficiency of energy release or conversion must be extremely high. A lot of physical information about GRBs was revealed after the discovery of afterglows[1], but we have still little knowledge about their emission mechanism. Theoretically, the prompt emissions and the following afterglows are thought to be generated by synchrotron radiation. In that case, we expect to detect a strong polarization (e.g. max 70%) especially during the prompt gamma-ray emission[2; 3]. There were earlier reports of measurement of GRB polarization[4],

X-ray Polarimetry: A New Window in Astrophysics, eds. R. Bellazzini, E. Costa, G. Matt and G. Tagliaferri. Published by Cambridge University Press.

Figure 50.1 A schematic view of the solar-powered sail spacecraft, IKAROS. Expanding a huge membrane of 20 m diameter, it translates the radiation pressure from the sun to the thrust of the spacecraft (from JSPEC/JAXA home page).

but not conclusive. Therefore, the direct measurement of the polarization degree of the prompt emission is a key to solving their emission mechanism.

50.2 Solar-powered sail mission – IKAROS

The solar-powered sail[5], named IKAROS, is a Japanese engineering verification spacecraft planned to launch in 2010. Figure 50.1 is an image of the IKAROS spacecraft. Expanding the huge membrane of 20 m in diameter, it transfers the radiation pressure from the Sun to the thrust of the spacecraft. IKAROS will cruise toward Venus and establish the solar-sail technology. During its cruising phase, we will observe GRBs and realize an interplanetary network system to determine the direction of GRBs with a few arcmin accuracy. For bright GRBs, we will measure the polarization degree of the prompt emission of GRBs.

50.3 Gamma-ray burst polarimeter – GAP

Figure 50.2 (left) is a schematic view of the GAP instrument[6]. The detection principle is the angular anisotropy of Compton scattering for the electric field vector of the gamma-ray photon. A large plastic scintillator with a super-bialkali type of photo-multiplier tube (PMT) is attached at the center, and 12 cesium iodide (CsI) scintillators are set around it. The central plastic works as the Compton scatterer, and the angular distribution of the scattered photons with coincidence are measured by surrounding CsI scintillators. This simple structure and the geometrical symmetry

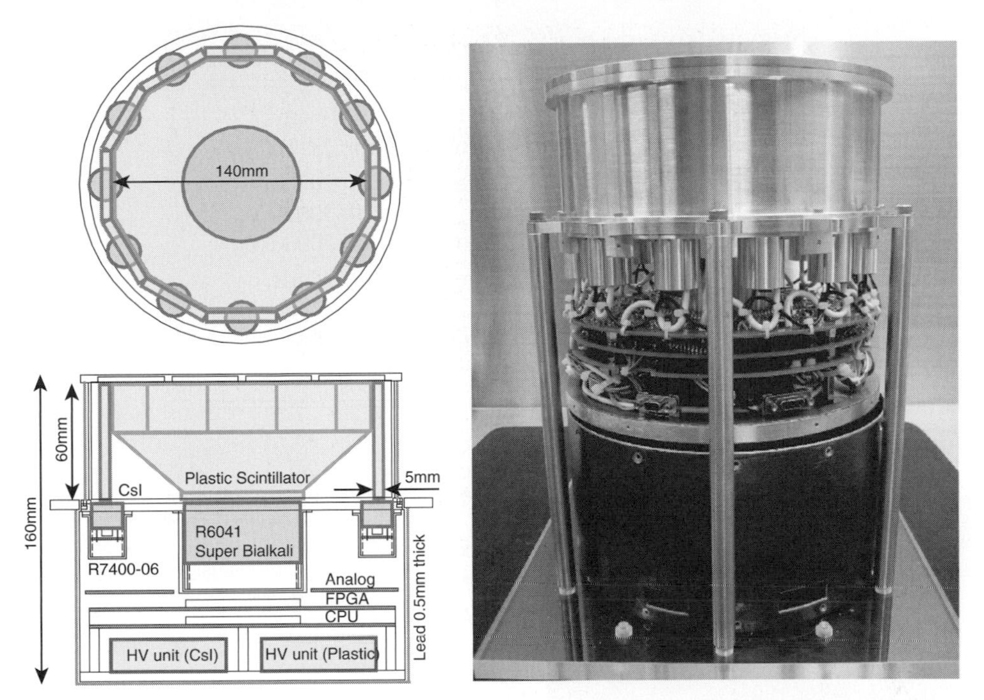

Figure 50.2 A schematic view of GAP (left) and a whole integrated system (right).

allow us to avoid a fake modulation. Even if the GRBs come from an off-axis direction, we may be able to estimate the asymmetry effect with the Monte-Carlo simulations. Several fundamental characteristics are listed in Table 50.1.

The low-energy threshold of the plastic scintillator should be set as low as possible to measure the coincidence events correctly. According to current experiments, we can set it as $E \sim 4$ keV. Therefore we can estimate that the energy range useful for polarimetry is $E > 50$ keV for 90° scattered events.

GAP will observe the direction of 45° off-axis from the anti-direction of the Sun. The instrument body, except for the detector surface, is protected by a thin layer of lead with 0.5 mm thickness for the purpose of shielding the cosmic X-ray background, low-energy cosmic rays,[1] and the solar flares.[2]

We have already completed the whole integration of the GAP flight model in April 2009. The total weight is very low, 4.0 kg including the electronics and high-voltage modules. Up to July 2009, we will perform several environmental tests, such as a vibration test for the HII-A rocket launching, temperature cycle tests, vacuum tests, and detailed software checks. We have already completed these kinds of tests using the prototype model, and their results were generally good. After July 2009, GAP will be integrated on the IKAROS spacecraft, and we will perform whole integration tests.

Table 50.1 *GAP characteristics*

Effective area	Geometry	176 cm^2
	Polarization	34 cm^2 @ 100 keV
Energy range	Lightcurve	10–300 keV
	Spectrum	10–300 keV
	Polarization	50–300 keV
Time resolution	Lightcurve	125 ms
	Spectrum	–
	Polarization	1 s
Field of view	Effective	π sr

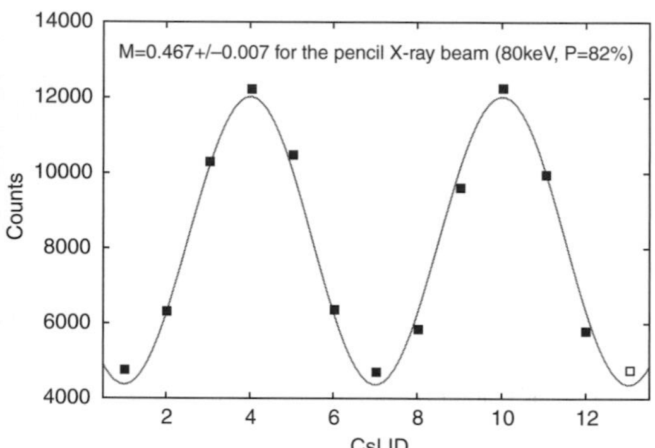

Figure 50.3 Experiment with the synchrotron facility at KEK (left). The monochromatic 80 keV X-ray beam had a polarization degree of 82% and the beam spread width was about 0.8 mm. An example of modulation curve irradiated at the center of GAP is shown (right).

50.4 Experiments with highly polarized X-rays

Hereafter we show the experimental results with the GAP prototype. Using highly polarized X-rays at the synchrotron facility of KEK photon factory, we estimate GAP performance for polarized X-rays. Figure 50.3 (left) shows the setup of the experiment. The monochromatic X-ray beam with 80 keV energy and 82% polarization degree comes from left to right. The beam size is about 0.8×0.8 mm, so we could only perform pencil beam experiments at KEK. The uniform irradiation for the entire surface can be operated in our laboratory using an X-ray generator which has about 15% polarization degree and the continuum spectrum up to 100 keV.

Figure 50.3 (right) shows an example of modulation curve with irradiation at the center of GAP. Each data point represents the coincidence events scattered from the central plastic, and the line is the best-fit function. From this experiment, we estimate the modulation factor of 0.446 for the pencil beam. On the other hand, using the EGS5/GEANT4 simulator, we succeeded in reproducing the experimental result within an uncertainty level of 2%.

Our GAP instrument has a wide field of view, and most GRBs come from a diagonal off-axis direction. So we estimate the detector response for the diagonal irradiation with the pencil beam. Comparing experiments and simulations, we can say that both results qualitatively describe each other. But we have not yet completed the detailed results. We recognize that the systematic uncertainty is extremely important for the polarimetry. We should make efforts to constrain the systematic uncertainty for various cases.

50.5 Detectability

50.5.1 Gamma-ray bursts and soft gamma-ray repeaters

In the previous section, we showed that the EGS5/GEANT4 simulations could represent the experimental results with the systematic uncertainty of about 2% level. Therefore, we consider that simulations can be adopted for various cases (energy, polarization degree, irradiation direction, etc.), and we estimate the GAP's basic capability. In Figure 50.4 (left), we show the modulation factor and the efficiency for the uniform irradiation case estimated by EGS5 simulator. The GAP performances are optimized around 100 keV, which generally corresponds to the energy at the maximum flux of νF_ν spectra. The efficiency is 0.22 and the modulation factor is 0.24, respectively.

According to the large amount of BATSE database, we estimate an expectation rate to detect positive polarization from GRBs. In Figure 50.4 (right), we show the expectation rate as a function of the minimum detectable polarization, which is equivalent to the GRB's polarization degree. The black line is the event rate in the

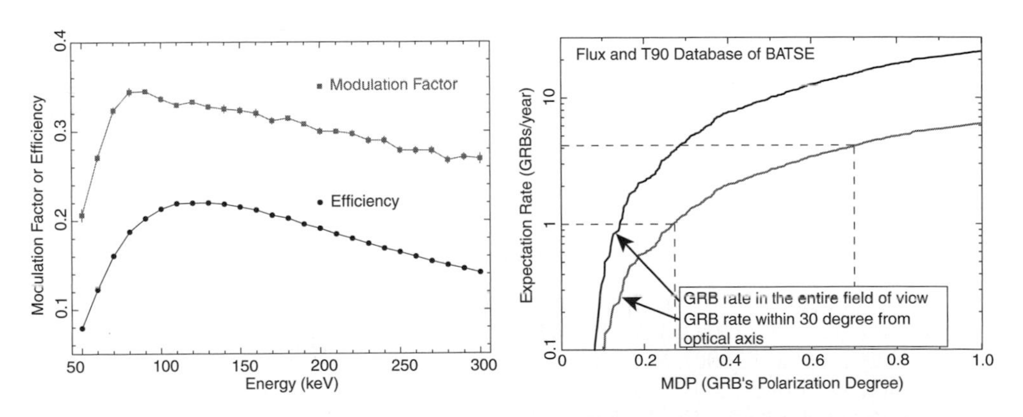

Figure 50.4 The energy dependence of the modulation factor and the efficiency (left), and the minimum detectable polarization for GRBs based on the BATSE database.

entire field of view, and the grey line is the one within 30° from the optical axis. If GRBs have 70% of polarization degree, we will detect about four events per year. In the case of 30% polarization degree, we may detect only one event with the significance of 3σ confidence level. This expectation rate is very small, but we will be able to measure the polarization degree with quite low systematic uncertainty, about 2%, because of the GAP's geometric symmetry. Moreover, if giant flares from soft gamma-ray repeaters are detected, we can perform polarization measurement on these as well as on GRBs.

50.5.2 Crab Nebula

We are planning to observe the Crab Nebula while it locates in front of the GAP instrument. Assuming the flux of the Crab Nebula and of cosmic X-ray background, as the optimistic estimation, we may detect 20% polarization with two days' exposure time. Even in the case of three times higher background rate and 10° of off-axis location, we may obtain a modulation with 3σ significance by ten days' exposure time. This observation also works as calibration for our system.

50.6 Summary

(i) We will install the GRB Polarimeter GAP on the solar-powered sail spacecraft which will launch in 2010.
(ii) We completed the integration of the flight model of GAP.
(iii) We will perform reliable polarimetry for GRBs, bright magnetar flares and the Crab Nebula with the systematic uncertainty of a few % level.
(iv) We require information about the GRB direction while we will contribute the interplanetary network system. We would be grateful if other teams would determine the GRB direction.

Notes

1. IKAROS is an interplanetary spacecraft, so the cosmic-rays environment is severe.
2. The period of 2010–2012 will be a maximum of solar activity, so we must avoid the huge number of solar flares.

References

[1] Costa, E., et al. (1997). *Nature* **387**, 783.
[2] Lazzati, D., et al. (2004). *MNRAS* **347**, L1.
[3] Lazzati, D., et al. (2006). *NJPh* **8**, 131.
[4] Coburn, W., Boggs, S. E. (2003). *Nature* **423**, 415.
[5] Mori, O. et al. (2004). *ESASP* **548**, 117.
[6] Yonetoku, D., et al. (2006). *Proc. SPIE* **6266**, 86.

51

Hard X-ray polarimeter for small-satellite missions

R. Cowsik

Washington University in St. Louis

B. Paul, R. Duraichelvan & P. V. Rishin

Raman Research Institute

We present here the design and estimated sensitivity of a hard X-ray polarimeter suitable for a small-satellite mission. Polarization fraction and direction of the X-rays from astronomical objects will be measured by recording the azimuthal anisotropy in the intensity of X-rays subsequent to their Thomson scattering on a circular disk of a light element like Li or Be. Extreme care is taken to make the instrument as azimuthally symmetric as possible and thus avoid spurious anisotropies from being seen. After describing the design of the present instrument, we present a semi-analytic calculation of the angular distribution of the scattered X-rays, taking explicitly into account the competing effects of photoelectric absorption. Thereafter, we derive the sensitivity of the instrument to measure the polarization fraction and the direction for astronomical sources with two possible spectral characteristics – a power law and thermal bremsstrahlung.

51.1 Introduction

The study of the polarization of X-ray sources may justifiably be called the next frontier in X-ray astronomy. Such measurements can reveal the physics and astrophysics under extreme conditions that are obtained in the presence of intense magnetic fields and strong gravity surrounding compact objects like pulsars and black holes and those encountered in high Mach number shocks. X-ray polarimetry of astronomical sources is a deep and sensitive probe of the astrophysical conditions that are obtained in these sources. Polarized X-rays are generated through a wide range of physical processes: synchrotron radiation, inverse-Compton scattering of low-energy polarized photons, curvature and thermal radiation in the intense magnetic

X-ray Polarimetry: A New Window in Astrophysics, eds. R. Bellazzini, E. Costa, G. Matt and G. Tagliaferri. Published by Cambridge University Press.

fields in the polar caps of pulsars, cyclotron emission in neutron star and magnetar magnetic fields, and Thomson scattering from anisotropic systems like jets, disks and plane parallel atmospheres. In the context of X-ray sources powered by accretion, the hard X-rays are produced close to compact objects like neutron stars and black holes and, accordingly, act as messengers of the conditions and phenomena associated with intense gravitational fields. Moreover, several theoretical studies have indicated that the percentage polarization is very high in the hard X-ray band and the direction of the polarization is expected to swing with energy in some compact sources[1]. Thus, hard X-ray polarimetry provides a new dimension to the astronomical studies of exotic objects and processes.

51.2 Design of the polarimeter and its response

The heart of the instrument is a circular disk (D) of Li or Be, of thickness ~1 cm and a diameter ~12.5 cm (Figure 51.1). Such a disk is azimuthally surrounded by a proportional counter (P) of height ~25 cm and diameter ~25 cm capable of measuring the azimuth in 24 15° bins. A single anode wire, interspaced with 24 resistors outside the common gas volume, is strung in a vertical raster pattern, to achieve uniformity of azimuthal response.

The astronomical X-rays enter the apparatus through a collimator (C) with a trapezoidal response function. The X-rays incident on the disk (D) are partially absorbed by the photoelectric effect and are also Thomson scattered, both in the forward and backward hemispheres. The azimuthal angular locations of the scattered X-rays are recorded by the position sensitive proportional counter (P). The intensity of the X-rays transmitted through the disk is recorded by the detector (T). Except perhaps in radio-astronomy, polarization studies are notoriously difficult – small azimuthal anisotropy in the telescope, beam transport system, or in the detector response can generate spurious signals. Accordingly, all the components noted above are made as axially symmetric as possible and mounted co-axially. What we have described thus far is one module of the instrument. We hope to deploy seven such units mounted as a tight triangular lattice on the pallet of a small satellite.

51.3 The response function and sensitivity of the detector

Let us define a coordinate system with the axis of the module as the z axis; the polar angle θ, the azimuthal angle ϕ with respect to any fiducial direction normal to $\hat{z}$, and thc radial distance r, are all measured with respect to the z axis. For an X-ray photon, the differential Thomson scattering cross section is given by

$$\frac{d\sigma_T}{d\Omega} = r_e^2(1 - \sin^2\theta\cos^2\phi) \equiv r_e^2 g(\theta,\phi). \tag{51.1}$$

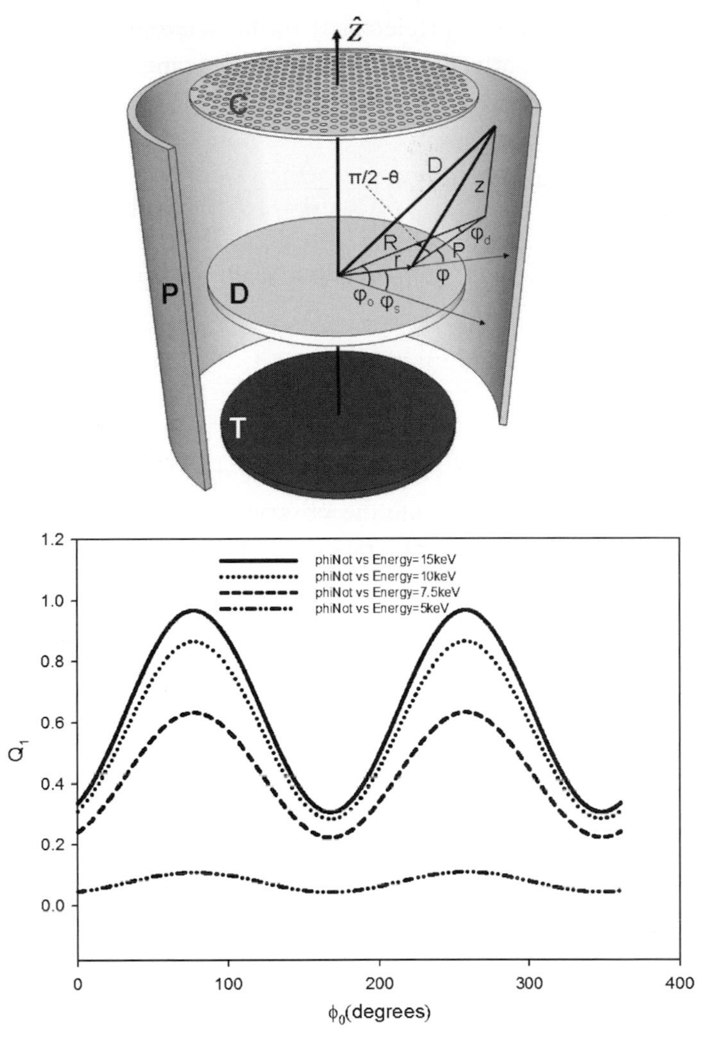

Figure 51.1 Sketch of the polarimeter showing the collimator (C), scattering disk made of Li or Be (D), through-put detector (T) and position-sensitive proportional counter (P) capable of measuring the azimuth of the scattered X-rays in 20° sectors. The geometrical reconstruction of the scattering event at $\hat{r}$ is also shown (top). Calculated azimuthal distribution of X-rays of various energies (bottom).

Defining λ_s as the Thompson scattering mean free path and $\lambda_a(E)$ as the photo-electric absorption mean free path in the material of the scatterer[2], the probability that the X-ray photon incident at $\theta = 0$ will be back-scattered at an angle (θ, ϕ) per unit solid angle is given by

$$P_1(\theta,\phi,E) = n_e r_e^2 g(\theta,\phi)\left(\frac{\Lambda}{1+\sec\theta}\right)\left\{1-\exp\left[-\frac{z_0}{\Lambda}(1+\sec\theta)\right]\right\}. \quad (51.2)$$

Here n_e is the number density of electrons in the scatterer, z_0 its thickness and $\Lambda^{-1} \equiv (\lambda_s^{-1} + \lambda_a^{-1})$. Similarly, the probability for the emergent scattered X-ray is given by

$$P_2(\theta,\phi,E) = n_e r_e^2 g(\theta,\phi) \left(\frac{\Lambda}{\sec\theta - 1}\right) \left\{e^{-\frac{z_0}{\Lambda}} - e^{-\frac{z_0 \sec\theta}{\Lambda}}\right\} \tag{51.3}$$

The probability of transmission, without either scattering or absorption is given by

$$P_0 = \exp\left(-\frac{z_0}{\Lambda}\right) \tag{51.4}$$

Because of the rapid decrease of the photoelectric cross section with energy as $E^{-\frac{7}{2}}, P_0, P_1$, and P_2 become essentially independent of energy beyond ~ 7.5 keV in Li and 10 keV in Be. To calculate the observed azimuthal distribution by the proportional counter, we should integrate over the area of the detector and the height of impact of the X-ray photons, always keeping track of the observed location ϕ_0 on the counter. Referring to Figure 51.1, which describes the geometry of the scattering process, the scattering probability to an azimuthal location per radian in the backward direction is given by

$$Q_1(\phi_0, E) = \int_{r=0}^{r_d} \int_{\phi_s=0}^{2\pi} \int_{z=0}^{z_x} P_1 \frac{R\cos\phi_d \sin\theta}{D^2} r dr d\phi_s dz \tag{51.5}$$

Figure 51.1 also gives the response function. When the energy-dependent angular distribution is integrated over the spectra from the sources, we obtain the appropriate count rates as a function of azimuth. For power law and bremsstrahlung spectra, we have

$$C_p(\phi) = \int_{E_{min}}^{E_{max}} Q(E,\phi) I_0 E^{-\beta} dE, \tag{51.6}$$

$$C_T(\phi) = \int_{E_{min}}^{E_{max}} Q(E,\phi) I_0 \exp\left(-\frac{E}{kT}\right) \frac{dE}{E}. \tag{51.7}$$

Assuming that the background in the proportional counter is dominated by veto inefficiency, unrelated to the source intensities, for an observation time of 10^4 s, we estimate the minimum detectable polarization fraction of hard X-rays by the instrument made of seven modules to be

$$p_\theta = \frac{0.2}{I_0} \left(\frac{10^4\,\text{s}}{t}\right)^{\frac{1}{2}}. \tag{51.8}$$

Here I_0 is in units of photons cm^{-2} s^{-1} keV^{-1} at 1 keV. For the Crab nebula we can detect about 5% polarization fraction in hard X-rays with 10^4 s of observation.

References

[1] Schnittman, J. D., Krolik, J. H. (2009). [arXiv:0902.3982].
[2] Berger, M. J., et. al. (2009). XCOM: Photon Cross Sections Database `http://physics.nist.gov/PhysRefData/Xcom/Text/XCOM.html`.

52

Performance of hard X-ray polarimeter: PHENEX

S. Gunji, Y. Kishimoto, Y. Tanaka, N. Fujita,
F. Tokanai, H. Sakurai & N. Toukairin
Yamagata University

K. Hayashida, M. Yamauchi, N. Anabuki & H. Tsunemi
Osaka University

T. Mihara
RIKEN

Y. Saito, M. Kohama & M. Suzuki
ISAS/JAXA

S. Kishimoto
KEK

In order to measure precisely the polarization of Crab Nebula and Cygnus X-1, we have been developing a hard X-ray polarimeter for balloon-borne experiments called PHENEX (Polarimetry of High ENErgy X-rays). It consists of several detectors called unit counters. The unit counter has a detection efficiency of 20% and a modulation factor of 53% at 80 keV. Up to now, we have finished the installation of eight unit counters to the polarimeter, that will be launched in Spring 2009 to observe the Crab Nebula. If the polarization of this source is more than 30%, the PHENEX polarimeter will be able to measure the degree and the direction of the polarization with errors less than 10% and 10°, respectively.

52.1 Introduction

X-ray astronomy has been much advanced by three observations: spectroscopy, timing, and imaging. Also in the hard X-ray region, these three observations will be realized by ASTRO-H and XEUS. However, the observation of the polarization is at the moment left out in spite of its potential usefulness. This is because of the difficulty of developing polarimeters with high sensitivity. Since the origin of the polarization is often due to nonthermal radiation processes such as synchrotron

X-ray Polarimetry: A New Window in Astrophysics, eds. R. Bellazzini, E. Costa, G. Matt and G. Tagliaferri.
Published by Cambridge University Press.

radiation, observations in the hard X-ray region are possibly more important than those in the soft X-ray region: it is expected that the degree of polarization in the hard X-ray region would be higher than that at lower energies. For the above reasons, we have been developing a hard X-ray polarimeter with high sensitivity, called PHENEX (Polarimetry for High ENErgy X-rays,[1]). In 2006, we observed the polarization of the Crab Nebula with PHENEX, using a geometrical area of 44 cm^2. Then the degree and the direction of the polarization were determined to be $33 \pm 26\%$ and $154° \pm 43°$[2]. However, more precise measurements are required to clarify the features of the magnetic field and of the emission region for the hard X-ray band, comparing with the past data for X-ray and γ-ray regions [3; 4]. Therefore, enlarging the geometrical area by a factor of two and decreasing the background level, we will observe the Crab Nebula in Spring 2009 and then Cygnus X-1 in 2010. Through these observations, we will open a new window for hard X-ray astronomy.

52.2 The PHENEX polarimeter

The PHENEX polarimeter consists of several unit counters which are Compton-scattering type polarimeters. As shown in the left panel of Figure 52.1, the unit counter consists of 36 plastic scintillators and 28 CsI(Tl) scintillators. The signals from the scintillators are read out by one multianode photomultiplier (H8500). Hard X-rays incident through the collimator (4.8° FWHM) enter one of the plastic scintillators and are scattered, then the scattered photon is absorbed by one of the CsI(Tl) scintillators. Detecting the scattering and the absorbing positions, the two-dimensional scattering direction is determined. So the information on the polarization of incident hard X-rays is obtained. The modulation curve for the injection of a polarized beam is shown in the right panel of Figure 52.1. The characteristics of a unit counter are summarized in Table 52.1.

In 2006, the PHENEX polarimeter had four unit counters. In 2009, it has eight unit counters and the geometrical area is doubled, as shown in Figure 52.2. Moreover, by reinforcement of passive shields and the fact that the neighboring unit counters work as a shield for each other, the ratio of signal to background goes up about 2.5 times and will be about 1:0.8.

52.3 Expected performances

In 2006, we carried out a balloon-borne experiment with the PHENEX polarimeter having four unit counters. However, the attitude control system did not function correctly and therefore the observation time for the Crab Nebula was only one hour. Now we have corrected the attitude control system problem and therefore we will obtain an observation time of six hours. Moreover, the performance of the PHENEX

Table 52.1 *Performance of a unit counter.*

Energy range	40 keV $\sim$ 200 keV
Geometrical area per one unit counter	11 cm^2
Modulation factor	53% at 80 keV
Detection efficiency	20% at 80 keV

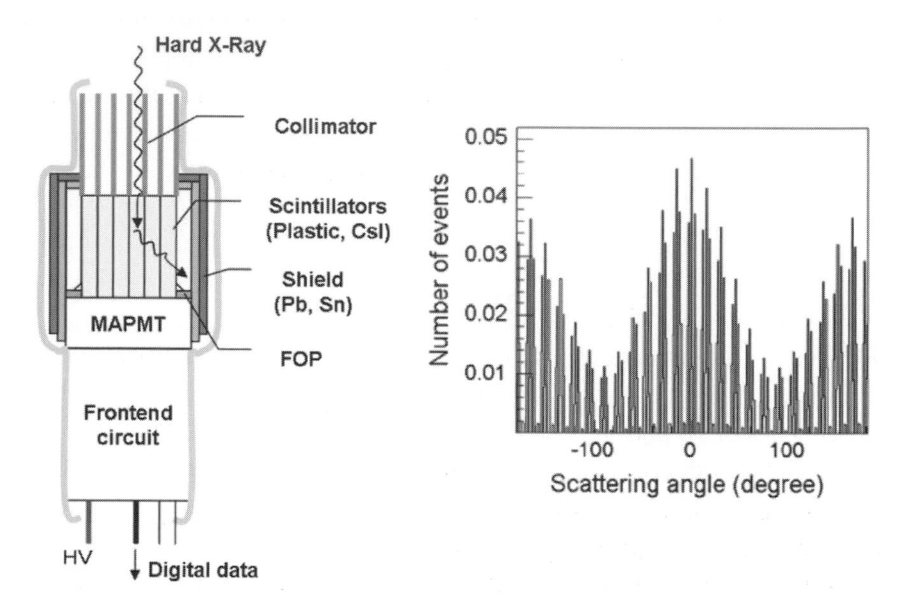

Figure 52.1 The left and right panels show the schematic view of the unit counter and the modulation curve for injection of polarized beam, respectively.

Figure 52.2 The left and right panels show the PHENEX polarimeter in 2006 and the enlarged one in 2009, respectively.

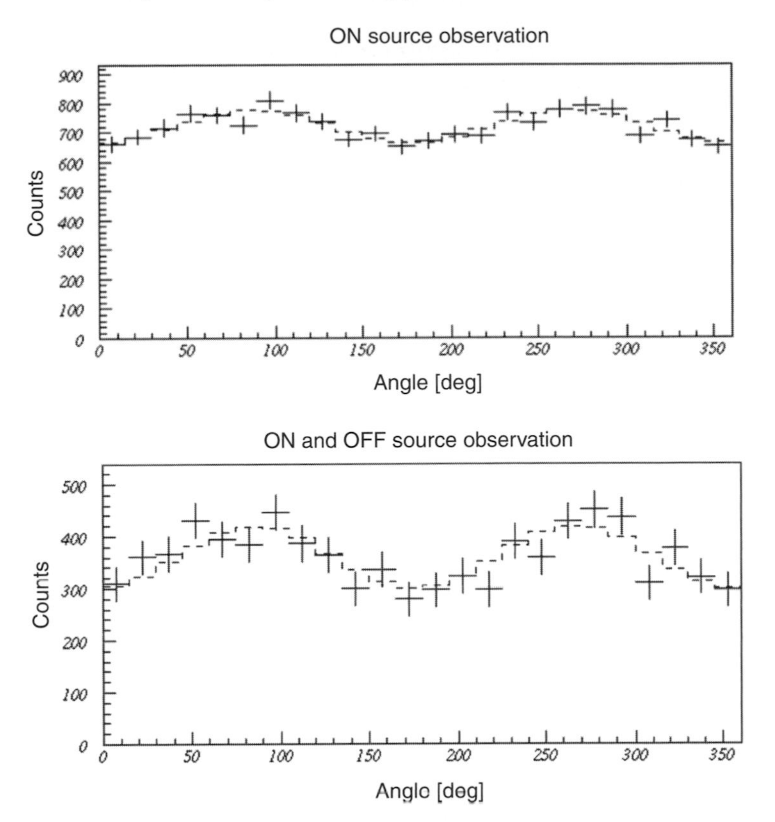

Figure 52.3 Using the data in 2006 and computer simulations, we investigated the expected modulation curve for the Crab Nebula with the PHENEX polarimeter in 2009. We assume four hours for the ON source observation and two hours for the OFF source one, respectively. The degree of polarization is assumed to be 30%.

polarimeter has been much improved. So, if the degree of the polarization is more than 30%, in 2009 we will be able to determine the degree and the direction of the polarization within less than 10% and 10°, respectively. Figure 52.3 shows the expected modulation curve for the Crab Nebula.

52.4 Conclusion

Now we are preparing the balloon launch to observe the Crab Nebula with the PHENEX polarimeter. The X day will be May 30th, 2009. We hope that this experiment will open a new window for hard X-ray polarimetry.

We appreciate the work of balloon launching by the balloon group in ISAS/JAXA. We wish to thank the Japan Society for the Promotion of Science for its support of the travel budget.

References

[1] Kishimoto Y., et al. (2007). *IEEE Trans. Nucl. Sci.* **54**, 561.
[2] Gunji, S., et al. (2007). *Proc. of SPIE* **6686**, 18:1.
[3] Weisskopf, M. C., et al. (1978). *ApJ* **220**, L117.
[4] Dean, A. J., et al. (2008). *Science* **321**, 1183.

53

GRB polarimetry with POET

M. L. McConnell

University of New Hampshire

on behalf of the POET collaboration[1]

POET (Polarimeters for Energetic Transients) represents a concept for a NASA Small Explorer (SMEX) satellite mission, whose principal scientific goal is to understand the structure of GRB sources through sensitive X-ray and γ-ray polarization measurements. The payload consists of two wide field-of-view (FoV) instruments: a low-energy polarimeter (LEP) capable of polarization measurements in the energy range from 2–15 keV and a high-energy polarimeter (Gamma-Ray Polarimeter Experiment or GRAPE) that would measure polarization in the 60–500 keV energy range. The POET spacecraft provides a rotating zenith-pointed platform for dealing with any residual systematic effects in the polarization response and for maximizing the exposure to deep space. POET would be capable of measuring statistically significant polarization (for polarization levels in excess of 20%) for ~80 GRBs in a two-year mission. High-energy polarization data would also be obtained for SGRs, solar flares, pulsars and other sources of astronomical interest.

53.1 Introduction

Gamma-ray bursts (GRBs) are amongst the most energetic events in the universe, and have stimulated intense observational and theoretical research. Theoretical models indicate that a refined understanding of the inner structure of GRBs, including the geometry and physical processes close to the central engine, requires the exploitation of high-energy X-ray and γ-ray polarimetry. To date, observations have been of limited sensitivity and subject to poorly understood systematics. POlarimeters for Energetic Transients (POET) is a SMEX mission concept that is capable of measuring the high-energy polarization of GRBs and other sources

X-ray Polarimetry: A New Window in Astrophysics, eds. R. Bellazzini, E. Costa, G. Matt and G. Tagliaferri. Published by Cambridge University Press.

of astronomical interest[1]. POET obtains measurements with two different polarimeters (both with wide fields of view) to provide observations over a broad energy range: LEP (Low-Energy Polarimeter) covers 2–15 keV and GRAPE (Gamma-RAy Polarimeter Experiment) covers 60–500 keV. In the context of studying GRBs, POET would address the following questions: What is the magnetic structure of the jets? What is the geometric structure of the jets? What is the prompt emission mechanism? Where does the emission originate? By measuring a distribution of GRB polarizations, POET could distinguish between jet models with ordered magnetic fields and those with random magnetic fields, each of which predicts different distributions of polarization values[2; 3].

53.2 Low-Energy Polarimeter (LEP)

LEP measures the direction of photoelectrons that are ejected after the photoelectric absorption of an incident photon. Photoelectrons tend to be ejected in a direction parallel to the electric field vector of the incident photon. The photoelectron tracks are measured with the innovative operation of a time projection chamber (TPC), providing unmatched polarization sensitivity over the 2–15 keV band-pass[4; 5]. The LEP polarimeter enclosure consists of four detector modules each with an isolated gas volume contained by a Be X-ray window. Each detector module contains two $6 \times 12 \times 24$ cm^3 TPCs. Each TPC is composed of a micropattern proportional counter, consisting of a shared drift electrode and a high-field gas electron multiplier (GEM) positioned 1 mm from a strip readout plane. When an X-ray is absorbed in the gas between the drift electrode and the GEM, a photoelectron is ejected in a preferential direction with a $\cos^2\phi$ distribution, where ϕ is the azimuthal angle measured from the X-ray polarization vector. (The photoelectron tends to be ejected in a direction parallel to the polarization vector of the incident photon.) As the photoelectron travels through the gas it creates a path of ionization that drifts in a moderate, uniform field to the GEM where an avalanche occurs. A track image projected onto the x-y plane is formed by digitizing the charge pulse waveforms. The coordinates are defined by strip location in one dimension, and arrival time multiplied by the drift velocity in the orthogonal dimension. The strips are smaller than the mean free path of the photoelectron and therefore an image of the photoelectron track can be reconstructed and the initial direction of the photoelectron determined. The magnitude and orientation of the source polarization can be determined from a histogram of the emission angles. LEP is sensitive to GRB emissions at off-axis angles up to at least 45°.

53.3 Gamma-Ray Polarimeter Experiment (GRAPE)

GRAPE[6; 7; 8] is designed to measure polarization in the range 60–500 keV and to provide spectroscopy over a broad energy range from 15 keV to 1 MeV. It relies on the fact that a Compton-scattered photon tends to scatter at right angles to the polarization vector of the incident photon. The GRAPE instrument is composed of 64 independent detector modules arranged in two identical assemblies that provide the associated electronics and the required mechanical and thermal support. Each polarimeter module incorporates an array of optically independent $5 \times 5 \times 50$ mm^3 scintillator elements aligned with and optically coupled to the 8×8 scintillation light sensors of a 64-channel MAPMT. Two types of scintillators are employed. Low-Z plastic scintillator is used as an effective medium for Compton scattering. High-Z inorganic scintillator (Bismuth Germanate, BGO) is used as a calorimeter, for absorbing the full energy of the scattered photon. An array of 6×6 plastic scattering elements is surrounded by 28 BGO calorimeter elements. Valid polarimeter events are those in which a photon Compton scatters in one of the plastic elements and is subsequently absorbed in one of the BGO elements. The azimuthal scatter angle is determined for each valid event by the relative locations of hit scintillator elements. To facilitate spectral measurements over a broader energy range (15 keV–1 MeV), GRAPE also includes two independent NaI(Tl) spectrometer modules. The magnitude and orientation of the source polarization can be determined from a histogram of the emission angles, where the polarization angle corresponds to the minimum in the modulation pattern. GRAPE will be sensitive to GRB emissions at off-axis angles up to at least 60°.

53.4 The POET MIssion

The mission concept involves a standard SMEX launch into a circular 600 km orbit. A zenith-pointed spacecraft (Figure 53.1) provides continuous exposure to

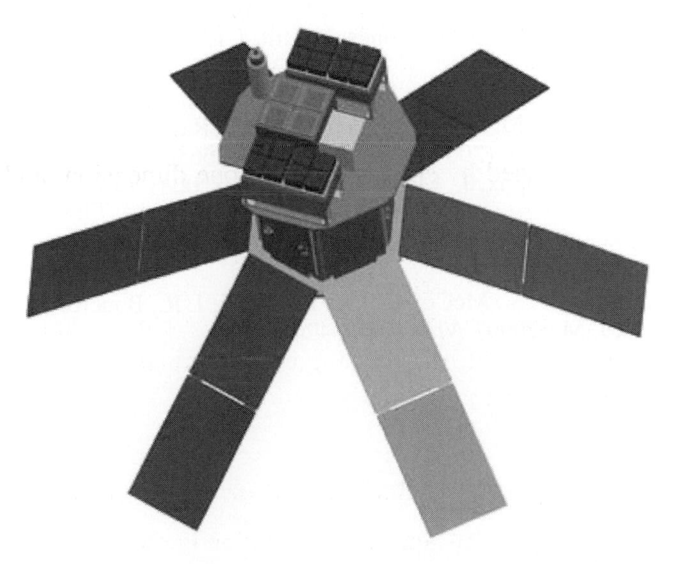

Figure 53.1 Spacecraft concept for the POET mission.

deep space. Coupled with the wide FoV of both instruments, this maximizes the probability of prompt GRB detection. Spacecraft rotation mitigates the effects of any residual systematic effects in the polarization response. The combined capabilities of LEP and GRAPE will allow us to measure the polarization over a broad energy range (2–500 keV) and also to measure E_p for a large number of GRBs. Simulations based on realistic GRB distributions indicate that both LEP and GRAPE will be capable of detecting about 40 GRBs per year with a polarization sensitivity of about 20%[1] in their respective energy bands. During a nominal two-year mission, POET would provide statistically significant polarization measurements for ~80 GRBs, providing diagnostic information on the GRB emission mechanism that cannot be obtained from currently available data[3].

Notes

1. POET Collaboration: L. Angelini, M. G. Baring, S. Barthelmy, J. K. Black, P. F. Bloser, B. Dennis, A. G. Emslie, J. Greiner, W. Hajdas, A. K. Harding, D. H. Hartmann, J. E. Hill, K. Ioka, P. Kaaret, G. Kanbach, D. Kniffen, J. S. Legere, J. R. Macri, R. Morris, T. Nakamura, N. Produit, J. M. Ryan, T. Sakamoto, K. Toma, X. Wu, R. Yamazaki, and B. Zhang.

References

[1] Hill, J. E., McConnell, M. L., Bloser, P., Legere, J., Macri, J., Ryan, J., Barthelmy, S., Angelini, L., Sakamoto, T., Black, J. K., Hartmann, D. H., Kaaret, P., Zhang, B., Ioka, K., Nakamura, T., Toma, K., Yamazaki, R. & Wu, X. (2008). *AIP Conf. Proc.* **1065**, 331–337.

[2] Waxman, E. (2003). *Nature* **423**, 388–389.

[3] Toma, K., Sakamoto, T., Zhang, B., Hill, J. E., McConnell, M. L., Bloser, P. F., Yamazaki, R., Ioka, K. & Nakamura, T. (2009). *ApJ* **698**, 1042.

[4] Black, J. K., Baker, R. G., Deines-Jones, P., Hill, J. E. & Jahoda, K. (2007). *NIM A* **581**, 755–760.

[5] Hill, J. E., Barthelmy, S., Black, J. K., Deines-Jones, P., Jahoda, K., Sakamoto, T., Kaaret, P., McConnell, M. L., Bloser, P. F., Macri, J. R., Legere, J. S., Ryan, J. M., Smith, B. R. & Zhang, B. (2007). *Proc. SPIE* **6686**, 66860Y.

[6] McConnell, M. L., Ledoux, J., Macri, J. & Ryan, J. (2004). *Proc. SPIE.* **5165**, 334–345.

[7] Legere, J., Bloser, P., Macri, J. R., McConnell, M. L., Narita, T. & Ryan, J. M. (2005). *Proc. SPIE.*, **5898**, 413.

[8] Bloser, P. F., Legere, J. S., McConnell, M. L., Macri, J. R., Bancroft, C. M., Connor, T. P. and Ryan, J. M. (2009). *NIM A* **600**, 424–433.

Author index

Subject index